BIRDS

Brain and Behavior

Contributors

Lynda D. Beazley
Jerram L. Brown
David H. Cohen
Irving J. Goodman
William Hodos
Harvey J. Karten
Masakazu Konishi
Richard H. McCollum
Wolfgang M. Schleidt
Laurence J. Stettner
D. M. Vowles
Hans Zeier
H. Philip Zeigler

BIRDS

Brain and Behavior

EDITED BY

Irving J. Goodman

Departments of Psychology and
Behavioral Medicine and Psychiatry
West Virginia University
Morgantown, West Virginia

Martin W. Schein

Department of Biology
West Virginia University
Morgantown, West Virginia

ACADEMIC PRESS NEW YORK SAN FRANCISCO LONDON 1974
A Subsidiary of Harcourt Brace Jovanovich, Publishers

ACADEMIC PRESS, INC.
111 Fifth Avenue, New York, New York 10003

United Kingdom Edition published by
ACADEMIC PRESS, INC. (LONDON) LTD.
24/28 Oval Road, London NW1

Library of Congress Cataloging in Publication Data

Lashley Memorial Conference, West Virginia University, 1971.
Birds: brain and behavior.

Includes bibliographies.
1. Birds–Behavior. 2. Brain. I. Goodman, Irving J., ed. II. Schein, Martin W., ed. III. Title. [DNLM: 1. Behavior, Animal–Congresses. 2. Birds–Congresses. QL698.3 L343b 1971]
QL698.3.L37 1971 598.2′5 73-18952
ISBN 0–12–290350–1

PRINTED IN THE UNITED STATES OF AMERICA

Contents

List of Contributors ix
Preface xi
Figure Credits xiii

1 THEORETICAL AND METHODOLOGICAL ISSUES

The Comparative Study of Behavior 3

Wolfgang M. Schleidt

I. One Bird 3
II. Two Birds 6
III. Many Birds 7
IV. Birds and Nonbirds 10
References 12

The Comparative Study of Brain–Behavior Relationships 15

William Hodos

I. Why Comparative Studies? 16
II. Which Animals Shall We Compare? 18
III. What Shall We Compare? 21
References 24

2 ORGANIZATION OF NEURAL SYSTEMS 29

The Structural Organization of Avian Brain: An Overview

David H. Cohen and Harvey J. Karten

I. Introduction 29
II. General Structural Features 31

III. Spinal Cord 37
IV. General Ascending Pattern 44
V. General Descending Pattern 56
VI. Telencephalic Organization 61
VII. Concluding Statement 67
References 68

3 NEUROBEHAVIORAL FINDINGS

Hearing and Vocalization in Songbirds 77

Masakazu Konishi

I. Auditory Mechanisms 77
II. Vocalization and Audition 83
References 85

Brain Stimulation Parameters Affecting Vocalization in Birds 87

Jerram L. Brown

I. Anatomical Considerations 87
II. Methodological Considerations 90
References 98

Feeding Behavior in the Pigeon: A Neurobehavioral Analysis 101

H. Philip Zeigler

I. Normative Studies of Feeding Behavior in the Pigeon 103
II. A Feeding Behavior "System" in the Pigeon 113
III. Feeding Behavior Mechanisms in the Pigeon: Neurobehavioral Mechanisms 118
IV. Conclusion: Implications for the Study of Feeding Behavior References 130

The Study of Sleep in Birds 133

Irving J. Goodman

I. Behavioral and Physiological Correlates of Sleep and Waking 133
II. Experimental Manipulation of Sleep 139
References 150

Behavioral Adaptation on Operant Schedules after Forebrain Lesions in the Pigeon 153

Hans Zeier

I. Hyperstriatal and Neostriatal Lesions 154
II. Archistriatal Lesions 156
III. Paleostriatal Lesions 162
IV. Conclusions 163
References 163

The Neural Basis of Avian Discrimination and Reversal Learning 165

Laurence J. Stettner

I. Introduction 165
II. Discrimination Procedures 167
III. Neurobehavioral Studies 172
References 198

Brain Perturbation and Memory Disruption: A Comparison between Classes 203

Richard H. McCollum and Irving J. Goodman

I. Findings in Mammals 204
II. Findings in Birds 209
III. Summary 217
References 218

The Neural Substrate of Emotional Behavior in Birds 221

D. M. Vowles and Lynda D. Beazley

I. Introducton 221
II. Methods 222
III. Results and Discussion 225
References 256

Author Index 259
Subject Index 266

List of Contributors

Numbers in parentheses indicate the pages on which the authors' contributions begin.

Lynda D. Beazley (221), Department of Psychology, University of Edinburgh, Edinburgh, Scotland

Jerram L. Brown (87), Department of Biology, University of Rochester, Rochester, New York

David H. Cohen (29), Department of Physiology, School of Medicine, University of Virginia, Charlottesville, Virginia.

Irving J. Goodman (133, 203), Departments of Psychology and Behavioral Medicine and Psychiatry, West Virginia University, Morgantown, West Virginia

William Hodos (15), Department of Psychology, University of Maryland, College Park, Maryland

Harvey J. Karten (29), Department of Psychology, Massachusetts Institute of Technology, Cambridge, Massachusetts

Masakazu Konishi (77), Department of Biology, Princeton University, Princeton, New Jersey

Richard H. McCollum (203), Department of Psychology, Allegheny College, Meadville, Pennsylvania

Wolfgang M. Schleidt (3), Department of Zoology, University of Maryland, College Park, Maryland

Laurence J. Stettner (165), Department of Psychology, Wayne State University, Detroit, Michigan

D. M. Vowles (221), Department of Psychology, University of Edinburgh, Edinburgh, Scotland

Hans Zeier (153), Department of Behavioral Biology, Swiss Federal Institute of Technology, Zurich, Switzerland

H. Philip Zeigler (101), Department of Psychology, Hunter College, City University of New York, *and* Department of Animal Behavior, American Museum of Natural History, New York, New York

Preface

What once seemed a wide and bottomless chasm between brain and behavior has within the past few decades been narrowed in at least a few places to gaps that are being tenuously bridged. There is no doubt that these initially tentative bridgeheads on both sides of the gap will someday be strengthened and broadened to the point where we will have a fairly clear picture of the mechanisms underlying behavior. At the present time, though, it behooves us to reassess our position if only to discover what we are going.

The present volume is an effort to contribute to the assessment, particularly with respect to birds. If we are to utilize efficiently the insights afforded by a comparative approach to brain–behavior problems, then the wealth of information presently available on mammals must be matched by comparable qualitative and quantitative information on other animals with functioning brains. Hence our concentration on birds, which are characterized by complex and well-integrated central nervous systems, but also by far more stereotyped behavioral repertoires than those exhibited by mammals.

The genesis of this volume was a 1971 Lashley Memorial Conference at Morgantown, West Virginia, which honored the memory of the late Professor Karl S. Lashley. Dr. Lashley was "a local boy," an alumnus of West Virginia University who went on to gain world renown as a neurobehaviorist. Much of the work reported in this volume is a logical extension and continuation of work that Lashley started many years ago; indeed, if Lashley were alive today, it is highly likely that we would have been a prime mover in the conference and in this volume.

The first section of the present volume deals with some theoretical and methodological questions involved in comparative studies of behavior and in brain–behavior relationships. These two areas of thought contain a number of ideas and viewpoints ranging from the controversial to the well accepted, and from the tenuously formulated to the moderately clear. Another section deals more directly with neuroanatomic information regarding sensory motor structure and connections. Some recent findings in this area have raised a number of questions about previous conceptualizations of avian neural organization. The last section examines and summarizes data pertaining to neural correlates or causes of various behavioral phenomena such as feeding, learning and memory, sleep, emotion, audition, vision, vocalization, and adaptation to various schedules of reinforcement.

The main purpose of the conference, and therefore of this volume, was to bring together a number of neurobehavioral scientists from several different disciplinary areas so that each could profit from the ideas and stimulation provided by the others. By the same token, is our hope that readers from many diverse areas of biology, psychology, and neurophysiology will find the materials as stimulating as did the participants.

We wish to thank the West Virginia University for its support and encouragement in this endeavor.

Figure Credits

Page 8: Reprinted by permission of the publishers from Konrad Lorenz, *Studies in Human and Animal Behaviour,* Volume II. Cambridge, Massachusetts: Harvard University Press, Copyright, 1971, by Konrad Lorenz.

Pages 32, 144, 181: From H. J. Karten and W. Hodos, *A Stereotaxic Atlas of the Pigeon (Columba livia).* Baltimore, Maryland: The Johns Hopkins Press, 1967.

Pages 33, 48, 174: From W. J. H. Nauta and H. J. Karten, A general profile of the vertebrate brain, with sidelights on the ancestry of cerebral cortex. G. C. Quarton, T. Melnechuk, and F. O. Schmitt (Eds.), *The Neurosciences: Second Study Program.* New York: The Rockefeller University Press, 1970.

Page 38: From C. U. A. Kappers, G. C. Huber, and E. C. Crosby, *The Comparative Anatomy of the Nervous System of Vertebrates, Including Man.* New York: Macmillan, 1936.

Page 53: From Boord, R. L. The anatomy of the avian auditory system. *Annals of the New York Academy of Sciences,* 1969, **167,** 186–198.

Page 79: From Boord, R. L., and Rasmussen, G. L. Projection of the cochlear and lagenar nerves on the cochlear nuclei of the pigeon. *Journal of Comparative Neurology,* 1963, **120,** 463–475.

Page 91: From Brown, J. L. The Integration of Agonistic Behavior in the Steller's Jay *Cyanocitta stelleri* (Gmelin). University of California Press Publications in Zoology, 1964, **60**. Originally published by the University of California Press; reprinted by permission of The Regents of the University of California.

Pages 105, 112, 128: From Zeigler, H. P., and Feldstein, R. A feedometer for the pigeon. *Journal of the Experimental Analysis of Behavior,* 1971, **16,** 181– 187. Copyright 1971 by the Society for the Experimental Analysis of Behavior, Inc.

Pages 106, 107, 113: From Zeigler, H. P., Green, H. L., Lehrer, R. Patterns of feeding behavior in the pigeon. *Journal of Comparative & Physiological Psychology,* 1971, **76,** 468–477. Copyright 1971 by the American Psychological Association and reproduction by permission.

Pages 108, 109: From Zeigler, H. P., Green, H. L., and Siegel, J. Food and water intake and weight regulation in the pigeon. *Physiology and Behavior,* 1972, **8,** 127–134.

Page 111: From Megibow, M., and Zeigler, H. P. Readiness to eat in the pigeon. *Psychnomic Science,* 1968, **12,** 17–18.

Page 119: From Zeigler, H. P., and Witkovsky, P. The main sensory trigeminal nucleus in the pigeon: A single unit analysis. *Journal of Comparative Neurology,* 1968, **134,** 255–264.

Page 120: Witkovsky, P., Zeigler, H. P., and Silver, R. A single-unit analysis of the nuclear basalis in the pigeon. *Journal of Comparative Analysis,* 1973, **147,** 119–128.

Page 157: Zeier, H., and Karten, H. J. Connections of the anterior commissure in the pigeon *(Columba livia). Journal of Comparative Neurology,* 1973, **150,** 201–216.

Pages 159, 160: From Hans Zeier, Archistriatal lesions and response inhibition in the pigeon. *Brain Research,* 1971, **31,** 327–339, Figs. 3, 4.

Page 195: From Gossette, R. L., Gossette, M. F., and Riddell, W. Comparisons of successive discrimination reversal performances among closely and remotely related avian species. *Animal Behavior,* 1966, **14,** 560–564.

1 THEORETICAL AND METHODOLOGICAL ISSUES

The Comparative Study of Behavior[1]

Wolfgang M. Schleidt
University of Maryland

I. One Bird

If we look at "the bird" and observe its behavior, we automatically initiate a comparative study: We compare the observed actions (and reactions) among themselves and form abstractions on the basis of similarities and differences. As observation continues, we assign these actions to established categories or we create new categories if existing ones are not appropriate. For example, "the bird" (Fig. 1) is at first motionless, its neck in upright position, its beak closed; then it bends the neck down and mandibulates food objects on the ground, and after a while, it bends its neck backward and mandibulates the feathers on its back. Customarily we pay great attention to the objects with which the bird interacts, but if no interactions with objects are initially noticed we say "the bird does nothing." The mandibulation of the food objects is labeled "feeding" and the mandibulation of the feathers "preening." In this way we build up a system of classification and obtain a list of behaviors, called an "ethogram."

A word of caution must be interjected here about the naming of behavioral acts: it is advantageous to choose purely descriptive terms and to avoid any premature assumption about the potential function of a behavior. A certain call of a bird might be labeled by an onomatopoetic term like "kooo," or called cooing, but preferably not "caressing call" since the latter

[1] In memory of a most respected investigator and dear friend, Peter A. Winter, who met an untimely death on March 10, 1972, through a skiing accident.

Fig. 1. The Bird, displaying three fundamental behaviors: head down, feeding; head up, doing nothing; head on back, preening.

implies motivational and functional properties that we cannot predict confidently at this early state of the investigation. Such implied connotations of a behavior's name (e.g., "fear" response; "alarm" call) can obscure other motivational tributaries or functions. In my own studies of the responses of the turkey to flying predators (Schleidt, 1961) "clucking" was labeled an "alarm call" during the exploratory phase of the research project, since it was evoked by a variety of predator models. Because of the "alarm call"label, it was not until we later took a census of the various behavioral acts occurring throughout the day that we realized that clucking also occurs in a number of nonthreatening situations. The new data showed that clucks occur concomitantly with a state of general arousal and suggest a classification such as "alert" rather than "alarm" call; however, as a label, I prefer the onomatopoetic term "cluck."

Another serious problem is to know when to split and when to lump categories of behavioral units. Take as an example a bird's song: Should we consider as a unit everything from the beginning of the first syllable to the end of the last syllable which precedes another activity of "no-song," like preening or pecking? Or should we accept as a unit any sequence of syllables up to the moment when the pause between two syllables exceeds a critical value? Or should we use the syllable itself as the ultimate unit? As Altmann (1965) has pointed out, the units for classification can be empirically determined: "One divides up the continuum of action wherever the animals do." This in itself can become a laborious and time-consuming task. Nevertheless, purely descriptive criteria for subdividing behavior are to be preferred over motivational or functional ones, because the latter might preclude the discovery of interactions between motivational or functional systems. For example, if in nest-building only one specific hormonal state of the individual is considered, or if only the acts which lead to the final product (the nest) are considered, the role of the nest-building acts in the context of establishing or maintaining a pair-bond might be overlooked or misjudged.

As we continue to observe "the bird" we see that certain behavioral acts are repeated in nearly identical fashion. The surprisingly high degree of stereotypy has been used as a diagnostic characteristic for a special type of behavior, the *fixed action pattern* (Lorenz, 1932; Tinbergen, 1951), and it is the delight of the operant psychologist who uses such stereotyped, unitary events as output of his (black) box. Strangely enough, only a few investigators have attempted to measure the stereotypy of particular acts (e.g., Dane & Van der Kloot, 1964; Schleidt, 1974) or to express the degree of stereotypy of a behavioral sequence in quantitative terms (Altmann, 1965). But, why are behavioral acts so frequently stereotyped and only rarely highly variable? Apparently there are three major functional properties that favor stereotypy:

(1) Any stimulus situation that presents itself to the bird in nearly identical fashion over and over again will elicit a similar response each time (e.g., a food object of a certain kind and size will most likely be handled in nearly the same way whenever it is encountered).

(2) Some parameters of a stimulus object or of the environment, in general, vary within a wide range. They are not responded to in a graded fashion but rather by a varying rate of repetition of a stereotyped element (e.g., a bird that shells seeds does so with a series of stereotyped strikes rather than with one carefully measured blow, and the speed of locomotion is within a certain range adjusted by the rate rather than by length).

(3) A behavioral act that serves the function of communication with another animal must be stereotyped so that its signal is clearly distinguishable from another act of a different meaning; the act must also be stereotyped so that it contrasts with the ambient noise of noncommunicative acts of the transmitting bird and with the environment of the receiving bird in general. Stereotyped acts tend to optimize the signal-to-noise ratio under a given set of conditions and thereby facilitate detection of the signal (Lorenz, 1935).

As we continue to observe the bird over an extended period of time, perhaps throughout its life span, we will note that some behavioral acts remain surprisingly constant, although size and other morphological characteristics might change drastically (e.g., the strutting of the male turkey, Schleidt, 1971; crowing in the male *Coturnix* quail, Schleidt & Shalter, 1973), whereas others will change in response to environmental, especially social, conditions (e.g., the call notes of finches, Mundinger, 1970). Various conceptual frameworks have been suggested to account for constancy or changes in individual behavioral acts during ontogeny. Among the suggested frameworks are the maturation of "innate" behavioral traits, innate meaning here "heritable and environment resistant" (Lorenz, 1965) and the "epigenetic" concept (Kuo, 1932; Aronson, Tobach, Lehrman, & Rosenblatt, 1970).

An additional set of temporal behavioral properties consists of the various cyclic changes in the probability of occurrence of individual acts during a day (circadian rhythms), a month (lunar cycle), or a year (annual cycle), or over periods that are less rigidly controlled by "*Zeitgeber*" from our planetary system, such as the egg-laying cycle of a bird.

Still dealing with a single bird, we can compare its behavior under the influence of various environmental conditions, and study the effect of a variety of natural and artificial stimuli. In this sense, any experiment that tests the influence of a treatment on behavior is based on a comparison between the conditions before and after the treatment is imposed.

II. Two Birds

Looking at more than one bird raises a gamut of problems for the conscientious scientist. First, the chances are about equal that the two birds will be of different sexes. Whereas some species show a marked sexual dimorphism in plumage and in other morphological characteristics (e.g., a fully grown male turkey is about twice as heavy as the adult hen), many other bird species are monomorphic. The observed behavior of an individual might be utterly misleading for recognition of its sex (except for egg-laying, the most reliable indicator of femaleness). In many species, even courtship behavior shows identical elements in both sexes. In some cases, the frequency of occurrence of these elements is sex specific, but in others either "role" can be performed by either sex. For example, turkey poults, a few days old, or even newly hatched, show "precocious" courtship and mating behavior in both sexes, but the characteristic male courtship behavior "strutting" occurs more frequently in male poults than in females (Schleidt, 1970). Two male pigeons in a cage can easily be mistaken for a happily mated pair. However, all this should not distract from the fact that there are often dramatic differences in the behavior of the two sexes.

Second, the two birds might not even belong to the same species and so the issue of species-specific behavior is raised. Since two birds are an insufficient sample to investigate the ramifications of species specificity, we must postpone the discussion.

Even if we can trust that the two birds are members of the same sex and species, are of the same age and in the same internal ("motivational") state, and are both exposed to the same environment, we must not expect that they will behave identically, and therefore we must not pool the data until it can be shown that we drew the sample from a homogeneous population of individuals or from repeats of the same type of behavior or both. What we can consider a homogeneous population will depend upon our concept of population. At an early state of our investigation, we might consider any animal that performs a behavioral act of a certain kind as a part of a population, independent of sex or age of the individual, and without concern for whether it comes from different individuals or is the *n*th repeat of that act performed by the same individual. Many field studies on unmarked animals deal with such vaguely defined samples. One must face similar problems in the laboratory with work done on rare species or on those that are difficult to keep. However, as a rule, one should consider intraindividual and interindividual variability separately, especially since it has been found that parameters of communication signals, which apparently are used by the bird for individual recognition, show a much greater variation among individuals than within any individual (e.g., Schleidt & Maiorana, 1974).

Finally two birds can interact with each other, and the observer thereby benefits from the opportunity to do simultaneous comparisons. To mention but one example, nearly any two birds put together in one cage will establish a "peck order" (e.g., Schjelderup-Ebbe, 1935; McBride, 1958). Before two birds encounter each other, it may be highly uncertain which one will achieve the top rank (some people bet on cock fights), but afterward the result is clear cut. This fact might explain in part the popularity of the concept of peck order among animal psychologists.

III. Many Birds

As the number of birds increases, the problems become more numerous and complex. In fact, one might see the concept of "the bird" seriously threatened, so wide is the range of variety in morphology and behavior among the gamut of different species and other taxonomic groupings. Looking at the morphology and at the behavior of a variety of bird species, we can find only a few characteristics that they all share and that distinguish them from other vertebrates: In essence, these characteristics relate to the ability to fly and are expressed in the features of wing and feather and the associated behaviors of flight and preening. Most other traits can be traced back to the bird's reptilian heritage.

It is important to remember that all species of birds are related through the bond of common ancestry. The closeness of this bond between any two species can be expressed by the number of characteristics they have in common; species of greatest similarity are gathered under one "genus,"

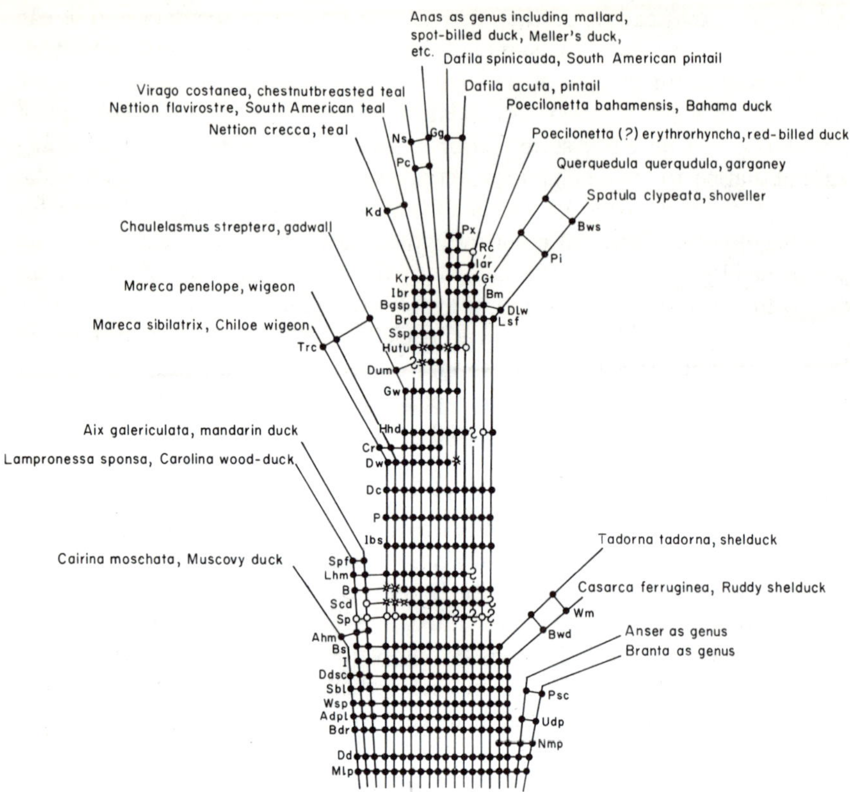

Fig. 2. A comparison of motor patterns of *Anatinae*. The vertical lines represent species; the horizontal lines characters common among them. A cross indicates the absence of a character in a species crossed at the point concerned by a character cross-line. A circle indicates special emphasis and differentiation of the character. A question mark indicates the author's uncertainty. [Adapted from Lorenz, K. *Studies in animal and human behaviour*. Vol. II. Cambridge, Massachusetts: Harvard University Press, 1971. Pp. 112–114.]

similar genera are united in one "family," families in one "order," all the orders in one "class" of birds (Aves), and the resulting hierarchical structure can be visualized as a phylogenetic tree. Often strikingly gross morphological characteristics are used to distinguish among members of larger taxonomic entities, like the structure or form of bones. Closely related species are often distinguishable only by their plumage, and as the number of distinguishing characteristics decreases the phylogenetic relations among their bearers become more tenuous. For further details on taxonomy and evolution see Mayr (1963).

The classic comparative study of the motor patterns of ducks and geese

CHARACTERS

Mlp	monosyllabic "lost-piping"
Dd	display drinking
Bdr	bony drum on the drake's trachea
Adpl	Anatine duckling plumage
Wsp	wing speculum
Sbl	Sieve bill with horny lamellae
Ddsc	disyllabic duckling social contact call
I	incitement by the female
Bs	body-shaking as a courtship or demonstrative gesture
Ahm	aiming head-movements as a mating prelude
Sp	sham-preening of the drake, performed behind the wings
Scd	Social courtship of the drakes
B	"burping"
Lhm	lateral head movement of the inciting female
Spf	specific feather specializations serving sham-preening
Ibs	introductory body-shaking
P	pumping as prelude to mating
Dc	decrescendo call of the female
Br	Bridling
Cr	chin-raising
Hhd	hind-head display of the drake
Gw	grunt-whistle
Dum	down-up movement
Hutu	head-up-tail-up
Ssp	speculum same in both sexes
Wm	black-and-white and red-brown wing markings of Casarcinae
Bgsp	black-gold-green teal speculum
Trc	chin-raising reminiscent of the triumph ceremony
Ibr	isolated bridling not coupled to head-up-tail-up
Kr	"Krick"-whistle
Kd	"Koo-dick" of the true teals
Pc	post-copulatory play with bridling and nod-swimming
Ns	nod-swimming by the female
Gg	*Geeeeegeeeee*-call of the true pintail drakes
Px	Pintail-like extension of the median tail-feathers
Rc	R-calls of the female in incitement and as social contact call
Iar	incitement with anterior of body raised
Gt	graduated tail
Bm	bill markings with spot and light-colored sides
Dlw	drake lacks whistle
Lsf	lancet-shaped shoulder feathers
Bws	blue wing secondaries
Pi	pumping as incitement
Dw	drake whistle
Bwd	black-and-white duckling plumage
Psc	polysyllabic gosling social contact call of Anserinae
Udp	uniform duckling plumage
Nmp	neck-dipping as mating prelude

(Lorenz, 1941) can serve as a representative example of the use of behavioral characteristics in the reconstruction of evolutionary pathways within a relatively close knit taxonomic group. Various behavioral characteristics of the young, and the courtship behavior patterns and their related morphological correlates in adults (such as the bony drum on the male's trachea or the male's plumage) are especially suited to taxonomic work. Differences in feeding habits and related morphological structures (e.g., form of beak, skull, or extremities) are apparently less useful in this group. Assuming that all species have descended from a common ancestral form which subdivided at various instances in the past, we can expect that characteristics shared by many species were acquired earlier than those that are found in only a few species. In Fig. 2, 17 species and 3 genera of ducks and geese are represented as vertical lines horizontally tied together by lines representing shared characteristics, following the general strategy that the more general a characteristic is, the lower the tie is placed in this scheme. This representation shows not only the applicability of the phylogenetic homology concept to behavioral characteristics, but it also illustrates well the relative specificity of certain characteristics in comparably small taxonomic groups.

With respect to brain and behavior, however, a grave disparity on the taxonomic levels of comparison must be pointed out. Modern ethology, as shown in the example of the ducks, centers on the species or on even smaller units such as subspecies or domestic strains, where behavioral differences are most striking. Brain research, morphological or physiological, deals with rather large taxonomic units, with phyla, classes, or orders (e.g., "insect," phylum Arthropoda; "snail," phylum Mollusca; "vertebrate," phylum Chordata; "fish," class Pisces; "frog," class Amphibia; "bird," class Aves; "mammal," class Mammalia). Comparative brain studies on a group of species are exceedingly rare (e.g., Winter, 1963), and very little is known about the morphological correlates of species-specific or taxon-specific behaviors in the central nervous system. We must recognize this void and start to collect the needed information. As with most gaps, it can be bridged from either side: "brain people" must realize that their experimental animal is not just a bird, or even worse, "the bird," but a distinct species, distinctly different from many other birds; at the same time, "behavior people" must turn their attention to behavioral acts which are shared by larger taxonomical units. Note that ethologists tend to overemphasize the idea of *species specificity* of particular acts by applying this term to cases where it is evident that the acts are common to a wider taxonomical unit; in such cases the term *taxon-specific* behavior in general and *genus-specific* behavior (or whatever the common level might be) should be used if the act is known to occur in all species of this genus.

IV. Birds and Nonbirds

By now it should be evident that I advocate the position that any kind of comparison can help us to understand systems or organisms so long as we find a "common level" on which to draw the comparison. It can be done on the basis of similar function or of common ancestry. For example, a comparison of the flight behavior of a swift and a bat, or of a swift and a nighthawk, or of two different species of swift, can yield new insights. However, if the level of comparison is vague or is based on excessive confidence that either species is completely representative of its class, little insight is gained though the differences observed might be striking. Consider for a moment, the speed with which wrens, swifts, rats and cheetahs, all forced to walk, might solve a multiple Y maze. Knowing the normal environment of each species, it is safe to predict that wren and rat will do very well on this type of problem and a comparison of the two species could reveal interesting differences in performances. However, a comparison of swift and cheetah performance in the same experimental situation is unlikely to further our understanding of the capacities of birds and mammals. The prospect for gaining insight does not improve even if there is a multitude of vague similarities between the two species that we are tempted to compare, e.g., the matched pair "bird" and "man."

"Through a certain anthropocentricity of interest we are likely to choose an organism as similar to man as is consistent with experimental convenience and control [Skinner, 1938, p. 47]." The rat, a long-time favorite of psychologists, is threatened in its leading role by "the bird," to some extent because we feel "the bird" is a better model for man ("the man"). On what evidence is this shift in fashion founded? Here is a list of some of the more striking similarities between man and bird: vision (colors; Purkinje shift), hearing (frequency range; rich communication system), olfaction (microsmatic), locomotion (bipedal), and social structure (peck-order; territoriality). However, a closer look at each of these properties shows us that the similarity holds only for gross functional aspects, and that structures of the systems and their dynamic properties are vastly different because of their different evolutionary histories. Color vision apparently has evolved several times independently within the animal kingdom and is achieved by different means; in birds, the spectral sensitivity of the individual cone is apparently set by an oil droplet which acts as a filter, while in man the same effect is produced by the visual pigments in the cone itself. The whole hearing system is vastly different in bird and man, down to the mechanics of the inner ear, which in birds is better equipped to handle fast transients. Whereas birds apparently are microsmatic, man and other primates are reasonably sensitive to olfactory cues. Bipedality is achieved by very different means, as closer inspection even by a layman's eye shows (find a bird's knee!). Of the behavioral traits, territoriality has become a popular issue, and it offers a good opportunity to point out where the anthropocentricity of our interest can get us. The human idea of territory is a strange mixture of agricultural efficiency, personal privacy in the realm of bed and hearth ("my home is my castle"), and national identification, leading to a specific concept of real estate where plots of land are demarcated by straight, fixed borders. When the concept of territoriality in birds was originally proposed (Howard, 1920), the defended border was an essential part of the concept and with it a rather anthropomorphic, real-estate-like idea of space nicely divided into parcels. This bias proved so strong that the actual distributions of essential behaviors, e.g., locations of feeding activity or of contests between neighbors, still have never been systematically mapped as a test of the validity of the "parcel concept."

Anthropomorphism and anthropocentricity, though at first sight acceptable from the point of view of relevance for the understanding of our own problems, can thus turn into severe handicaps. I believe that the only way one can achieve relevance is to uncover the principles which govern the behavior of a variety of species first, and then test whether or not they apply to human behavior.

References

Altmann, S. A. Sociobiology of rhesus monkeys. II. Stochastics of social communication. *Journal of Theoretical Biology*, 1965, **8**, 490–522.

Aronson, L. R., Tobach. E., Lehrman, D. S. & Rosenblatt, J. S. *Development and evolution of behavior*. San Francisco: Freeman, 1970.

Dane, B. & Van der Kloot, W. G. An analysis of the display of the goldeneye duck (*Bucephala clangula* L.). *Behaviour*, 1964, **22**, 282–328.

Howard, E. *Territory in bird life*. London: Collins, 1920.

Kuo, Z. Y. Ontogeny of embryonic behavior in Aves. IV. The influence of embryonic movements upon the behavior after hatching. *Journal of Comparative Psychology*, 1932, **14**, 109–122.

Lorenz, K. Betrachtungen über das Erkennen arteigener Triebhandlungen der Vögel. *Journal für Ornithologie*, 1932, **80**, 50–98.[2]

Lorenz, K. Der Kumpan in der Umwelt des Vogels. *Journal für Ornithologie*, 1935, **83**, 137–213 289–413.[2]

Lorenz, K. Vergleichende Bewegungsstudien an Anatiden. *Journal für Ornithologie*, 1941, **89**, 194–293.[2]

Lorenz, K. *Evolution and modification of behavior*. Chicago: Univ. of Chicago Press 1965.

Mayr, E. *Animal species and evolution*. Cambridge, Massachusetts: Harvard Univ. Press, 1963.

McBride, G. Relationship between aggressiveness and egg production in the domestic hen. *Nature*, 1958, **181**, 858.

Mundinger, P. Vocal imitation and individual recognition of finch calls. *Science*, 1970, **168**. 480–482.

Schjelderup-Ebbe, T. Social Behavior of birds. In C. Murchison (Ed.), *A handbook of social psychology*. Worcester, Massachusetts: Clark Univ. Press, 1935.

Schleidt, W. M. Reaktionen von Truthühnern auf fliegende Raubvögel und Versuche zur Analyse ihres AAM's. *Zeitschrift für Tierpsychologie*, 1961 **18**, 534–560.

Schleidt, W. M. Precocial sexual behaviour in turkeys (*Meleagris gallopavo* L.). *Animal Behaviour*, 1970, **18**, 760–761.

Schleidt, W. M. *Meleagris gallopavo domesticus* (Meleagrididae), Elemente des Sexualverhaltens bei Küken nach Injektion von Testosteron, Filmbeiheft der *Encyclopaedia Cinematographica*, 1971, **E 488** Göttingen.

Schleidt, W. M. How "fixed" is the fixed action pattern? *Zeitschrift für Tierpsychologie*, in press, 1974.

Schleidt, W. M., & Maiorana. V., Intra-individual and interindividual variability in communication signals. In preparation. 1974.

---

[2] Note: English translations of these articles are available in: Lorenz, K. *Studies in animal and human behavior*. Cambridge, Massachussetts: Harvard Univ. Press, Volume I, 1970; Volume II, 1971.

Schleidt, W. M. & Shalter, M. Stereotype of a fixed action pattern during ontogeny in *Coturnix coturnix coturnix. Zeitschrift ffür Tierpsychologie*, 1973, **33**, 35–37.
Skinner, B. F., *The behavior of organisms*. New York Appleton, 1938.
Tinbergen, N. *The study of instinct*. Oxford: Oxford Univ. Press, 1951.
Winter, P. Vergleichende qualitative and quantitative Untersuchungen an der Hörbahn von Vögeln. *Zeitschrift für Morphologie und Ökologie der Tiere*, 1963, **52**, 365–400.

The Comparative Study of Brain–Behavior Relationships

William Hodos
University of Maryland

"The step from neural structure to an understanding of the details of behavior is . . . obscure, but I believe that the general principles of organization are within our grasp. Progress toward an understanding of the evolution of behavior depends on our ability to analyze the properties of the nerve net and to discover the phylogenetic differences in its structure [Lashley, 1949, p. 475]." With these words, Karl Lashley ended his thoughtful and challenging paper, "Persistent Problems in the Evolution of Mind." In the twenty five years that have followed the publication of this paper, we have made great strides in analyzing the electrical and chemical "properties of the nerve net" and some of the phylogenetic differences in its structure, but the chasm in our understanding of the relationship between brain structure and behavior yawns as wide for us as it did for Lashley. It is fortunate that the persistence of the problem has been matched by the persistence of the scientist. This volume is yet another attempt to draw some insight into the mystery from our combined pool of knowledge.

A view held by most of the authors in this book is that an approach based on comparative studies and an evolutionary perspective may untimately provide a breakthrough in our search for a deeper understanding of the complex relationship between brain and behavior. The purpose of this chapter is to suggest a number of strategies for experimentation and conceptualization.

I. Why Comparative Studies?

The first point that I would like to discuss is the value of the comparative approach as a research strategy. First, the comparative approach allows the experimenter to investigate the diversity of a phenomenon in nature and to determine the generality of his or her conclusions. From a strictly behavioral point of view, the comparative approach can lead to a behavioral taxonomy similar to the conventional taxonomy based on morphological similarity. Furthermore, behavioral characteristics can be used as a means of differentiating species that cannot be otherwise discriminated on the basis of morphology (Welty, 1963). However, a more important value of the comparative approach lies in the unique advantages that it provides for relating structure and function. For example, Cobb (1963, 1964) was able to establish a relationship between vocalization ability and the differentiation of the mesencephalic component of the auditory system. Similarly, Bang and Cobb (1968) suggested a relationship between the olfactory ability of various birds and the development of their olfactory bulbs. Donner (1951) was able to show a relationship between the visual acuity of various passerine birds and morphological characteristics of the retina. Although studies of this type suffer from the inherent weaknesses of any correlational study, they nevertheless can form the basis of powerful working hypotheses that can guide direct attacks on the structure–function question. Thus, Cobb's speculations on which subdivision of the avian mesencephalon should be regarded as comparable to the mammalian inferior colliculus were in large measure confirmed by the subsequent silver impregnation studies of Karten (1967, 1968).

A closely related benefit of the outcomes of comparative studies is their implication for understanding evolutionary trends. These implications may be of two sorts: first for the understanding of general mechanisms of ecological adaption and survival; second for the reconstruction of evolutionary trends in specific lineages. Since there has been some misunderstanding in the literature of the difference between these two interpretations of comparative data, I will take this opportunity to point out that the "adaptive value" interpretation is usually based on the observation that a rank order of the species studied according to morphological complexity is correlated with a rank order of those same species according to behavioral complexity or ability. This is a reasonable approach. However, a number of investigators have also interpreted such an ordinal arrangement as representing the evolutionary history of that particular brain–behavior relationship.

Whereas a progression from simple to complex may seem to be an appropriate way for nature to have gone about the business of evolution, the field of paleontology is replete with examples of historical progressions from complex to simpler (Simpson, 1967; Romer, 1966). When we think of evolution we often have in mind the progression from protozoans to metazoans or the development of complex, specialized organ systems and cell types from simpler, more generalized types. However, one should consider that the earliest vertebrates were already quite complex organisms, and a feature that seems to have been common to a number of vertebrate lineages has been a trend in the direction of simplification of some systems. Thus we should not assume, in the absence of other evidence, that a particular ordinal arrangement of behavioral and/or morphological features represents an historical progression. On the other hand, if the animals that have been studied have been selected because they are fairly direct descendents of a common ancestral lineage, and if they are ranked in order of the appearance in the fossil record of their immediate precursor, then the observed morphological or behavioral ranking might be considered as representing something like an historical sequence. However, even under these circumstances, great caution is required in drawing conclusions, because one cannot determine from a single species whether a given character has been retained in a relatively unmodified form from ancestral times, or whether the character represents an adaptive change peculiar to that particular species and not present in ancestral forms. This means that quasi-evolutionary sequences, in which living animals are intended to represent an historical progression, are extremely tenuous if only a single species is used to represent each stage of development. Moreover, no single living animal can be regarded as representing a stage in the development of any other living animal (Nelson, 1969). Thus we must infer what the brains and behaviors of the ancestors were probably like from the study of as many different descendent species as possible. The more species studied within a given taxon, the greater will be the liklihood that we will be able to differentiate between the ancestral characters that will serve as the basis of our inference and the derived characters that were not present in the ancestors. An important aid in discriminating ancestral from derived characters is the study of descendents of other lineages derived from the same stem. I will return to this point later.

The net result of this "phylogenetic arithmetic" (holding ancestral characters constant and subtracting derived characters) is a hypothetical creature that may be described as a "morphotype" (Nelson, 1969, 1970). A morphotype may not closely resemble any actual animal now living or that ever lived, although it would clearly be recognized as a relative of an actual animal. It is the embodiment of the defining characters of a given taxon. Since much of modern taxonomy is based on phyletic relatedness, we assume that the morphotype of the extant animals will be rather similar to the morphotype of the ancestors, i.e., the archetype.

As an illustration of how this process might work, one could attempt to formulate the morphotype of the modern automobile. Such features as fenders, wheels, engine, doors, steering wheel, etc., would emerge as common features of all automobiles and would be included among the ancestral characters. Fender shape, front grill design, hub caps, etc., would clearly fall into the category of derived characters. At the conclusion of our survey, we would want to compare our automobile morphotype with the Model-T Ford. The morphotype would certainly not look like a Model-T, even though it would have many of the characteristics of this ancestral automobile. To push the analogy somewhat further, consider how much greater the resemblance between the morphotype and the Model-T would be if, in addition, we were to base our reconstruction on other groups of vehicles derived from the same ancestral stem such as trucks.

While this analogy is not perfect, it does illustrate both the surprising degree of precision as well as the disconcerting degree of uncertainty that exist in evolutionary reconstruction. It also points up the need to have data on as wide a variety of species as possible. On the other hand, no single investigator may be able to study a sufficient number of species in his or her own laboratory. Thus the development of a morphotype must be the synthesis of activities in a number of laboratories.

II. Which Animals Shall We Compare?

A fundamental question in any comparative investigation is which animals should be studied? Should they be selected at random, or for practical reasons, or according to theoretical considerations? Random selection, or selection of an "interesting" animal, would probably result in considerable misdirection of effort, although some worthwhile data could accrue. Practical reasons must always be viewed seriously, but should not be permitted to strangle innovation. Far too often my question, "Why are you studying rats?" has been answered with, "Well, we already had this rat apparatus in the lab . . . ." In my opinion, theoretical considerations should be primary in the selection of species. If the orientation of the research is one of adaptation, then the species should be selected on a basis of their position along the particular behavioral, morphological, or physiological dimension that is the independent variable in the research. Animals that show extensive development of some morphological feature should be behaviorally compared with closely related forms that do not show such hypertrophy. Animals that exhibit a highly developed form of behavior should be morphologically or physiologically compared with animals that display an inferior capacity for the behavior. On the other hand, if the experimenter is interested in a phylogenetic perspective of his comparisons, the choice of species must be dictated by the conclusions of paleontologists and systematic biologists as to the genealogies and phyletic affinities of the various groups of extant forms.

In fields as inexact as systematics and paleontology, differences of opinion among experts are common, as are changes of opinion. The student of brain–behavior phylogeny must be prepared to act accordingly in the face of these uncertainties and to be flexible in reinterpreting data as new light is shed on the vertebrate family tree. The brain–behavior evolutionist must also periodically check the paleontological and systematic literature to keep abreast of current thought. For example, the relationship of the modern amphibians to the amphibian ancestors of reptiles, birds and mammals is very uncertain and is continually being reevaluated (Ørvig, 1968; Hecht, 1969).

This brings us to the question of which specific groups of animals are appropriate for use in various types of phyletic comparisons. The four major radiations of vertebrates are the cyclostomes (lampreys, hagfishes), the chrondrosteans (sharks, rays), the actinopterygians (ray-finned fishes including bony fishes) and the sarcopterygians (fleshy-finned fishes and tetrapods). There is no strong evidence that any of these groups should be regarded as representing stages in the development of any other, since they all first appear in the fossil record at about the same time (Romer, 1968). A reasonable possibility is that all four of these groups are descended from a common ancestral stock, of which we have no direct knowledge. Nevertheless, comparisons among these major lineages are of great importance since features common to all four lines may be presumed to have been present in the very earliest vertebrates. Of these, the actinopterygians have been the most neglected considering the vast number of species and the wide range of adaptations that the bony fishes have developed. They are an ideal group for the study of the relationship between structure and function. Because descendents of various stages of actinopterygian development have survived to the present time, the possibility exists for the selection of species according to the time of appearance of the lineage in the fossil record or probable phyletic affinity as determined by morphological similarity.

As I mentioned previously, the relationship between modern amphibians and the rest of the tetrapod lineage is very uncertain. They do not closely resemble the amphibian ancestors of reptiles and their use as representatives of these ancestors is very hazardous. Similarly, the earliest reptiles have left no direct, relatively unmodified descendents; perhaps their closest living relatives are the turtles (Hecht, 1969). The remaining groups of living reptiles are rather more remote from the ancestral reptilian lineage, but closely related to these latter are the birds. Mammals are derived from a different branch off this early reptilian stock. Thus many mysteries about the development and organization of the nervous system and behavior of the amniotes would be unraveled if we were to have accurate information about what these characters were like in the earliest reptiles. If sufficient data were available, we could construct a morphotype of modern reptiles to represent a reptilian archetype. However, the image of this hypothetical creature would be made much sharper if its characters were also based on a large group of closely related descendents of the stem reptiles—the birds. Thus common features of the brains and behavior of living reptiles, birds and mammals can provide some insight into what the brains and behavior of the common ancestral reptiles were like.

On the negative side of the ledger, so far as birds are concerned, the systematics of modern birds have been relatively neglected and the fossil record is rather poor (Bock, 1969). Thus, we have no clear idea as to which of the many orders of living birds, if any, can be regarded as direct, relatively unchanged descendents of the earliest modern birds (Bock, 1969). On the other hand, this seeming disadvantage can have its positive aspects, for the narrow, monophyletic origin of birds and the high degree of morphological uniformity that makes the systematics of birds so difficult, may put us on somewhat safer ground when we make general statements about "the avian brain" than when we talk about "the mammalian brain" or "the reptilian brain."

Before I turn from the question of which animals to compare, I want to mention the concept of a "phylogenetic scale" or "animal series." As I have pointed out in detail elsewhere (Hodos, 1970; Hodos & Campbell, 1969), the concept has persisted for so long because of its seductive simplicity and its superficial similarity to the phylogenetic trees of the paleontologists and systematic biologists. We are all familiar with pronouncements such as, "As we ascend the phylogenetic scale, such and such events take place." In the pre-Darwinian era of biological thought, each animal was regarded as having some sort of natural rank in the hierarchy of living things (Lovejoy, 1936; Wightman, 1950). As the idea of evolution gained acceptance, the hierarchy gradually became transformed into a historical sequence. Thus the familiar rat–cat–monkey–man hierarchy became an evolutionary sequence. The powerful influence of this idea can be seen in the prevalence of the opinions that some animals are "the higher animals" while others are "lower." Monkeys are often described as "subhuman primates" and birds are included with the "submammalian" forms. The concept of a unilinear, nondivergent phylogenetic scale has no scientific status and is only coincidentally related to the known facts of evolution. Unfortunately the widespread acceptance of this way of thinking about the relationships among animals has had a damaging effect on theoretical development in comparative psychology, comparative neuroanatomy, and comparative physiology (Hodos & Campbell, 1969). As students of the comparative approach, we must make every effort to avoid this conceptual trap.

III. What Shall We Compare?

The final theoretical point that I will address is the matter of which are the appropriate features of the brain and behavior to be compared. At the simplest level of analysis, one could rightly argue that any features may be compared for almost any purpose. The retina of an eagle could be compared meaningfully to the liver of a codfish, if the chemistry of Vitamin A were the subject of the investigation. However, in most comparative studies, we are interested in a comparison of equivalent features. But simply asking for a statement of equivalence is not sufficient; we must also know the basis of the equivalence.

A term closely associated with the idea of equivalent features is "homology." This term has undergone a dramatic change in meaning during the past century. It was originally conceived in the pre-Darwinian era by Owen (1843), who later became one of Darwin's bitterest opponents. In its original conception, the term "homologue" was used to refer to "the same" organ in different species. The basis of this equivalence was structural similarity. As the evolutionary views of Darwin and Wallace began to gain acceptance in the scientific world, a new meaning of the term homology began to emerge. This meaning was based on the derivation of a particular structure in two species from its precursor in a common ancestor.

Let me take a moment to point out the differences between the two meanings of homology. The original structuralist definition has its roots in the pre-evolutionary doctrine of nonmutability of species and in typological anatomy, which were the prevalent theoretical positions in comparative anatomy in the mid-nineteenth century. This definition holds that structures that are morphologically similar are "the same," irrespective of how they came to be the same. The phylogenist definition of homology is also based, in part, on structural correspondence, but limits the term to those structures that are the result of inheritance from a common ancestry (Haas & Simpson, 1946). The phylogenists hold that structural correspondences, not based on inheritance from common ancestors, should be regarded as examples of "homoplasy"; i.e., the result of similar environmental pressures producing similar structural adaptations in more or less unrelated species. Among the processes that produce homoplasy are convergence and parallelism (Ghiselin, 1969).

One of the causes of considerable ambiguity in the use of the term homology stems from the fact that adherents of both the structuralist and phylogenist usages of the term are active contributors to the scientific literature today (Boyden, 1943, 1969). Since both of these definitions are based to some extent on structural correspondence, the reader often may not be able readily to discern the intended meaning of the term. Moreover, a number of scientists, particularly in the neuroanatomical and behavioral fields, seem to be unaware of the controversy and thus fail to indicate whether the term is used in a structural or phyletic sense. I should add that both the structuralists and the phylogenists are each asking legitimate, meaningful questions, but the questions are different, as frequently are the answers.

Campbell and Hodos (1970) have suggested a definition of homology that could be applied to the nervous system and behavior. Our bias toward the phylogenetic meaning of homology is obvious in the definition:

Homology: *Structures or other entities are homologous when they could, in principle, be traced back through a genealogical series to a stipulated common ancestral precursor, irrespective of morphological similarity.*

Structures that are homologous and also morphologically similar are regarded as "homogenous." Structures that are morphologically similar, but that cannot in principle be traced back to a common precursor are "homoplastic." These definitions suggest that merely describing two characters as homologous is an insufficient description (Ghiselin, 1966). For example, the wings of a crow, a sparrow and a bat are all homologous, but for different reasons. The crow and sparrow wings are homologous as derivatives of the *wings* of their common bird ancestors. Whereas, the birds' wings are homologous with the bat's wings as derivatives of the *forelimbs* of their common ancestors in the stem reptiles. The term "analogy" refers to similarities in *function* independent of common phyletic origin.

The problem of behavioral homology is considerably more difficult than that of morphological homology. In his recent paper on behavioral homology, Atz (1970) has clarified the nature of the problem by pointing out that homology is basically a morphological concept and thus the degree to which the idea of homology is applicable to behavior depends on the extent to which behavior can be characterized in morphological terms. Yet one cannot deny the existence of homologous behavior since we know that behavior is a character of organisms that is responsive to the pressure of natural selection and survival. Since behavior does not exist independently of structure, the resolution of this paradox may result from efforts similar to those of the participants of this conference, i.e., the attempt to relate behavior to its morphological substrate. Thus, homologous behaviors would be those that could be related to specific homologous morphological entities.

Similarity of appearance of the behavior would not be a necessary condition for behavioral homology, just as it is not a necessary condition for morphological homology. Similar appearing behaviors, but not related to homologous morphological entities would be regarded as homoplastic. Thus the behavior of lifting a food object and inserting it in the mouth would be homologous in a rhesus monkey and a chimpanzee, since the hands of these animals can be traced to the hands of ancestral prosimians. The comparable behaviors of a monkey and a raccoon would be homoplastic, according to the preceding definition. Although the hands of monkeys and raccoons are homologous as forelimb derivatives from ancestral insectivores, their structure as hands are homoplastic since hands were not a feature of their common ancestor. The ingestive behavior of an elephant would be regarded as analogous to that described for the monkey and raccoon, since it serves the same function for the animal, but does not have a similar behavioral topography.

An illustration closer to the topic of this volume may be found in the studies that Karten and I have carried out on the visual system of pigeons (Hodos & Karten, 1966, 1970; Hodos, 1969; Karten, 1969; Karten & Revzin, 1966; Karten & Hodos, 1970). We have found that pigeons possess two major ascending visual projection pathways to the telencephalon. These pathways seem to be comparable from a number of points of view to the dual pathways described for mammals (Schneider, 1969; Diamond & Hall, 1969). The electrophysiological observations of Revzin (1969) have pointed to one type of functional correspondence between pigeons and mammals. Our behavioral observations of the similarity of lesion effects in comparable cell groups of mammals and pigeons, suggests another type of functional correspondence.

An important question for understanding the phyletic development of this dual mechanism is: Did ancestral reptiles possess this dual mechanism, or did these pathways evolve independently in birds and mammals? This question opens a Pandora's box of additional questions.

> *Question:* Is the condition found in pigeons typical of all or most birds?
> *Answer:* We do not know, since owls are the only other birds that have been studied to date using this battery of techniques. However, in view of the owl data and the general homogeneity of birds, we have reason to suspect that the answer to this question probably is yes.
> *Question:* Is the condition found in pigeons the same as in reptiles?
> *Answer:* A tectofugal component thus far has been identified in turtles and portions of a thalamofugal component have been reported in several reptilian orders. However, the survey is far from complete. Nor have the higher order neurons been traced as far in reptiles as they have in pigeons. From the data at hand, the condition found in reptiles seems to be quite similar to the condition found in pigeons.

These various observations have led us to propose, as a working hypothesis, that at least the tectofugal components of these dual pathways of birds and mammals may be homologous as derivatives of the same cell populations in ancestral reptiles that are represented by the tectofugal pathway in living reptiles. If verified, this hypothesis would suggest that a retino–

tectal–thalamic–telencephalic visual pathway has existed at least since the middle Carboniferous, which is the presumed time of separation of the reptilian ancestors of mammals from the stem reptilian stock. Our suspicion is that the thalamofugal pathway is at least as old as the tectofugal pathway. If correct, these conclusions would be of great importance in understanding the origins of neothalamus, neocortex, and the behavioral functions associated with these cell groups.

I shall close with another quotation from Lashley's 1949 paper: "The evolution of mind is the evolution of nervous mechanisms, but only the simpler of these can yet be analyzed directly. Comparative studies of the brain and behavior are, therefore, still largely separate in method and problems [p. 461]." I think that Lashley would agree that a volume such as this clearly represents a recognition that comparative studies of brain and behavior have common problems that can be solved by common methods. I think that he would also agree that we may be on the right road to a greater understanding of the enormous complexities that confront us.

References

Atz, J. W. The application of the idea of homology to behavior. In L. R. Aronson, E. Tobach, D. S. Lehrman & J. S. Rosenblatt (Eds.), *Development and evolution of behavior*. San Francisco: Freeman, 1970.

Bang, B. G. & Cobb, S. The size of the olfactory bulb in 108 species of birds. *The Auk*, 1968, **85**, 55–61.

Bock, W. J. The origin and radiation of birds. *Annals of the New York Academy of Sciences*, 1969, **167**, 147–155.

Boyden, A. Homology and analogy: A century after the definitions of "homologue" and "analogue" of Richard Owen. *Quarterly Review of Biology*, 1943, **18**, 228–241.

Boyden, A. Homology and analogy. *Science*, 1969, **164**, 455–456.

Campbell, C. B. G. & Hodos, W. The concept of homology and the evolution of the nervous system. *Brain, Behavior, and Evolution*, 1970, **3**, 353–367.

Cobb, S. Notes on the avian optic lobe (*tectum and nucleus mesencephalicus lateralis*) *Brain*, 1963, **86**, 363–372.

Cobb, S. A comparison of the size of an auditory nucleus (*n. mesencephalicus lateralis, pars dorsalis*) with the size of the optic lobe in twenty-seven species of birds. *Journal of Comparative Neurology*, 1964, **122**, 271–280.

Diamond I. T. & Hall, W. C. Evolution of neocortex. *Science*, 1969. **164**, 251–262.

Donner, K. O. The visual acuity of some passerine birds. *Acta Zoologica Fennica*, 1951, **66**, 3–40.

Ghiselin, M. T. An application of the theory of definitions to systematic principles. *Systematic Zoology*, 1966, **15**, 127–130.

Ghiselin, M. T. The distribution between similarity and homology. *Systematic Zoology*, 1969, **18**, 148–149.

Haas, O. & Simpson, G. G. Analysis of phylogenetic terms with attempts at redefinition. *Proceedings of the American Philosophical Society*, **1946**, **90**, 319–245.

Hecht, M. K. The living lower tetrapods: their interrelationships and phylogenetic position. *Annals of the New York Academy of Sciences*, 1969, **167**, 74–79.

Hodos, W. Color discrimination deficits after lesions of the nucleus rotundus in pigeons. *Brain, Behavior & Evolution*, 1969, **2**, 185–200.

Hodos, W. Evolutionary interpretation of neural and behavioral studies of living vertebrates. In F. O. Schmitt (Ed.), *The neurosciences: Second study program*, New York: Rockefeller Univ. Press, 1970, 26–39.

Hodos, W. & Campbell, C. B. G. *Scala naturae*: Why there is no theory in comparative psychology. *Psychological Review*, 1969, **76**, 337–350.

Hodos, W. & Karten, H. J. Brightness and pattern discrimination deficits in the pigeon after lesions of nucleus rotundus. *Experimental Brain Research*, 1966, **2**, 151–167.

Hodos, W. & Karten, H. J. Visual intensity and pattern discrimination deficits after lesions of ectostriatum in pigeons. *Journal of Comparative Neurology*, 1970, **140**, 53–68.

Karten, H. J. The organization of the ascending auditory pathway in the pigeon (*Columba livia*). I. Diencephalic projections of the inferior colliculus (*nucleus mesencephalicus lateralis pars dorsalis*). *Brain Research*, 1967, **6**, 409–427.

Karten, H. J. The organization of the ascending auditory pathway in the pigeon (*Columba livia*). II. Telencephalic projections of the nucleus ovoidalis thalami. *Brain Research*, 1968, **11**, 134–153.

Karten, H. J. The organization of the avian telencephalon and some speculations on the phylogeny of the amniote telencephalon. *Annals of the New York Academy of Sciences*, 1969, **167**, 164–179.

Karten, H. J. & Hodos, W. Telencephalic projections of the nucleus rotundus in the pigeon (*Columba livia*). *Journal of Comparative Neurology*, 1970, **140**, 35–51.

Karten, H. J. & Revzin, A. M. The afferent connections of nucleus rotundus in the pigeon. *Brain Research*, 1966, **2**, 368–377.

Lashley, K. S. Persistent problems in the evolution of mind. *Quarterly Review of Biology*, 1949, **24**, 28–42. [Reprinted in F. A. Beach, D. O. Hebb, C. T. Morgan & H. W. Nissen (Eds.) *The neuropsychology of Lashley*, New York: McGraw Hill, 1960.]

Lovejoy, A. O. *The great chain of being*. Cambridge, Massachusetts: Harvard Univ. Press, 1936.

Nelson, G. J. Origin and diversification of teleostean fishes. *Annals of the New York Academy of Sciences*, 1969, **147**, 18–30.

Nelson, G. J. Outline of a theory of comparative biology. *Systematic Zoology*, 1970, **19**, 373–385.

Ørvig, T. *Current problems of lower vertebrate phylogeny*. Fourth Nobel Symposium. Stockholm: Almqvist & Wiksell, 1968.

Owen, R. *Lectures on the comparative anatomy and physiology of the invertebrate animals*. London: Longman, 1843.

Revzin, A. M. A specific visual projection area in the hyperstriatum of the pigeon (*Columba livia*). *Brain Research*, 1969, **15**, 246–249.

Romer, A. S. *Vertebrate paleontology*, 3rd ed. Chicago: Univ. of Chicago Press, 1966.

Romer, A. S. *The procession of life*. Cleveland, Ohio: World, 1968.

Schneider, G. E. Two visual systems. *Science*, 1969, **163**, 895–902.

Simpson, G. G. *The meaning of evolution*, New Haven, Connecticut: Yale Univ. Press, 1967.

Welty, J. C. *The life of birds*. New York: Knopf, 1963.

Wightman, W. P. D. *The growth of scientific ideas*. Edinburgh: Oliver & Boyd, 1950.

2 ORGANIZATION OF NEURAL SYSTEMS

The Structural Organization of Avian Brain: An Overview

David H. Cohen[1]
University of Virginia

and

Harvey J. Karten[2]
Massachusetts Institute of Technology

I. Introduction

> Among morphological characters there appear to be very great differences in phylogenetic stability; some basic structural patterns have remained relatively unchanged throughout mammalian and even vertebrate evolution, while others have run the scale of imaginable changes. . . . Progress toward an understanding of the evolution of behavior depends upon our ability to analyze the properties of the nerve net and to discover the phylogenetic differences in its structure. . . . We must seek the clue to behavioral evolution in the number and interconnections of the nerve cells or in their biochemical characteristics, not in their gross structural arrangement [Lashley, 1949, p. 475].

These statements of Karl Lashley in his 1949 paper, "Persistent Problems in the Evolution of Mind" are perhaps interpretable as a mandate for intensive investigation of connectional neuroanatomy in a comparative context. It may be of some significance that comparative neurology was in a rather dormant period at the time of that publication, and the literature prior to the

[1] This work was supported by National Institutes of Health, Research Career Development Award No. HL-16579.

[2] This work was supported by National Institutes of Health, Research Career Development Award No. HD-29979.

period strongly emphasized species differences in "gross structural arrangement" with respect to both the external topography and internal structure of the brain. The fortunate development in recent years of histochemical methods and the powerful techniques for the selective staining of degenerating fibers and boutons now enables a rigorous and systematic comparative neuroanatomy, and in all likelihood this is providing the impetus for the present resurgence of interest in comparative neurology. In view of this, it is conceivable that within a few years Lashley's prediction of a quarter of century ago will be realized.

In this spirit, the principal emphasis in this chapter will be upon the morphological organization of the avian brain and in particular its connectional anatomy; functional data will be included on a highly selective basis and generally within the context of an anatomical argument. This clearly reflects the authors' bias that a comprehensive view of the structure of the nervous system is an essential substrate for functional studies and in itself has the potential of elucidating certain basic features of functional organization.

At the outset it is necessary to point out that no attempt will be made to present an encyclopedic description of the avian brain. The best approximation to this is the extensive review provided in Pearson's (1972) excellent volume, *The Avian Brain*. Thus, in no sense does this chapter constitute a textbook of avian neuroanatomy; such an effort would be premature at this time as evidenced by the lack of emphasis on connectional details in Pearson's comprehensive work. Rather, a strategy has been adopted in which specific regions or systems of the avian brain are described as illustrations of general organizational principles. Furthermore, the results to be presented are based mainly upon studies of a single family of birds, *Columbidae*, and more specifically the pigeon; where other species are involved specific note will be made.

Although there have been attempts to review the structure of the avian nervous system in a comprehensive manner (e.g., Jungherr, 1969; Kappers, Huber, & Crosby, 1936; Papez, 1929; Portmann & Stingelin, 1961), these are generally limited, since none is based on data obtained with the more contemporary methods. Cajal's (1952) classic Golgi studies should be singled out as having sustaining value, particularly his descriptions of the cerebellum, retina, spinal cord, and optic tectum of the bird, since results based on the Golgi techniques have been less susceptible to historical limitation. There are reviews of more limited scope that the reader may find of value, including descriptions of the avian spinal cord (Huber, 1936; Nieuwenhuys, 1964; Van den Akker, 1970), cerebellum (Dow & Moruzzi, 1957; Larsell, 1967), optic tectum (Cajal, 1889; LaVail & Cowan, 1971), and telencephalon (Haefelfinger, 1958; Karten, 1969). Furthermore, for a useful overview of the pigeon brain including the most contemporary nomenclature for the major nuclear groups and fiber tracts the reader is referred to the stereotaxic atlas of Karten and Hodos (1967).

To preview briefly the organization of this chapter, the first section deals with the more prominent gross anatomical features of the avian brain. This is followed by a section treating the spinal cord. Relying on selected systems, the subsequent two sections describe the general patterns of ascending and descending projections in the avian brain. Finally, telencephalic organization is discussed at some length, since it is the telencephalon that shows the greatest phylogenetic variation in internal structure and consequently may provide the most provocative information concerning the evolution of the nervous system.

II. General Structural Features

A. External Topography

In external topography the avian nervous systems most closely resemble those of the modern reptiles, a similarity that is particularly striking with respect to head structure and cranial nerve distribution. The olfactory bulb is rudimentary in many avian species, corresponding to the frequently microsmatic nature of birds, and the prominent and highly developed cerebral hemispheres are lissencephalic (Fig. 1). Other notable external characteristics are the median cerebellum and in particular the large, laterally displaced optic lobes; such impressive tectal development undoubtedly reflects the predominance of the visual system in many birds. The brainstem topography and the spinal cord are both representative of the general vertebrate pattern with the exception of the large rhomboid sinus and glycogen body found in the lumbosacral region of the cord. Finally, the paravertebral sympathetic chain also resembles those of the mammals with respect to general characteristics, despite some tendency for a greater variation in the number of ganglia.

B. Internal Structure

1. Telencephalon

The greatest volume of the avian telencephalon is constituted by what has traditionally been described as a striatal complex (e.g., Kappers *et al.*, 1936). This is composed primarily of five major nuclear masses located lateral to the ventricle, which in birds is displaced toward the dorsomedial hemispheric margin (Fig. 2). Thin hippocampal and entorhinal areas occupy

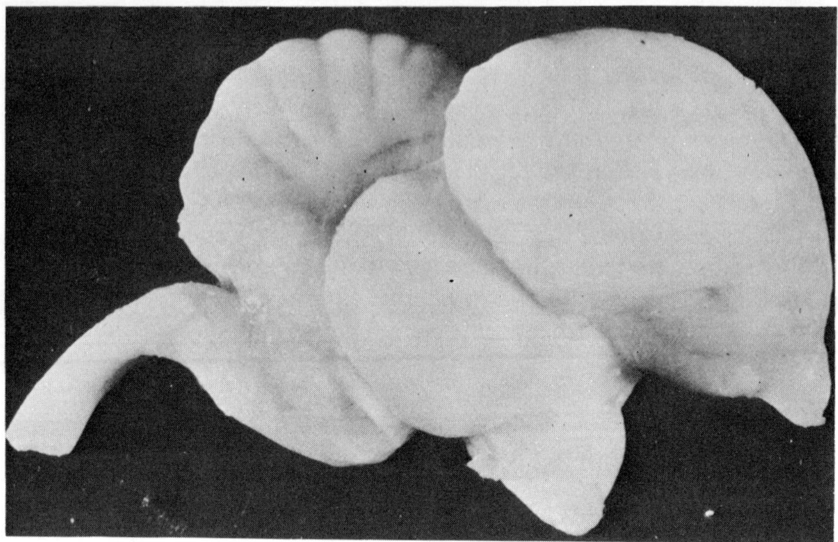

Fig. 1. Lateral view of the pigeon brain. Note the large, laterally displaced optic lobes, median cerebellum and lissencephalic hemispheres. [From Karten & Hodos (1967).]

the medial wall of the telencephalon, and a superficial region of dorsolateral corticoid tissue caps the caudal aspect of the hemisphere. The basal telencephalon is formed principally by the septum and the parolfactory lobe.

The cellular aggregates constituting the five major striatal masses are differentiated largely on a cytoarchitectonic basis and are the hyperstriatum, neostriatum, paleostriatum, ectostriatum, and archistriatum (Fig. 2). The hyperstriatum occupies a dorsomedial position in the hemisphere and may be further differentiated into the hyperstriatum accessorium, largely corresponding to the sagittal Wulst, and a group of ventral hyperstriatal nuclei situated below the lamina frontalis superior. The neostriatum, clearly the largest hemispheric nuclear mass, extends to the caudal pole of the telencephalon and is separated from the more dorsal hyperstriatal complex by the lamina hyperstriaticus. The neostriatum has classically been differentiated into the neostriatum frontale, intermediale and caudale with several subfields notable at each level. The neostriatum is greatest in extent in the caudal portion and gradually decreases in size in the direction of the rostral hemispheric pole. The third cellular mass, the paleostriatum, may be readily divided into augmentatum and primitivum segments; the latter is composed of a distinct group of large neurons, the paleostriatum primitivum proper, and a ventral smaller-celled zone, the nucleus intrapeduncularis. The entire paleostriatal

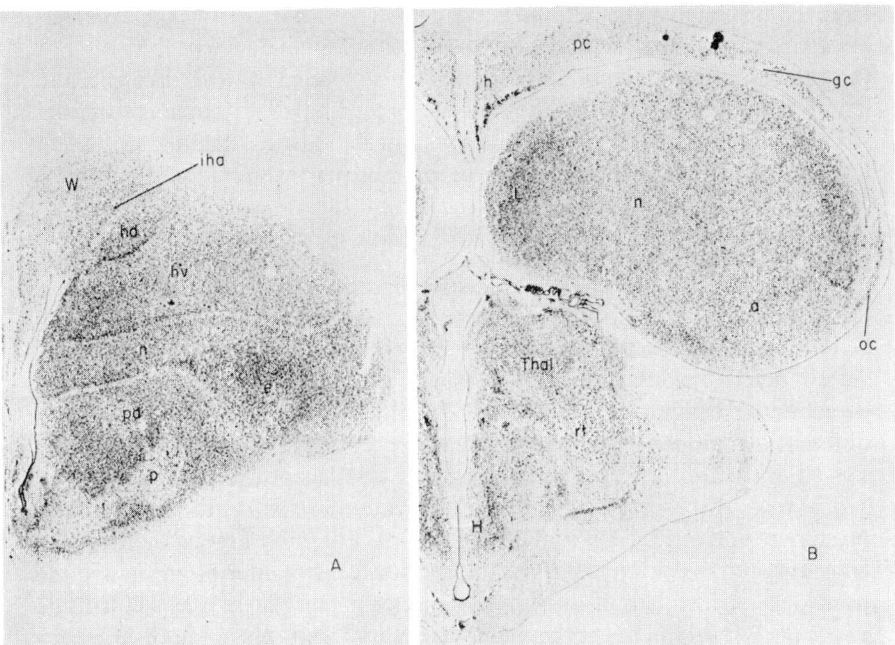

Fig. 2. Nissl-stained transverse sections of the pigeon brain with the section in panel A being more rostral than that in panel B. Abbreviations: a, archistriatum; e, ectostriatum; gc, general cortex; h, hippocampus; H, hypothalamus; hd, hyperstriatum dorsale; hv, hyperstriatum ventrale; iha, nucleus intercalatus of hyperstriatum accessorium; L, Field L of neostriatum; n, neostriatum; oc, olfactory cortex; p, paleostriatum primitivum; pa, paleostriatum augmentatum; pc, parahippocampal cortex; rt, nucleus rotundus; Thal, thalamus; W, Wulst. [From Nauta & Karten (1970).]

complex is separated from the more dorsally situated neostriatum by the lamina medullaris dorsalis. The fourth mass, the ectostriatum, is found embedded in the anteroventral portion of the neostriatum, immediately dorsal to the lamina medullaris dorsalis; finally, the archistriatum is located in the caudal third of the telencephalon ventral to the neostriatum caudale.

It is essential to point out, before proceeding to more caudal regions of the avian brain, that the preceding brief description of the telencephalic internal structure is founded on the more classical views that prevailed for many years. However, of all the divisions of the avian nervous system, our thinking regarding telencephalic organization has undergone the most radical revision in recent years (Karten, 1969; Nauta & Karten, 1970). Not only have the constituent cell groups been subjected to more detailed analysis, but, perhaps more important, the study of the fundamental afferent and efferent connections has generated major reinterpretations of avian telencephalic organization and its relationship to that of mammalian telencephalon. For this reason a separate section of this chapter is directed toward this topic, and it is emphasized at this point that the preceding description of internal structure is intended only to provide a view of the classical landmarks with no intended implications as to basic telencephalic organization.

2. *Diencephalon*

As with the telencephalon, the classic descriptions of the avian thalamus did not have the benefit of detailed connectional data and consequently can be relied upon only to establish a general structural framework, particularly since much of the interpretation of thalamic organization is tightly coupled to an understanding of the telencephalon. The avian thalamus is highly differentiated, much more so than that of the reptiles (Powell & Kruger, 1960; Powell & Cowan, 1961). There are a large number of thalamic nuclei, and rather than discussing each individually, attention is drawn to the more prominent and intensively studied cell groups. Foremost among these are the nucleus rotundus and nucleus ovoidalis of the central inferior group and the principal optic nucleus of the central superior group. These nuclei constitute particularly prominent diencephalic landmarks, but, more important, they are specific relays in the major lemniscal pathways. As will be discussed in greater detail in a subsequent section, the principal optic nucleus and the nucleus rotundus constitute the thalamic relays for the thalamofugal and tectofugal visual pathways, respectively, and the nucleus ovoidalis is the specific thalamic cell group for the ascending auditory pathway.

With regard to the hypothalamus, unfortunately it remains poorly understood. Huber and Crosby (1929) differentiate a number of nuclear groups, but the absence of distinct boundaries in many instances renders such subdivisions somewhat arbitrary without more detailed information concerning the afferent and efferent projections of the hypothalamus.

3. *Mesencephalon*

The most striking feature of the avian mesencephalon is, of course, the extensively developed optic lobes, their supraventricular portion constituting the most elaborately laminated structure of the avian nervous system, namely the optic tectum (Fig. 3). Although the so-called tectal nuclei are not constituents of the optic tectum proper, they do represent important features of the avian midbrain. The most prominent of these are the nucleus mesencephalicus lateralis, pars dorsalis, associated with the lemniscal auditory pathway, the isthmic complex consisting of the isthmi pars principalis magnocellularis and isthmi pars principalis parvocellularis, and the isthmo-

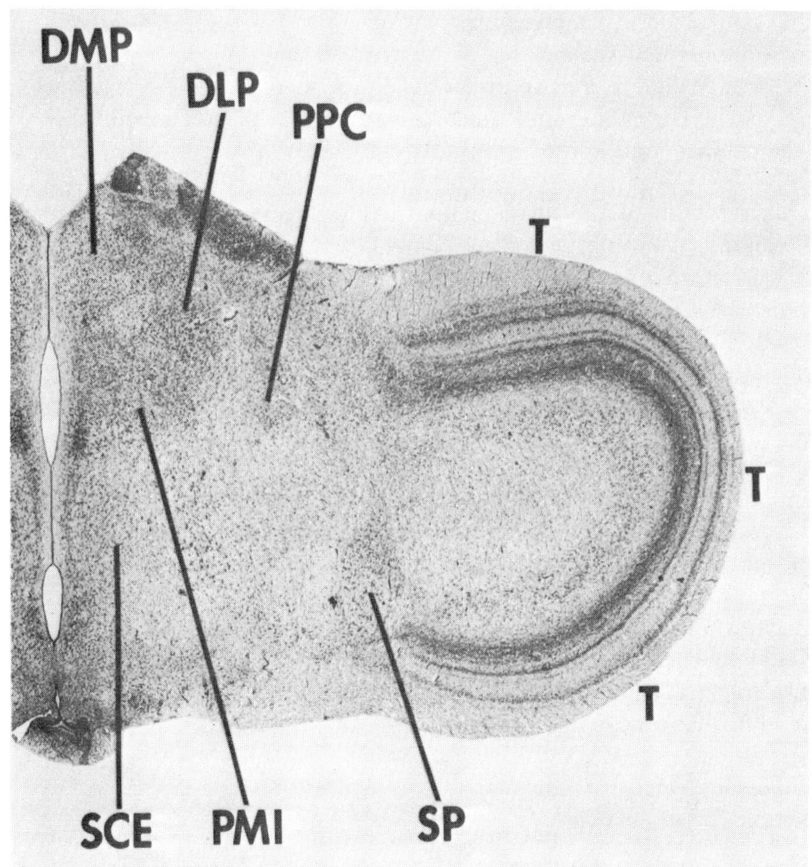

Fig. 3. Nissl-stained transverse section through the pigeon mesencephalon showing the highly laminated optic tectum. Abbreviations: DLP, nucleus dorsolateralis posterior thalami; DMP, nucleus dorsomedialis posterior thalami; PMI, nucleus paramedianus internus thalami; PPC, nucleus principalis precommissuralis; SCE, stratum cellulare externum; SP, nucleus subpretectalis; T, optic tectum.

optic nucleus containing the cells of origin of the centrifugal fibers to the retina. The nucleus mesencephalicus lateralis, pars dorsalis is a large cell mass situated just below the optic ventricle. It comprises the central gray of the interior of the tectum and is partially enclosed within a capsule formed by the lateral lemniscus. The isthmic complex is located in the caudal portion of the optic lobe and receives a massive and topographically arrayed input from the optic tectum. The rostral limit of the magnocellular portion is at the level of the third cranial nerve, while caudally it lies ventral to the nucleus mesencephalicus lateralis, pars dorsalis. The pars parvocellularis of the complex is found ventral to the magnocellular portion and lies in its hilus. Gradually the parvocellular portion elongates, forming a narrow band ventromedial to the pars magnocellularis and ventrolateral to the medial aspect of the nucleus semilunaris. The isthmo-optic nucleus is situated at the dorsomedial margin of the optic tectum approximately at the level of the fourth cranial nerve nucleus.

In addition to the optic tectum and the tectal nuclei, the midbrain contains a group of tegmental nuclei including the oculomotor complex, red nucleus, ectomammillary nucleus and tegmental reticular structures. The red nucleus is found in the medial tegmentum and is almost certainly homologous to the red nucleus of mammalian brain. The ectomammillary nucleus, located just medial to the ventromedial aspect of the tectum, has been clearly identified as the recipient of the basal optic root, a direct projection from the retina.

Before leaving the internal structure of the avian mesencephalon, mention should be made of the pretectal nuclear complex forming a line between the mesencephalon and diencephalon. According to the description of Kappers *et al.* (1936) this complex includes the nuclei pretectalis, subpretectalis, spiriformis medialis and lateralis, and the tectal gray. Although the nature of these nuclei is poorly understood, recent evidence to be discussed later clearly indicates that at best a limited part of the complex can be considered truly pretectal in character. For example, the spiriform nuclei should probably be eliminated as constituents of the pretectal complex.

4. *Rhombencephalon*

The cerebellum is reasonably well developed in birds, being a median structure with distinct transverse folia. According to Larsell (1948), anterior and posterior lobes, separated by a primary fissure, may be clearly distinguished. In the midsagittal plane at its ventral aspect, the folds of the lingula and nodulus are evident. More laterally the nodulus continues into the flocculus on each side, while the uvula merges laterally into the parafloccular lobes. Larsell (1948) has provided a useful nomenclature in which each primary folium is designated by a Roman numeral with secondary folia being identified by letters. The avian cerebellar nuclei are well developed and generally correspond to the mammalian arrangement. Also, the cerebellar cortex follows the well-described vertebrate pattern, and all the major cell types seen in the mammalian cerebellar cortex are present.

In keeping with the generally advanced development of the avian rhombencephalon, the bird, in contrast to the reptile, has a rudimentary pontine homologue (Brodal, Kristiansen, & Jansen, 1950). This forms a band over the ventral surface of the bulb at its rostral extent. The auditory and vestibular structures are well developed, and both superior and inferior olivary nuclei may be identified. Eight of the twelve cranial nerves are found in this region, and, as mentioned previously, their disposition is much the same as that found in reptiles and in mammals. Finally, the pontine and medullary reticular formations are extensive, and the patterning of the reticular nuclei is similar to that observed in mammalian brains (Karten & Hodos, 1967).

III. Spinal Cord

A. External Topography

In external topography the avian spinal cord is rather representative of the general vertebrate pattern, and bears many similarities to the mammalian spinal cord. Although there is marked species variability, one can reasonably propose the generalization that birds tend to have elongated cervical and reduced thoracic regions relative to mammals. For example, the pigeon has 15 cervical, 15 sacro-coccygeal and only 6 thoracic segments (Huber, 1936). Thus, the total number of spinal nerves in birds generally exceeds that of mammals, there being, for example, 39 in the pigeon (Huber, 1936) and 51 in the ostrich (Streeter, 1904). Relevant to this is the absence of a cauda equina in birds, since as in reptiles the avian cord occupies almost the entire length of the vertebral column.

The characteristic cervical and lumbosacral enlargements are present in all birds, and these, of course, correspond to the spinal segments innervating the wings and legs respectively. By way of illustration, in the pigeon the brachial plexus arises from spinal segments 11–15 (C11-T1) and the lumbosacral plexus from segments 21–27 (L1–S3) (Huber, 1936). Although the cervical enlargement generally exceeds the lumbosacral in size, the reverse is the case for cursorial species such as the ostrich.

Perhaps the most unique structural feature of the avian spinal cord is the expansion of the dorsomedian fissure at the lumbosacral enlargement to form the rhomboid sinus (Fig. 4). This sinus contains the distinctive glycogen body whose function is still obscure. Also unique to the avian lumbosacral cord are ventrolateral protuberances designated the accessory lobes of Lachi (Fig. 4). These apparently contain neurons of the marginal paragriseal cell column which is found throughout the cord, but protrudes in the lumbosacral region at the junction of the lateral and ventral funiculi.

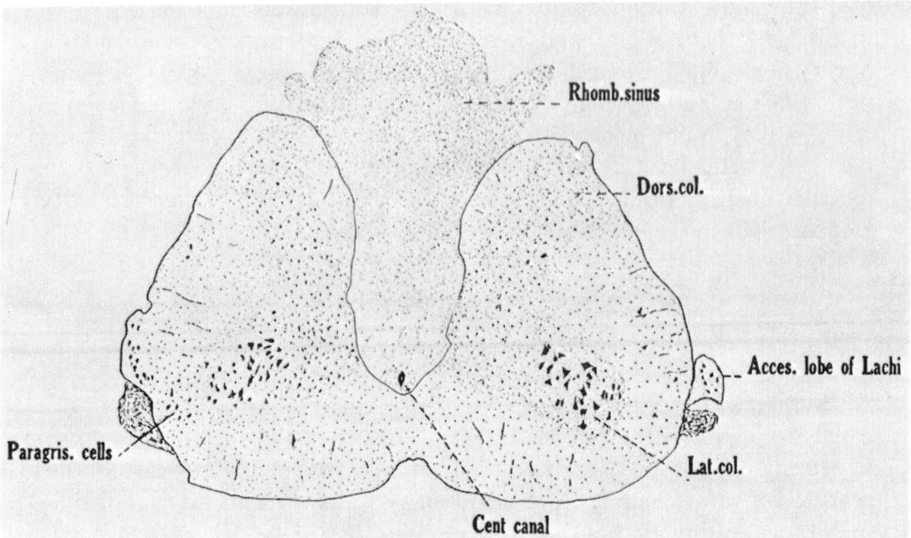

Fig. 4. Transverse section through the lumbosacral enlargement of the pigeon spinal cord. Abbreviations: Acces. lobe of Lachi, accessory lobes of Lachi; Cent, canal; central canal; Dors. col., dorsal horn; Lat. col., lateral column of motoneurons; Paragris. cells, paragriseal neurons; Rhomb. sinus, rhomboid sinus. [From Kappers, Huber, & Crosby (1936).]

B. Peripheral and Segmental Innervation

1. Peripheral Innervation

As in all vertebrates, birds possess numerous free nerve endings which ramify into various types of terminals. The skin and feathers are richly endowed with sensory endings, and, while still moot, it has been claimed that mammalian-like Merkel discs, Krause end bulbs and Meissner corpuscles are present, in addition to the various classes of free nerve endings. Physiological studies of avian cutaneous receptors (Dorward, 1970a) suggest that they may be classified into functional groups that have some correspondence to those described in the mammalian literature. For example, receptors associated with down feathers in the duck appear physiologically to have much in common with the hair receptors of the cat.

Birds also possess two unique and particularly interesting receptors, the Herbst lamellar and Grandry corpuscles (Botezat, 1906; Pearson, 1972), beautifully illustrated by Quilliam and Armstrong (1963). The Herbst lamellar corpuscles are apparently an avian specialization of the Pacinian corpuscle, though they are histochemically distinct (Winkelman & Myers, 1961). These receptors are widely distributed and occur in featherless skin, the feet, at the base of contour feathers, between muscles and connective tissue surrounding the bones of the leg, and most prominently in the skin of the bills of aquatic birds. Functionally they have been shown to transduce vibrational stimuli and to share many physiological properties with the mammalian Pacinian corpuscle (Dorward & McIntyre, 1971). Grandry corpuscles are most frequently found in the tongue and palate and seem to be topographically associated with Herbst corpuscles. While they have been hypothesized to function as touch receptors, both their function and precise relationship to the Herbst corpuscles remain uncertain.

Finally, although little is known with respect to proprioception in birds, it has been demonstrated in the duck that tendon organs behave as in mammals and that muscle spindle receptors with rather typical "in-parallel" behavior are present (Dorward, 1970b).

2. *Segmental Innervation*

Concerning the course of the peripheral nerves in birds, the most detailed descriptions are with respect to the cranial nerves such as the trigeminal (Barnikol, 1954) and the vagus (Cohen, Schnall, Macdonald & Pitts, 1970; Malinovsky, 1962). In this context a particularly useful series of papers dealing with the topographic anatomy of the fowl has been appearing over the years in the *Japanese Journal of Veterinary Science* (e.g., Watanabe, Isomura, & Yasuda, 1967).

As regards the segmental innervation, information on the myotomes of birds would be most helpful but with a few exceptions is not generally available. The avian dermatomal organization is somewhat better described, primarily in the pigeon (Kaiser, 1924) and to a lesser extent the chicken (Yasuda, 1964). We would refer the reader to the paper of Kaiser (1924), since this is a particularly comprehensive report.

3. *Autonomic Nervous System*

We shall conclude this section with a brief discussion of the autonomic nervous system; as in mammals the craniosacral (parasympathetic) and thoracolumbar (sympathetic) divisions are present in birds. Although detailed information is, in most instances, limited, there is no reason to believe that the craniosacral system differs in any significant respect from that of mammals. The cranial portion arises from cell bodies in the motor nuclei of the third, seventh, tenth, and possibly ninth cranial nerves, and the sacral component consists of visceral efferent neurons in the spinal gray which send their axons through the pelvic nerve. The parasympathetic preganglionic axons then synapse in terminal ganglia in close proximity to the target organ. For example, the cardiac portion of the vagus nerve in the bird arises from neurons located in the dorsal motor nucleus of the vagus (Cohen & Schnall, 1970; Cohen *et al.*, 1970) and projects upon neurons located in the intrinsic cardiac ganglion (Ssinelnikov, 1928), as in mammals.

Similarly, the sympathetic or thoracolumbar division of the autonomic nervous system generally resembles that of mammals (Langley, 1904; Macdonald & Cohen, 1970). The cells of origin of the preganglionic fibers are found in a cell column dorsal to the central canal, the column of Terni (Macdonald & Cohen, 1970; Terni, 1923). In the pigeon this column extends from the most caudal cervical intersegment to the most rostral lumbar intersegment, corresponding to the rostrocaudal distribution of mammalian preganglionic neurons. The preganglionic fibers leave the cord through the ventral roots to enter the paravertebral ganglia via one of the rami communicantes; however, their ramification is more restricted than in mammals. Moreover, birds tend to have a greater number of rostral and, to a lesser extent, caudal paravertebral ganglia than mammals (Huber, 1936). For example, in the pigeon there is a ganglion associated with each cervical spinal nerve. A final difference that might be noted is the absence of gray rami in the bird, since all postganglionic sympathetic fibers appear to be myelinated (Langley, 1904).

C. Internal Structure

The shape and size of the spinal gray relative to the funiculi vary at different levels of the cord in rather characteristic fashion, and as in mammals it is possible to identify the spinal level on the basis of these parameters. The major cell groups of the spinal gray resemble those of the mammal in many respects and are more highly differentiated than those of the reptilian cord. Furthermore, Golgi material supports this contention in indicating that the dendrites of neurons in the avian cord are confined primarily to the spinal gray and have a rather restricted distribution.

With respect to the major cell groups, the most comprehensive description is that of Huber (1936) for the spinal cord of the pigeon. He defined a number of major cell columns. First, the motoneurons of the ventral horn are divided into medial and lateral columns, the medial column extending throughout the entire cord and being concentrated in the ventromedial aspect of the ventral horn. These neurons are presumed to innervate the trunk musculature. The lateral column is divided into medial and lateral divisions, and, because of its prominence at the cervical and lumbosacral enlargements, it is presumed to innervate the muscles of the extremities. The visceral efferent neurons of the thoracolumbar system have already been mentioned and appear to be located in the preganglionic column of Terni; this column straddles the midline just dorsal to the central canal and extends from caudal cervical through upper lumbar levels. The visceral efferent neurons of the sacral component of the parasympathetic system are thought to be located in a similar region of the sacral spinal gray.

Another presumptive source of motor fibers is the column of von Lenhossék which is located in the lateral part of the upper cervical gray at the level of the central canal. These neurons are considered to be cells of origin of motor fibers exiting via the dorsal roots. Running throughout the entire cord at the surface of the lateral funiculus at its junction with the ventral funiculus is the marginal paragriseal cell column, and at lumbosacral levels this cell group protrudes into the white matter to form the accessory lobes of Lachi. These neurons are suggested as one source of commissural fibers in the avian spinal cord, another possible source being certain of the scattered paragriseal cells found in the white matter ventrolateral to the anterior horn from caudal cervical to sacral levels. However, in all likelihood, this cell group also includes displaced motoneurons. Finally, a prominent nucleus at the base of the dorsal horn, the dorsal magnocellular cell column, extends from midcervical through sacral levels. Whereas its identification is uncertain, it may well correspond to the mammalian column of Clarke.

More recent analyses of the avian spinal gray have involved a cytoarchitectonic division into zones which in some areas appear to constitute a laminar pattern (Leonard & Cohen, in preparation; Van den Akker, 1970). The scheme of Leonard and Cohen (in preparation), although still considered tentative, is the more detailed of the two and is illustrated in Fig. 5. Along the dorsal margin of the dorsal horn and extending ventrally on its medial and lateral aspects is an area containing small fusiform neurons resembling the cells of Waldeyer; this has been designated as region I. Ventral and parallel to this is an area of cells resembling those of region I but larger; this is designated as region II. Region III is a large area in the head of the dorsal horn containing small ovoid cells. At the base of the dorsal horn and extending into its neck is an elliptical area of medium-sized multipolar neurons designated as region IV. Regions V and VI lie below region IV and contain a heterogeneous collection of multipolar neurons. There is a rather continuous dorsoventral distribution with respect to cell size, the large neurons being situated more ventrally, and this gradual increase in cell size makes it somewhat difficult to delineate regions V and VI with great confidence.

In the ventral horn three major cellular regions may be distinguished. Region VIII is an area of moderate packing density containing numerous large neurons, particularly at the medial aspect. In contrast, region VII contains almost exclusively small cells that are more loosely packed. The gradual transition between these two regions makes their precise delineation difficult, analogous to the problem in distinguishing regions V and VI. The motoneuronal cell groups, region IX, contain characteristically large, deeply staining multipolar neurons, particularly in the lateral cell groups.

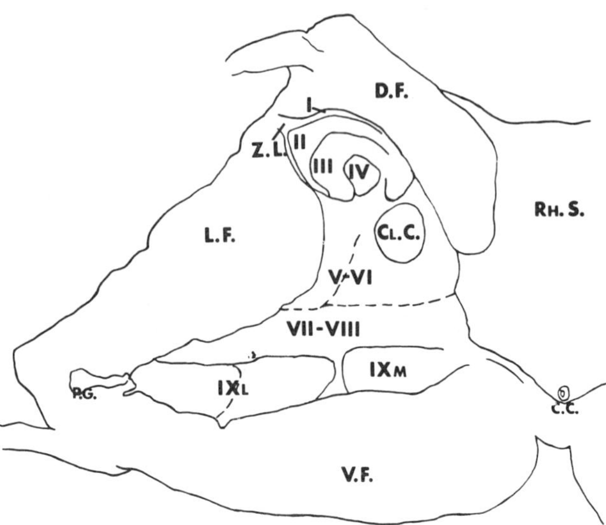

Fig. 5. Schematic diagram of the cytoarchitectonic zones of the pigeon lumbar spinal cord. See text for the definition of the zones. Abbreviations: C.C., central canal; CL.C., column of Clarke; D.F., dorsal funiculus; L.F., lateral funiculus; P.G., paragriseal cells; RH.S., rhomboid sinus; V.F., ventral funiculus; Z.L., zone of Lissauer. [From Leonard & Cohen (in preparation).]

There are variations in this zonal pattern at different levels of the cord, but in the main, the basic pattern is maintained. Also, it should be recognized that certain distinct cell groups have not been included in this description, such as the column of Terni and the marginal paragriseal cell column. For this reason it is necessary that no single approach to describing the avian spinal organization be relied upon entirely at this time, since the nuclear approach is of greater value for certain aspects of the spinal gray and the zonal analysis for other aspects, such as the dorsal horn.

D. Pattern of Dorsal Root Termination

One particular advantage of a zonal analysis is that it provides a convenient foundation for describing the terminal fields in the spinal gray, such as the pattern of dorsal root distribution. This has been studied by both Leonard and Cohen (in preparation) and Van den Akker (1970), and on most points, the two descriptions are in general agreement. For the present discussion the material of Leonard and Cohen (in preparation) will be used.

As in the mammal, the large afferent fibers enter the cord medially, with a lateral bundle consisting of thinner fibers. In experimental material, degenerating fibers can be seen sweeping across the dorsal funiculus posterior to the cap of the dorsal horn. Fascicles enter the spinal gray primarily along its dorsomedial and medial aspects. Degenerating terminals are prominent throughout regions I, II and III, but they tend to be sparser laterally. Similarly, dense degeneration is present in regions IV, V, VI, and the column of Clarke with sparing of their lateral aspects. Region VIII is free of degeneration, whereas in region VII, a few degenerating fibers can be seen passing ventrolaterally toward the lateral motoneurons. Although there are variations as a function of segmental level, this fundamental terminal pattern is maintained.

Dorsal root terminations are, of course, densest in the segment corresponding to the level of the root section. The rostral and caudal terminations are rather restricted, and by three segments rostral to the root section degeneration is confined entirely to the dorsal magnocellular cell column of Huber (1936). The distribution in the caudal direction is in general less extensive than that rostrally.

E. Long Fiber Systems

Unfortunately, information concerning the major ascending and descending fiber systems of the avian spinal cord is rather limited. With respect to descending pathways, the avian rubrospinal tract, a crossed pathway descending in the dorsolateral funiculus, is directly comparable to that of mammals with regard to its cells of origin, course, and area of terminal distribution in regions V and VI (Fig. 6). Also, a component of the ipsilaterally descending occipitomesencephalic tract and a crossed projection from the anterior Wulst descend to the spinal cord, traveling in the dorsal funiculus to terminate at the base of the dorsal horn (Karten, 1971; Wallenberg, 1902; Zecha, 1962; Zeier & Karten, 1971).

Although not described in detail, birds do appear to have prominent interstitiospinal, reticulospinal, and vestibulospinal pathways all traveling in the ventral funiculus to terminate in the ventral horn. There is a limited tectospinal projection which does not extend beyond the first few cervical segments, and cerebellospinal projections, while reported, are not reliably documented.

Concerning ascending fiber systems, the dorsal column system of the bird is present but not large. Dorsal root fibers do ascend to terminate in the gracile and cuneate nuclei of the medulla in an apparently somatotopic fashion (Karten, 1963; Leonard & Cohen, in preparation; Van den Akker, 1970). There is general agreement that both dorsal and ventral spinocerebellar pathways are present and travel in the lateral funiculus (e.g., Karten, 1963; Van den Akker, 1970); however, Oscarsson, Rosen, and Uddenberg (1963)

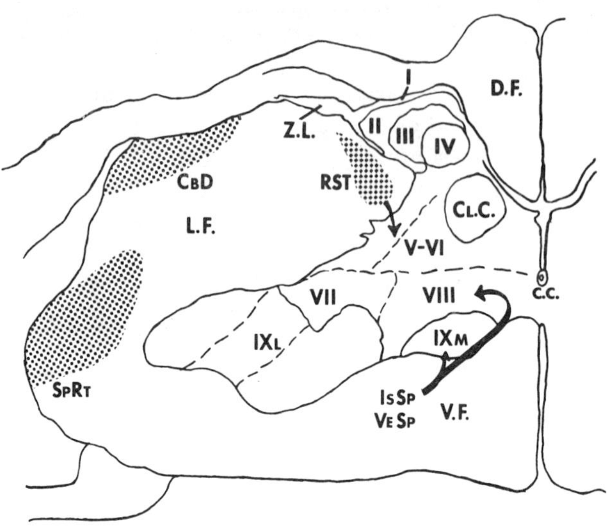

Fig. 6. Schematic diagram of the courses of some major pathways in the pigeon brachial spinal cord. Abbreviations: CBD, dorsal spinocerebellar tract; C.C., central canal; CL.C., column of Clarke; D.F., dorsal funiculus; IsSP, interstitiospinal tract; L.F., lateral funiculus; RST, rubrospinal tract; SPRT, spinoreticular tract; VESP, vestibulospinal tract; V.F., ventral funiculus; Z.L., zone of Lissauer.

claim only limited physiological similarity to the corresponding mammalian spinocerebellar pathways. Karten (1963) has also described a number of regions of termination following cervical hemisection. Although no terminal fields were described in the telencephalon, a variety of fields were found in medulla, pons and mesencephalon, including a spino-olivary path, projections to the solitary complex, spinoreticular projections to the lateral pontine and mesencephalic reticular formations, and a spinotectal pathway. Of particular interest is the description of terminations in the thalamus, demonstrating the existence of an avian spinothalamic tract. However, it is not yet clear as to whether this arises from all spinal levels or has its cells of origin restricted to the more rostral segments of the spinal cord.

IV. General Ascending Pattern

A. Introduction

Having briefly described the major gross neuroanatomical features of the avian brain and our present knowledge regarding the organization of the avian spinal cord, this section is directed toward characterizing the general pattern of sensory projections. With few exceptions (e.g., Karten, 1969), treatment of the avian lemniscal pathways within a framework of connectional neuroanatomy has been lacking, and even the most recent texts (e.g., Pearson, 1972) minimally emphasize the recent developments and discoveries concerning these pathways. Yet, their comprehensive description is fundamental to any general understanding of the avian nervous system and particularly to specifying its relationship to the brains of other vertebrates. By using the visual, auditory and trigeminal systems as illustrative cases, in this section we shall try to provide some overview regarding the organization of the avian lemniscal systems relative to those of other vertebrates, the emphasis being upon the more recent anatomical findings and their influence upon associated behavioral studies.

B. Visual System

The organization of the avian eye is well treated in several recent reviews, including those of Pearson (1972) and Sillman (in press). However, the details concerning central projections of the retina are less extensively described and are discussed in what follows.

As in all amniotes and most anamniotes studied within the past decade (see Ebbesson, 1970), the distribution of avian retinal projections has proven strikingly similar to those of the more widely studied mammalian forms. Primary retinal projections terminate contralaterally in the anterior dorsolateral thalamic complex, ventral geniculate nucleus, suprachiasmatic nucleus, posterodorsal pretectal nucleus, area pretectalis, ectomammillary nucleus, mesencephalic lentiform nuclei (pars parvo- and magnocellularis), and most extensively within the optic tectum. Despite the years of concern with the visual system, surprisingly little is known in any class of vertebrates regarding the second-order projections of many of these structures. This is particularly true of the ventral geniculate, suprachiasmatic, pretectal, ectomammillary and mesencephalic lentiform nuclei. Furthermore, it is only within the past 5 years that any substantial data concerning features of the avian visual system other than the projection upon the optic tectum have become available. Despite our gross ignorance of the totality of subsequent central ramifications of the primary retinal projections, recent experimental investigations of the avian visual system have considerably extended our knowledge and have dramatically emphasized their detailed similarity to comparable pathways recognized in mammalian forms. Such studies have been mainly concerned with those pathways arising from the dorsal thalamus and the optic tectum, and the majority of work has involved the pigeon with several critical studies in the owl as well.

1. Optic Tectum and the Tectofugal Pathways

As described in the section on the gross morphology of the avian brain, birds possess large optic lobes (Fig. 1). As also alluded to earlier, however, the term "optic lobes" is a misnomer, as only a limited portion of the lobe is actually related to the visual system. The superficial laminated rind constituting the surface of the lobe contains the primary retinal input and associated interneuronal and efferent zones. This is designated as the optic tectum (Fig. 3) and is separated from the underlying portions of the optic lobe by the tectal ventricle and its ependymal lining. The laminae comprising the tectum proper are readily traced laterally and ventrally and may be easily distinguished from the remaining, though contiguous, portions of the lobe, which include the central gray substance, nucleus mesencephalicus lateralis, pars dorsalis (central nucleus of the inferior colliculus), lateral mesencephalic reticular formation, and the very prominent isthmic nuclei (pars parvo- and magnocellularis). These last cell groups, the isthmic nuclei, are clearly related to central visual pathways, but receive their input from the optic tectum.

The failure to recognize the fundamental distinction between the optic lobe and optic tectum has led to considerable confusion, particularly as regards the relationship of avian and mammalian brains. Properly speaking, the optic lobe of birds is equivalent to the superior colliculus of mammals, and the optic tectum of birds is considered equivalent to the superficial cap of the superior colliculus, as defined by the ventral margin of the incoming axons from the striate cortex. The distinction is often not as apparent in mammals as in birds and reptiles, because of the fusion of the tectal ventricle during the late stages of mammalian embryogenesis.

Cajal (1952) has provided a detailed study of the cytoarchitecture and Golgi architecture of the avian optic tectum, as well as demonstrated the pattern of ramification of the primary retinal axons within its superficial portion. These retinotectal axons form the outermost tectal lamina, the stratum opticum (Fig. 3), and course through the upper 7 of the 14 tectal laminae to terminate in layers 2–7. This distribution has been more recently confirmed by Cowan, Adamson, and Powell (1961) using degenerating terminal methods.

The efferent connections of the optic tectum were initially studied by Münzer and Wiener (1898) using the Marchi method for degenerating myelin and more recently by Karten using the Nauta–Gygax (Karten, 1965) and Fink–Heimer (Karten, unpublished observations) methods. The tectum projects topographically upon the isthmo-optic nucleus which in turn projects topographically upon the retina (Cowan, 1970). The optic tectum also projects ipsilaterally upon the nuclei isthmi magnocellularis and parvocellularis, the nucleus semilunaris, the lateral pontine nucleus, the lateral mesencephalic reticular formation, parts of the trapezoid body and the contralateral paramedian nuclei via the so-called tectospinal tract. We are virtually totally ignorant of the significance of the majority of these projections in either birds or mammals. The tectal projections that have received the greatest attention within the past several years are the rostrally directed connections with the pretectum, diencephalon, and opposite optic tectum and pretectum via the tectal commissure.

Karten and Revzin (1966) demonstrated the existence of a massive ascending projection from the optic tectum, via the brachium of the "superior colliculus," to the ipsilateral nucleus rotundus. Using experimental neuroanatomical methods, they were unable to confirm a variety of other reported inputs to this nucleus, particularly of somatosensory origin. Rotundus appears to be dominated by the visual input from the tectum, and Hodos and Karten (1966) and Karten and Revzin (1966) concluded on the basis of anatomical, physiological and behavioral studies that the nucleus rotundus represented a major lemniscal terminus of an ascending visual system. In view of the lack of direct retinal contributions to rotundus and its massive input from the optic tectum, Karten (1969), Karten and Hodos (1970), and Nauta and Karten (1970) designated this as a tectofugal pathway to indicate its basic affiliation with the tectum. Additional projections of the optic tectum include the mesencephalic lentiform nuclei, area pretectalis, ventral geniculate nucleus, and portions of the dorsal thalamus proper.

In view of the prominent size of the nucleus rotundus and its massive afferentation from the optic tectum, Hodos and Karten (1970), Karten and Hodos (1970), and Revzin and Karten (1966) directed their major attention to further investigation of the rotundal component of the tectofugal system. The nucleus rotundus was found to project massively, and in a seemingly topographic manner, upon the ipsilateral telencephalon to end selectively within a distinct nuclear mass, the ectostriatum (Revzin & Karten, 1966). Rotundal efferents appear to end within a restricted portion of the ectostriatum, namely the central core zone. This is a region of generally similar neurons and is well separated from the surrounding smaller cells designated the periectostriatal belt. The significance of this seeming isolation of the ectostriatal core from the periectostriatal belt became apparent with subsequent studies of the so-called thalamofugal path described later. As with nucleus rotundus, the ectostriatal core seems to receive only a single major input, that from the nucleus rotundus, and this emphasizes the lemniscal "labeled line" quality of the pathway (Fig. 7). The roles of these various structures in visual discriminative performance have been confirmed in several recent studies (Hodos, 1969; Hodos & Karten, 1966, 1970).

Until the past several years, this tectofugal system as already outlined appeared without a clear mammalian parallel. The major recognized pathway to the mammalian telencephalon was the so-called geniculostriate system,

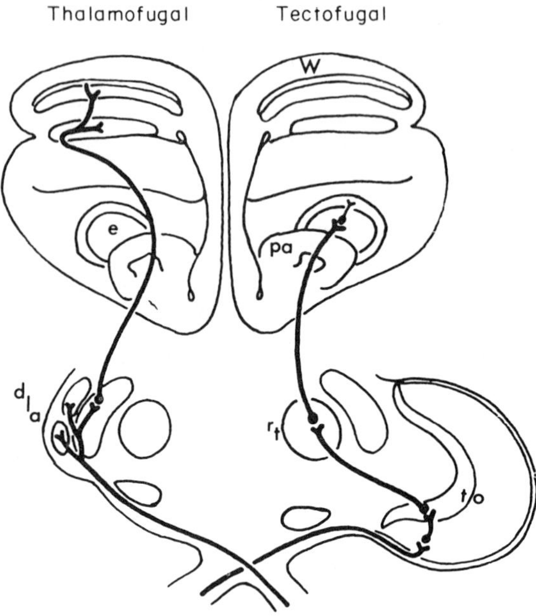

Fig. 7. Schematic illustration of the two principal pathways from the retina to the telencephalon of the owl. The right half of the figure shows the path over the optic tectum (to) and nucleus rotundus (rt) to the ectostriatum (e). The left half of the figure shows the conduction route over the nucleus dorsolateralis anterior thalami (dla); this pathway is distributed to various cell layers of the Wulst (W). pa, paleostriatum augmentatum. [From Nauta & Karten (1970).]

and the possible presence of a distinct tectofugal system was overlooked. However, we have now come to recognize the important contributions of such a tectofugal system in several mammalian forms (e.g., Altman & Carpenter, 1961; Diamond & Hall, 1969; Schneider, 1969). Though the gross morphology of the telencephalic projection field differs markedly between birds and mammals, there is surprisingly little difference in the sequence of anatomical projections, electrophysiology or behavioral functions of these systems. The nature of the differences in the gross morphology of the hemispheric fields will be dealt with in a subsequent section concerned with telencephalic organization.

2. *Thalamofugal Pathways*

Whereas the existence of a primary retinal projection upon the dorsal thalamus has been recognized for some time (Cowan *et al.*, 1961), a re-examination of this projection in the pigeon and owl using the Fink–Heimer methods for terminal degeneration indicated a far more extensive and dense terminal field than described previously (Karten & Nauta, 1968). In the owl (*Speotyto conicularis*) the dorsal thalamic terminal fields were greatly enlarged in comparison to those of the pigeon, and there was a distinct pattern of lamination in both cell types and character of the axons and terminals. This projection appears completely crossed, and Karten, Hodos, Nauta, and Revzin (1973) have designated the dorsal thalamic terminal field as the principal optic nucleus of the thalamus. This is composed of several nuclear clusters, including those nuclei of the pigeon previously designated as lateralis anterior, dorsolateralis anterior, dorsolateralis anterior pars magnocellularis, and suprarotundus. In the owl, the nuclear complex in receipt of the retinal afferents corresponds to that of the pigeon, although its greater size, topological rearrangement and displacement relative to the pigeon renders precise subnuclear comparisons difficult. With the exception of the virtually totally crossed nature of the retinothalamic projections, the principal optic nucleus bears several similarities to the dorsal nucleus of the lateral geniculate of mammals. This suggested similarity is strikingly enhanced when we consider the subsequent projections of the principal optic nucleus and the character of its zone of termination within the telencephalon.

In tracing the efferent connections of the principal optic nucleus, Karten and Nauta (1968) and Karten *et al.* (1973) found that in both the owl and pigeon it projects to the telencephalon, bypassing the ectostriatum to enter the overlying Wulst. This undoubtedly accounts for the short latency, retinotopically organized visual projection upon the dorsal surface of the pigeon telencephalon described electrophysiologically by Revzin (1969).

Massive terminal fields have been found in the Wulst, and these are particularly dense in the lateral division of the hyperstriatum dorsale and the broad bilaminate band of granule cells of the nucleus intercalatus hyperstriatum accessorium. This restricted distribution has been confirmed electrophysiologically by Cohen and Dooley (in preparation). A smaller number of axons and terminals reached the contralateral Wulst via the dorsal supraoptic commissure, ending in homotopic regions. In parallel with the greater degree of development of the principal optic nucleus of the owl, the Wulst is greatly hypertrophied and bears a striking resemblance to the striate cortex (area 17) of many mammalian forms, We should like to emphasize, however, that the "visual" portion of the Wulst does not include its totality, as recognized by many investigators. Rather, the Wulst appears to be a broad cortical zone with several afferent sources and different efferent projections.

Regarding these efferent projections, they have been described over the years by many investigators (e.g., Adamo, 1967; Kalischer, 1905; Kappers *et al.*, 1936; Karten, 1971; Wallenberg, 1902). One of the most notable features that is consistently observed is the presence of a substantial descending projection upon the optic tectum, rendering the similarity of the visual Wulst to the mammalian striate cortex even more likely. Moreover, as in mammals this hyperstriatotectal projection is topographically organized and extends to the more superficial tectal layers in the owl. Also noteworthy here is the presence of a topographically arrayed intratelencephalic projection from the Wulst onto the periectostriatal belt, providing an anatomical substrate for interactions at the telencephalic level between the tectofugal and thalamofugal visual systems (Fig. 7).

C. Auditory System

Extremely informative reviews of the avian auditory system have recently been provided by Erulkar (1972), Pearson (1972), Schwartzkopf (in press), and Smith and Takasaka (1971). Although less well developed than in some mammalian forms, many of the specializations emphasized in mammals are evident in birds, as for example their discriminative ability with respect to species-specific calls and auditory mechanisms adapted for prey catching (Payne, 1962). Perhaps less well known is the specialization of certain avian auditory systems for echolocation (e.g., Medway, 1959; Novick, 1959). The comparative studies of Cobb (1964), Ilyichev (1960), Sanders (1929), and Winter (1963) have documented particular species differences within the central auditory path, yet as in the visual and trigeminal systems, these variations occur only in the relative degree of development of individual nuclei that are common to all avian forms rather than in the appearance of nuclear groups unique to a given species.

The peripheral auditory apparatus of birds, as in mammals, consists of an external, middle, and inner ear. The external ear is a duct patent to the outside, but since this duct is frequently covered by feathers it is not always immediately apparent. In contrast to mammals, there is no external mobile pinna. Yet birds are clearly capable of binaural localization, and the organization of their central auditory connections suggests that the relevant nuclei are similar to those involved in mammals for the localization of sound in space (Erulkar, 1972).

The avian external auditory canal ends bluntly at the tympanic membrane, and the middle ear contains only a single major ossicle, the columella. This ossicle abuts on the inner ear at the oval window to excite the papilla basilaris, which is similar in all major details to the mammalian organ of Corti with the single exception that it is straight rather than coiled. Takasaka and Smith (1971) have recently published an excellent ultrastructural description of the papilla basilaris which clearly confirms the presence of both inner and outer hair cells with a distinct kinocilium. In mammals (*Eutheria*) the kinocilium is absent and only a basal body remains. At the distal end of the papilla basilaris is the macula lagena whose function is uncertain, although it is speculated to be concerned with low-frequency vibration or possibly otolith-type functions.

The studies of Boord (1968), Boord and Rasmussen (1963), Cajal (1952), and Karten (1967, 1968) have been instrumental in outlining the major features of the auditory lemniscal system of birds. The cochlea-lagena has been shown to project topographically upon the nuclei angularis and magnocellularis (Boord & Rasmussen, 1963), and in combination with electrophysiological data (Stopp & Whitfield, 1961), it is suggested that these nuclei may well compare, respectively, to the dorsal and ventral cochlear nuclei of mammals. Although there is no direct projection of the papilla basilaris upon the nucleus laminaris, this nucleus does receive inputs from the nuclei magnocellularis of each side, the ipsilateral and contralateral afferent projections terminating on opposing polarized dendrites of laminaris neurons. This agrees with Cajal's (1952) early suggestion that the nucleus laminaris is most directly comparable to the medial accessory superior olivary nucleus of mammals both in patterns of dendritic orientation and preferential afferentation. In addition, the macula lagena projects directly upon the nucleus quadrangularis parvocellularis, an ostensible component of the vestibular complex. However, this affiliation is based largely upon its proximity to other nuclei more obviously related to the vestibular apparatus.

The efferent projections of the cochlear nuclei contribute to an ascending lateral lemniscal system that decussates dorsally in the dorsal acoustic stria and ventrally through the corpus trapezoideum. These crossed fibers terminate in part upon the nuclei of the lateral lemniscus, but the majority project to the central core nucleus of the torus semicircularis, also known as the nucleus mesencephalicus lateralis, pars dorsalis. This structure is directly comparable to the central nucleus of the mammalian inferior colliculus, and it projects in turn via the brachium of the inferior colliculus to the nucleus ovoidalis of the thalamus (Karten, 1967). A small but significant number of fibers cross the midline through the dorsal supraoptic decussation to terminate in the opposite nucleus ovoidalis. The considerations leading to the comparison of the avian nucleus ovoidalis, reptilian nucleus reuniens posterior and mammalian lobus inferior of the medial geniculate body are discussed at length by Karten (1967).

The nucleus ovoidalis projects upon a well-delineated cell group within the caudal neostriatum, previously designated as Field L by Rose (1914). Despite its ostensible location within the neostriatum, Karten (1968) has proposed that Field L bears the same relation to the avian ascending auditory pathway as do the cells of lamina IV of mammalian auditory neocortex.

Further support of these various hypotheses regarding both the nucleus ovoidalis and Field L have been provided by the electrophysiological investigations of Biederman-Thorson (1967, 1970a,b), Erulkar (1955) and Harman and Phillips (1967), and the principal projections comprising the ascending auditory system in birds are illustrated in Fig. 8.

D. Trigeminal System

Avian species demonstrate as great a variation in the degree of development of their trigeminal systems, as may be found in any other class of vertebrates, largely reflecting extreme differences in bill and, to a lesser extent, head size. For example, the marsh and shore birds generally show the greatest development of their bills and, expectedly, an attendant expansion of their trigeminal systems. Yet, to the extent of our current knowledge of avian neuroanatomy, these trigeminal variations consist only of the greater or lesser development of trigeminal components that are common to all birds.

The trigeminal, or fifth cranial, nerve is a mixed nerve including a small motor branch innervating the muscles of mastication and a large sensory component. The nerve derives its name from the three characteristic branches arising from the Gasserian, or trigeminal, ganglion, namely the opthalmic, maxillary and mandibular branches in dorsoventral order. It is well developed in pigeons, chickens, owls, and many passerine birds, but the trigeminal nerve reaches truly massive proportions in the large billed birds (Stingelin, 1965). Regarding its sensory components, the variety of sensory endings of the avian trigeminus have not been fully explored, but the Herbst and Grandry corpuscles are undoubtedly important constituents of the system (see Section III.B.1).

The Gasserian ganglion contains the cells of origin of the sensory trigeminal fibers and is derived from a mixed population during embryogenesis which receives contributions from cells of both neural crest and placodal origin. The central axons of these sensory ganglion cells then form two major tracts upon entering the brainstem, the ascending and descending trigeminal tracts. According to Cajal (1952), the entering axons are of three types; those with only ascending branches, those with only descending branches, and bifurcating axons with both ascending and descending branches. The ascending axons terminate in the nucleus of the tractus trigemini ascendans (principal sensory nucleus of the trigeminus), a nucleus consisting of separate though adjacent subdivisions, the pars dorsale and pars ventrale. These entering axons end topographically to form an inverted mapping of the facial region with the mandibular branch terminating most dorsally and the opthalmic branch most ventrally (Dubbeldam & Karten, in preparation; Zeigler & Witkovsky, 1968). Furthermore, Zeigler and Witkovsky (1968) report that the receptive fields of the neurons of the prin-

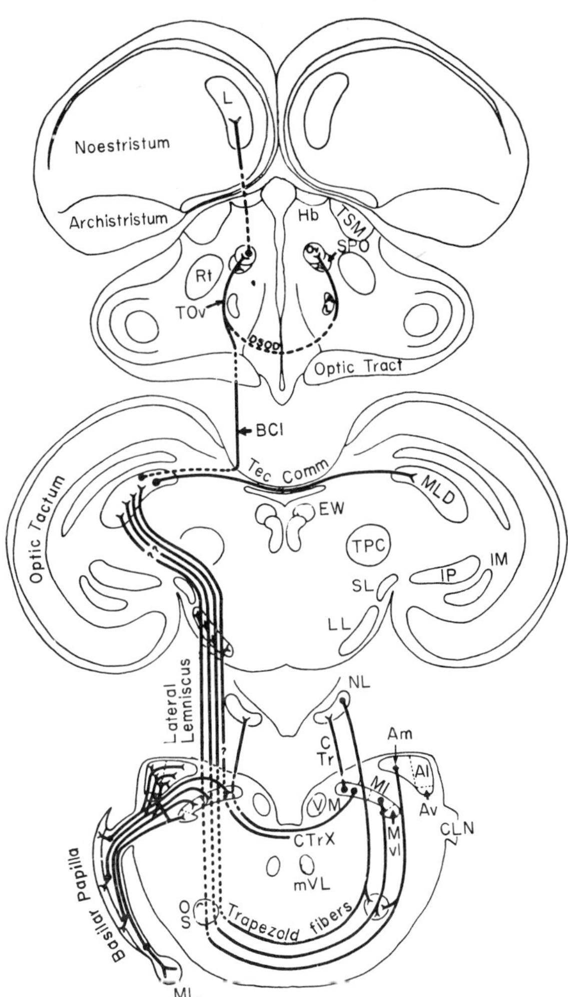

Fig. 8. Schematic illustration of the principal ascending auditory pathways of the pigeon. Abbreviations: Al, nucleus angularis pars lateralis; Am, nucleus angularis pars medialis; Av, nucleus angularis pars ventralis; BCI, auditory fibers in brachium of inferior colliculus; CLN, cochlear and lagenar nerves; CTr, uncrossed dorsal cochlear tract; CTrX, crossed dorsal cochlear tract; DSOD, auditory fibers in dorsal supraoptic decussation; EW, Edinger–Westphal nucleus; Hb, habenular nuclei; IM, nucleus isthmi pars magnocellularis; IP, nucleus isthmi pars parvocellularis; L, auditory area of neostriatum; LL, nucleus of lateral lemniscus; ML, macula lagenae; Ml, nucleus magnocellularis pars lateralis; MLD, nucleus mesencephali lateralis pars dorsalis; Mm, nucleus magnocellularis pars medialis; Mvl, nucleus magnocellularis pars ventrolateralis; NL, nucleus laminaris; n III, oculomotor nucleus; n IV, abducens nucleus; OS, superior olive; Ov, nucleus ovoidalis; Rt, nucleus rotundus; SL, nucleus semilunaris; SPO, nucleus semilunaris parovoidalis; Tec Comm, intertectal commissure; TSM, tractus septomesencephalicus; TOv, tractus nuclei ovoidalis; TPC, nuclei tegmenti pedunculopontinus pars compacta; VM, medial vestibular nucleus. [From Boord (1969).]

cipal nucleus are generally small and restricted to the region of the mouth and beak. The descending axons, caudal to their point of entry, terminate in the nucleus of the descending tract, some continuing into the rostral spinal cord to end in laminae I–IV. In parallel with the pattern of mammalian organization, the descending tract and nucleus are divisible into a pars oralis, pars interpolaris, pars caudalis and pars spinalis.

One of the seemingly most curious aspects of the subsequent organization of the trigeminal system pertains to the efferent projections of the principal nucleus. In 1903 Wallenberg described a bilateral ascending projection from this nucleus which, after a partial decussation in the isthmus, passes through the diencephalon and paleostriatum primitivum without apparent termination to end in a rostrobasal telencephalic nucleus; Wallenberg (1903) designated this the quintofrontal tract. Karten (in preparation) has confirmed the original observations of Wallenberg (1903) and further demonstrated that the ipsilateral component of the quintofrontal tract arises mainly from the pars dorsale of the principal nucleus, whereas the contralateral component arises from the pars ventrale. The term "nucleus prosencephali trigeminalis" has been substituted for "nucleus basalis" to designate this telencephalic terminal nucleus, since the latter name has had variable and confusing usage. A further finding is that the nucleus prosencephali trigeminalis appears to be laminated with alternating projections from the ipsilateral and contralateral principal nuclei. Subsequently, this nucleus projects caudally within the telencephalon via the tractus frontoarchistriaticus, terminating within the caudolateral telencephalon in a region lateral to the tractus dorsoarchistriaticus, dorsal to the archistriatum, and dorsomedial to the olfactory cortex (Karten, in preparation; Zeier & Karten, 1971). Figure 9 summarizes these projections of the ascending trigeminal system.

The discovery of the quintofrontal tract posed a major conceptual problem for neuroanatomy, in view of the generally accepted dictum, based on studies of mammalian forms, that all sensory systems, except the olfactory, relay in the thalamus before projecting upon the telencephalon. Whether this is even necessarily true in mammals may be questioned; yet the notable exception of the quintofrontal tract has enforced the notion in many people's minds that the nonmammalian brain is in some ways peculiar and different from the mammalian brains. However, the seemingly unusual nature of the quintofrontal system relates only to the relative locus of the nucleus trigeminalis prosencephali, since it is otherwise similar in organization to the mammalian trigeminal lemniscal system. The paradoxical nature of the quintofrontal system may perhaps be resolved if we consider that strict boundaries between diencephalon and telencephalon are arbitrary and that the precise position of a cell group within the broader zone of the prosence-

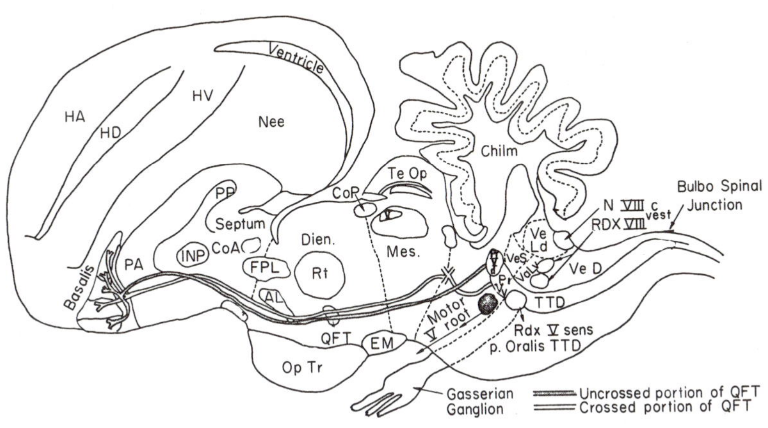

Fig. 9. Schematic summary of the trigeminal system of the pigeon. Abbreviations: AL, ansa lenticularis; Cbllm, cerebellum; CoA, anterior commissure; CoP, posterior commissure; Dien., diencephalon; EM, ectomammillary nucleus; FPL, fasciculus prosencephali lateralis; HA, hyperstriatum accessorium; HD, hyperstriatum dorsale; HV, hyperstriatum ventrale; INP, nucleus intrapeduncularis; Mes., mesencephalon; Neo, neostriatum; N VIIIc, cochlear division of eighth nerve; OpTr, optic tract; PA, paleostriatum augmentatum; PP, paleostriatum primitivum; PrVd, nucleus principalis nervi trigemini pars dorsale; PrVv, nucleus principalis nervi trigemini pars ventrale; QFT, quintofrontal tract; RDX VIIIvest, radix nervus octavus pars vestibularis; RDX Vsens., radix nervi trigemini; Rt, nucleus rotundus; TeOp, optic tectum; TSM, tractus septomesencephalicus; TTD V p. Oralis, nucleus tractus descendens nervi trigemini pars oralis; TTD V ip, nucleus tractus descendens nervi trigemini pars interpolaris; TTd V cd, nucleus tractus descendens nervi trigemini pars caudalis; TTD V sp, nucleus tractus descendens nervi trigemini pars spinalis; VeD, nucleus vestibularis descendens; VeLd, nucleus vestibularis lateralis pars dorsalis; VeLv, nucleus vestibularis lateralis pars ventralis; VeS, nucleus vestibularis superior.

phalon may not be fully predictable on the basis of its absolute or relative location in any single class of vertebrates.

Regarding possible functions of this quintofrontal system, Edinger (1908) was the first to propose its role in feeding behavior. However, this was never specifically investigated until recently (Zeigler, this volume; Zeigler & Karten, 1973; Zeigler, Karten, & Green, 1969). Numerous investigators of the avian brain had observed that following extensive bilateral telencephalic lesions pigeons stopped eating and drinking (Flourens, 1824; Rogers, 1922), the observed deficits often being permanent and leading to death in the absence of forced feeding. Rogers (1922) placed small lesions in the telencephalon and concluded that the critical locus of this effect was within the paleostriatum primitivum. In retrospect, the critical lesion probably involved the quintofrontal tract as it passes through the paleostriatum primitivum rather than the paleostriatum itself. Zeigler and Karten

(1973) have studied feeding behavior in the pigeon and confirmed this latter interpretation, as well as validating Edinger's (1908) insightful proposals on the broader problems of the relationship of trigeminal sensory mechanisms and feeding behavior. Although studies to date have concerned themselves largely with pigeons, the wide range of variation in the development of both the peripheral and central trigeminal system in birds provides a rich ground for subsequent investigation. For example, the relationship of the trigeminal system to vocalization has yet to be explored. It is perhaps more than slightly amusing that the enigmatic aphasia (loss of vocalization, stammering, etc.) of parrots as described by Kalischer (1905) was consequent to lesions in a region now recognized as the projection field of the nucleus trigeminalis prosencephali.

E. Concluding Remarks

In conclusion, the similarities of particular sensory pathways in birds and mammals are becomingly increasingly apparent, a point we have attempted to generate inductively. A reasonable inference might then be that such anatomically comparable paths subserve *generally* similar functions and that the accurate definition of the functional role of a given neural unit in any single form, be it bird or mammal, will lead to valid generalizations as to its role in all forms. Despite the frequent emphasis upon the divergent evolutionary histories of birds and mammals, they clearly seem to have elaborated central sensory mechanisms that unquestionably are manifestations of structural arrangements already present in lesser degree in reptilia. Though this may suggest a return to "preformism" as theory in evolutionary biology, it may actually indicate the constraints within which central sensory structures have developed. Recognizing the evident similarities among the sensory pathways of birds and mammals may well establish the soundest foundation for revealing the most significant points of difference and for experimentally approaching the detailed species variations in synaptic arrangements and terminal densities that may account for the observable differences in their behaviors.

V. General Descending Pattern

A. Introduction

The investigation of the detailed neuronal architecture of the forebrain of any avian form has generally been limited; experimental efforts have been most intensive with respect to describing the terminal fields of the ascending projections (see Section IV) and somewhat less vigorous as regards defining the cells of origin and distributions of the telencephalic efferents. Research to date has clearly emphasized that lemniscal terminal fields within the telencephalon are several synapses removed from the cells of origin of the major efferent bundles; yet, our present knowledge regarding the intrinsic telencephalic connections is minimal. This is unfortunate, since the organization and architecture of intratelencephalic projection systems bear on some of the more important morphological and functional questions concerning the forebrain.

Among the more striking features emerging from the investigations of the telencephalic efferent systems is once again their striking similarities to certain of the descending projections and their distributions in mammals (e.g., Adamo, 1967; Edinger, Wallenberg, & Holmes, 1903; Kalischer, 1905; Karten, 1971; Karten & Dubbeldam, 1973; Zecha, 1962; Zeier & Karten, 1971). The major novel findings pertain to those constituents of the forebrain that have been found to have axonal distributions identical to those of mammalian neocortex; that is, projections upon the thalamus, optic tectum, red nucleus, tegmentum, pontine nuclei, medial and lateral reticular formations, nuclei cuneatus and gracilis, and spinal cord. An important feature of these various projections is their differential origins from distinct telencephalic cell groups, since this provides potential insights as regards the nature and origin of certain mammalian efferent systems, such as the pyramidal tract.

The subtelencephalic efferent systems are perhaps even more poorly defined than those arising from the forebrain. Yet, their full description, including the cells of origin and spinal terminal fields, are clearly fundamental to an eventual understanding of the avian motor system and its relationship to mammalian motor organization. Again, as with the telencephalic efferents, the minimal available information suggests rather remarkable similarities to the major mammalian pathways projecting upon the spinal cord.

B. Telencephalic Efferents

The loci of the cells of origin of the efferent telencephalic projections are summarized schematically in Fig. 10. It may be noted that there are only five major regions from which axons arise to leave the telencephalon, and these fibers exit as distinct, compact, and easily identifiable bundles.

The occipitomesencephalic tract proper arises from the various components of the anterior two-thirds of the archistriatum (Zeier & Karten, 1971). It has both ipsilateral and contralateral components, and along its descending course terminal fields are found in the dorsal thalamus, nucleus spiriformis medialis, nucleus subrotundus, nucleus principalis precommis-

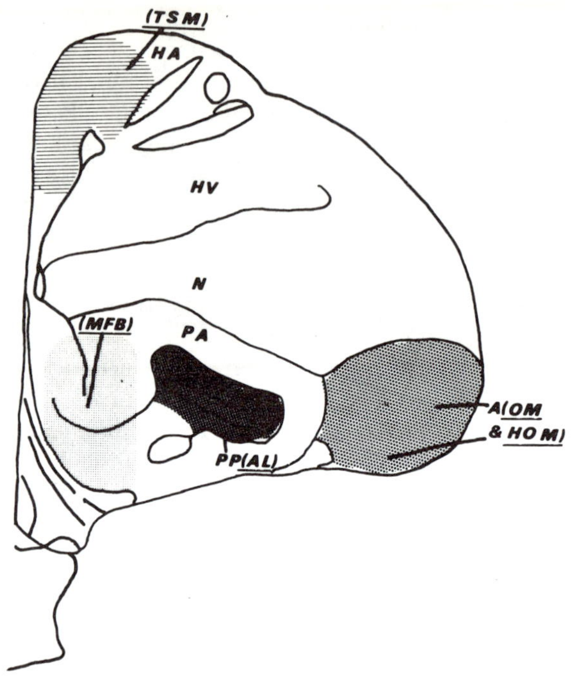

Fig. 10. Summary of loci of origin of telencephalic projections and the bundles arising from each. Abbreviations: A, archistriatum; AL, ansa lenticularis; HA, hyperstriatum accessorium; HOM, tractus occipitomesencephalicus pars hypothalami; HV, hyperstriatum ventrale; MFB, medial forebrain bundle; N, neostriatum; OM, tractus occipitomesencephalicus; PA, paleostriatum augmentatum; PP, paleostriatum primitivum; TSM, tractus septomesencephalicus. [From Karten & Dubbeldam (1973).]

suralis, lateral reticular formation, nucleus intercollicularis, the deeper layers of the optic tectum, locus coeruleus, nucleus subcoeruleus dorsalis and ventralis, and the lateral pontine nucleus. Additional projections are found to end in the nuclei reticularis parvocellularis, subtrigeminalis, descendens nervi trigemini and gracilis and cuneatus. The crossed component of the pathway can be followed to corresponding nuclei in the brainstem and upper cervical cord. Its trajectory in the spinal cord is through the dorsal columns, and the terminations are largely restricted to the base of the dorsal horn. This bundle, the occipitomesencephalic tract, is quite similar in its course and terminal distribution to the neocortically derived bundle of Bagley described in ungulates (Haartsen & Verhaart, 1967; Zecha, 1962), a bundle which appears to be equivalent to a component of the primate pyramidal tract that terminates in the lateral reticular formation.

Also arising from neurons in the archistriatum is another pathway which is quite distinct from the occipitomesencephalic tract proper both in its cells of origin and terminal field; Zeier and Karten (1971) have designated this the tractus occipitomesencephalicus, pars hypothalami. These fibers arise from the medial and posterior aspects of the archistriatum to terminate upon the medial and lateral hypothalamus in a manner quite reminiscent of certain amygdalar projections in mammals. However, at present it is unclear as to what relationship this projection may bear to the mammalian stria terminalis and ventral amygdalofugal pathways.

A third distinct pathway, the septomesencephalic tract, originates in the dorsal telencephalon, predominantly the Wulst, and it exits along the medial telencephalic wall (Adamo, 1967; Karten, 1971; Zecha, 1962). This is a rather complex pathway, since its more medial components may include the avian fornix bundles (Kappers *et al.*, 1936). The terminal fields of the lateral components are extensive and include the lateral neostriatum and periectostriatal belt. Extratelencephalic projections are found to the internal lamella of the ventral geniculate nucleus, pretectal nucleus and the optic tectum. Furthermore, fibers arising from the anterior portion of the Wulst collect on the ventral aspect of the ansa lenticularis and terminate in the nucleus intercalatus thalami, spiriformis medialis and red nucleus. More caudally, the septomesencephalic tract can be traced along the ventral surface of the brainstem with terminal fields in the medial reticular formation, pontine nuclei and cuneate and gracile nuclei. A distinct group of fibers enters the contralateral dorsal funiculus to terminate in the base of the dorsal horn, overlapping only minimally with the spinal terminal field of the occipitomesencephalic tract.

The medial forebrain bundle or fasciculus prosencephali medialis constitutes a fourth system of efferents from the telencephalon. Crosby and Woodburne (1940), Huber and Crosby (1929), Karten and Dubbeldam (1973), and Zeier and Karten (1971) have all defined its course. In contrast to the other telencephalic efferent systems, all of which appear to contain predominantly if not exclusively descending projections, the medial forebrain bundle is composed of a variety of long and short projection systems of both an ascending and descending nature. The precise details concerning the loci of origin and the distributions of the individual fascicles of this pathway are unfortunately lacking, although they are generally known to arise from cells in the medial and basomedial telencephalic wall and the lobus parolfactorius to project mainly upon the hypothalamus (Zeier & Karten, 1971). Clearly, further information is needed regarding this important limbic pathway, as is equally true for the avian fornix system.

The final major telencephalic projection to be discussed is the ansa lenticularis (Karten & Dubbeldam, 1973). The paleostriatum augmentatum appears to receive a topographic projection from the overlying regions of the dorsal ventricular ridge, neostriatum and hyperstriatum ventrale, and it projects in turn upon the paleostriatum primitivum and nucleus intrapeduncularis. Cells in these latter regions then give rise to the ansa lenticularis, a bundle similar to the tract of the same name in mammalia. This pathway terminates in the nuclei ansae lenticularis anterior and posterior, dorsalis intermedius posterior thalami, spiriformis lateralis, and the various components of the nucleus tegmenti pedunculopontinus. Unfortunately, the subsequent projections of these nuclei are unknown and this severely limits our understanding of the available discharge pathways of the basal ganglia; this is true for mammals as well as birds. Considering the evolutionary stability of the basal ganglia and their significant volume within the telencephalon of all vertebrates, the exploration and definition of this total system in any single species of vertebrate would undoubtedly prove of substantial significance for our understanding of the functions of this system in all vertebrates.

At this point in the discussion a few words regarding the intrinsic telencephalic connections are relevant. After we exclude the terminal fields of the projections ascending to the telencephalon and the loci of the cells of origin of the telencephalic efferents, it may be concluded that the neostriatum and hyperstriatum ventrale contain only intrinsic neurons with axons ramifying within the telencephalon, save for those portions of the neostriatum in receipt of thalamic efferents (e.g., Field L). Mammalian neocortex is obviously replete with similar intrinsic neurons; yet, not all of the remaining intrinsic neurons of the avian neostriatum and hyperstriatum ventrale can be assuredly compared to the intrinsic neocortical cells. It is possible that some components may be without mammalian equivalents, although detailed study of the intrinsic populations in the bird may provide characteristic criteria pertinent to their identification in mammalia. Perhaps such an analysis would provide the distinction between those features of neocortex that are derived from lamination *per se* versus those inherent to that particular population of neurons, and might once again demonstrate the enormous value of the comparative approach even to a structure so seemingly uniquely mammalian as neocortex.

C. Subtelencephalic Efferents

As already mentioned, surprisingly little data have been obtained regarding the major fiber systems originating subtelencephalically and terminating in the spinal cord. The presence of an avian rubrospinal tract, comparable to that pathway in mammals, is reasonably well established. This pathway, after decussating, assumes a lateral course in its descent and in the spinal cord travels in the dorsolateral funiculus to terminate in laminae V and VI. Further supporting the comparability of this pathway to the mammalian rubrospinal tract are the findings that the avian red nucleus receives projections from the lateral nucleus of the cerebellum (Karten, 1964) and the anterior Wulst (Karten, 1971). Tectobulbar and limited tectospinal pathways are apparently present in the bird, although the tectospinal tract does not terminate beyond the first few cervical segments of the cord. Birds apparently also have prominent interstitiospinal, reticulospinal and vestibulospinal pathways all traveling in the ventral funiculus to terminate in the ventral horn. However, the precise origins, courses, and terminations of these pathways have not been well described. Moreover, poorly documented cerebellospinal projections have been reported but not confirmed by recent studies.

Consequently, the detailed description of the subtelencephalic projections to the avian spinal cord provides a fertile area for future investigation. Little is known at present about the details of their organization; such information will be necessary for a complete understanding of the influences playing on the avian spinal cord. Furthermore, a precise definition of the loci of the cells of origin of these various pathways may well be quite helpful in understanding the nature of other descending systems not projecting as far caudally as the spinal cord.

VI. Telencephalic Organization

Despite the numerous morphological surveys of the avian telencephalon and the various attempts to compare it with other vertebrate forebrains (see e.g., Haefelfinger, 1958; Kappers *et al.*, 1936), in the main such efforts have been restricted to normal material. In the absence of experimental material, the gray matter of the lateral wall, which constitutes the major portion of the avian telencephalon, was necessarily subdivided on a cyto-architectonic basis, and unfortunately a nomenclature was applied in which the suffix describing each major cell group was "striatum." This largely descriptive approach has provided little information regarding the possible differences between these various cell masses of the telencephalon, and in fact it has been frankly misleading both with respect to avian telencephalic organization and its relationship to that of other vertebrates.

The recent availability, however, of the powerful silver techniques for demonstrating degenerating fibers and boutons, as well as the development of the histochemical methods, have now permitted considerably more informative analyses of comparative telencephalic organization that are based on detailed connectional neuroanatomy. It is a brief review of such contemporary information for the avian brain that constitutes the basis of this section. The discussion will rely heavily on the previous descriptions of the general structural features of the avian brain, the general ascending pattern and the review of the descending pattern, the topics of three previous sections.

A. The Paleostriatal Complex

As described in the section dealing with the general structural features of the avian brain, the paleostriatal complex consists of two major cell groups: the paleostriatum primitivum and the paleostriatum augmentatum (Fig. 2). Based upon its position, constituent cell type, iron concentration (Kappers *et al.*, 1936), acetylcholinesterase distribution (Karten & Iversen, unpublished data), succinic dehydrogenase distribution (Baker-Cohen, 1968), and comparative electron microscopic observations (Fox, Hillman, Siegesmund, & Sether, 1967), the paleostriatum primitivum has been viewed as homologous to part of the mammalian globus pallidus (Karten, 1969). Furthermore, the paleostriatum primitivum gives rise to a major projection system that is identical in many of its features to the classical ansa lenticularis bundle of mammalian brain, the terminal fields in some instances overlapping in either their afferent or efferent projections with the cerebellum. One major zone of termination of the ansa lenticularis is the nucleus tegmenti pedunculopontinus, pars compacta, and in both birds and mammals this structure has been demonstrated to be rich in catecholamines (Bertler, Falck, Gottfries, Ljunggren & Rosengren, 1964).

The paleostriatum augmentatum may possibly be considered as homologous to at least part of the mammalian caudate-putamenal complex, and this may be the only region of avian brain corresponding to that complex (Karten, 1969). It has been shown that various telencephalic cell groups project topographically upon the paleostriatum augmentatum in a manner strikingly similar to that found in mammals, and the paleostriatum augmentatum in turn projects upon the paleostriatum primitivum. Furthermore, the paleostriatum augmentatum demonstrates an intensely positive reaction for both acetylcholinesterase and dopamine, and the dorsal boundary of the distribution of these substances sharply coincides with the morphological dorsal boundary of the paleostriatum augmentatum (Juorio & Vogt, 1967; Karten & Iversen, unpublished data). In mammals, the telencephalic dopamine is almost entirely confined to the caudate and putamen (Fuxe, 1965; Koelle, 1954), further suggesting their homology with the paleostriatum augmentatum.

Thus, there is rather compelling evidence that the avian paleostriatum primitivum and the mammalian globus pallidus are homologous. Also, the evidence is highly suggestive of an homology between the avian paleostriatum augmentatum and at least part of the mammalian caudate-putamenal complex. This then raises the provocative question as to what the nature of the remaining "striatal" nuclei might be.

B. The Neostriatal and Hyperstriatal Complex

The more dorsal regions of the avian "striatum" have classically been designated as the neostriatum, hyperstriatum ventrale, hyperstriatum dorsale, hyperstriatum intercalatus superior, and hyperstriatum accessorium (Fig. 2). The hyperstriatum dorsale, intercalatus superior, and accessorium constitute the Wulst or sagittal elevation. The nature of these telencephalic areas has been somewhat elusive, but it is rapidly being clarified through the analysis of their afferent and efferent projections, particularly as regards the lemniscal systems (see the section on the general ascending pattern of the avian brain).

Briefly, specific sensory receiving areas have now been identified in the avian telencephalon, and these appear to receive their inputs over pathways that are generally homologous to the mammalian lemniscal systems. First, an auditory area, receiving projections from the nucleus ovoidalis, has been localized to a sharply circumscribed population of granule cells in the medial portion of the caudal neostriatum (Karten, 1968). This corresponds to the Field L of Rose (1914), and evoked auditory field potentials have been demonstrated in this restricted region (Biederman-Thorson, 1970a,b; Erulkar, 1955).

Although little is known regarding a telencephalic somatosensory representation, a definitive representation of the beak region has been established in the nucleus prosencephali trigeminalis (Zeigler, this volume). It has been known for some time that cells of the principal nucleus of the trigeminal nerve give rise to an unusual fiber tract that projects directly to the telencephalon to terminate in the nucleus prosencephali trigeminalis (Wallenberg, 1903), and more recently it has been demonstrated that this latter nucleus projects to the anterior archistriatal nucleus and overlying neostriatum via the frontoarchistriatal tract (Zeier & Karten, 1971). The evidence is unfortunately rather inconclusive concerning a more general somatosensory representation in the telencephalon. On the basis of electrophysiological experiments, Delius and Bennetto (1972) report responsive areas in both the anterior hyperstriatum and medial caudal neostriatum, the latter corresponding to the region where Erulkar (1955) described somatosensory evoked potentials. Moreover, Delius and Bennetto (1972) localize thalamic somatosensory evoked responses to the nuclei superficialis parvocellularis and dorsolateralis posterior, a region thought to receive fibers of the medial lemniscus (Wallenberg, 1904) and the spinothalamic tract (Karten, 1963;

Karten & Revzin, 1966). Since, in each instance where a specific sensory thalamic nucleus has been identified in the bird, it has been shown to project upon a specific telencephalic sensory field, we would anticipate the existence of a general somatosensory representation in the hemisphere.

There is fortunately considerably more information available regarding the visual projections to the telencephalon. It is now well established anatomically that there are at least two major sources of visual input from the thalamus to the telencephalon in birds: the thalamofugal and tectofugal systems (Karten, 1969). The thalamofugal system consists of a topographically organized retino–thalamo–hyperstriatal (accessorium and intercalatus superior) projection (Karten & Nauta, 1968). Electrophysiological studies of this pathway confirm the highly circumscribed nature of the thalamo–hyperstriatal projection (Cohen & Dooley, in preparation; Revzin, 1969), and Revzin (1969) suggests that the physiological properties of the responsive hyperstriatal neurons are strongly reminiscent of those of the mammalian geniculo-striate system.

The tectofugal system consists of a retino–tecto–rotundo–ectostriatal series of projections, with maintenance of a certain degree of topography (Karten & Hodos, 1970; Karten & Revzin, 1966; Revzin & Karten, 1967). The telencephalic component of this system, the ectostriatum, consists of two distinct zones: a central core containing vertically oriented, heavily myelinated axons with round, medium-sized neurons and a peripheral belt sweeping around the dorsal surface of the core region with somewhat thinner myelinated axons and small fusiform neurons. The rotundal fibers terminate exclusively in the core region (Karten & Hodos, 1970) whose neurons project to the periectostriatal belt, the lateral portion of the neostriatum intermediale and a laminated population of neurons on the dorsolateral surface of the hemisphere in a region described physiologically by Belekhova (1968). Cohen and Dooley (in preparation) have electrophysiologically confirmed the existence of visually evoked field potentials in an area coincident with the ectostriatum, although Parker and Delius (1972) were unable to demonstrate this. Finally, as information on the organization of the ascending visual projections accumulates, there is increasingly suggestive evidence that there may exist another major source of visual input to the ectostriatum (Cohen, in press; Cohen & Dooley, in preparation; Kimberly, Holden, & Bamborough, 1971).

Although descending projections from the ectostriatum have not been described, the hyperstriatum appears to project back upon many structures receiving retinal input, such as the ventral geniculate (internal lamella), visual areas of the dorsal thalamus, and the optic tectum (Adamo, 1967; Karten, 1971). There is also a projection to the periectostriatal belt (Karten & Hodos, 1970), providing a pathway for possible telencephalic interaction between the thalamofugal and tectofugal systems.

In partial summary, there is cogent evidence for highly specific auditory and visual sensory fields in the avian telencephalon and suggestive evidence of a somatosensory representation. Furthermore, there are rather striking similarities between these and the mammalian lemniscal pathways. For example, the avian thalamofugal and tectofugal visual pathways are remarkably similar to the mammalian geniculostriate and tecto–thalamo–circumstriate systems, respectively. Consequently, the so-called "striatum" of the avian brain has been shown to contain well-defined, cytologically distinct regions receiving specific sensory projections in a manner reminiscent of mammalian sensory cortex, necessarily casting further doubt on the presumptive homology of much of the avian telencephalon to corpus striatum.

C. The Archistriatum

The archistriatum, a large and heterogeneous nuclear mass in the caudal and ventrolateral portion of the telencephalon, has for a number of years been considered as homologous to the mammalian amygdala (e.g., Haefelfinger, 1968). However, this putative homology has been based on minimal information regarding the cytoarchitecture, nuclear subdivisions, and origins and terminations of the afferent and efferent systems of the archistriatum. A study by Zeier and Karten (1971) contradicts the simple generalization that the entire archistriatum corresponds to the mammalian amygdala and is therefore limbic in nature.

Zeier and Karten (1971) suggest a division of the archistriatum into four major regions: anterior, intermedium, posterior, and mediale. The posterior and mediale divisions are clearly limbic, since they project to medial hypothalamus via the tractus occipitomesencephalicus, pars hypothalami. This tract may be the avian homologue of the stria terminalis; its terminal field is quite distinct from that of the medial forebrain bundle which arises largely from the lobus parolfactorius and anterior preoptic area and terminates in lateral hypothalamus. Consistent with these anatomical findings are results suggesting that the cardiovascular involvements of the archistriatum are largely restricted to its medial and caudal aspects (Cohen, in press; Macdonald & Cohen, 1973).

In contrast, the anterior and intermedium divisions, the anterior two-thirds of the archistriatum, do not project to any limbic structures. They give rise to the nonhypothalamic component of the occipitomesencephalic tract, a bundle consisting of thicker fibers and distributing to the thalamus, optic tectum, tegmentum, lateral reticular formation, lateral pontine nuclei, sensory nuclei of the brainstem, and to some extent rostral levels of the spinal cord. This tract had been previously described (Edinger,

1908; Herrmann, 1922; Zecha, 1962), but its exclusive origin in the archistriatum, and particularly its anterior two-thirds, had not been clearly established.

In its course and distribution, the nonhypothalamic component of the occipitomesencephalic tract resembles the bundle of Bagley in the goat, and this similarity suggests that the anterior two-thirds of the archistriatum may well be somatic rather than limbic in function, and as previously proposed by Zecha (1962) may be comparable to the pericentral cortex of the goat and the sensorimotor cortex of primates. Further suggesting a sharp demarcation between the anteriorintermedium and posteriormediale archistriatal divisions is the mutually exclusive nature of their afferents (Zeier & Karten, 1971), the anterior two-thirds receiving inputs from the frontoarchistriatal tract and anterior commissure.

D. General Comments

As histochemical and hodological data become available for the avian forebrain, the external striatum is yielding discrete populations of neurons that have striking parallels to particular neuronal populations of mammalian brain. In fact, recent anatomical findings indicate that the mammalian neocortex is perhaps the most comparable brain structure to the nonmammalian external striatum with respect to afferent and efferent connections (Nauta & Karten, 1970). The major neural connections of the avian external striatum are quite similar to those of the mammalian neocortex. As already described, specific sensory fields are present in the avian forebrain, at least for the auditory and visual systems. A pyramidal tract arises from at least a part of the Wulst, and a homologue of the ansa lenticularis originates in the paleostriatum primitivum. Another long descending system of a "somatic" nature stems from neurons of the anterior archistriatum, whereas the posterior and medial archistriatum give rise to a pathway possibly homologous to the stria terminalis. Together, findings of this sort certainly indicate that much of the avian external striatum can in no sense be considered "striatal" in nature, and the only clear striatal homologue at present appears to be the paleostriatal complex.

As stated earlier, it is perhaps more reasonable to view the external striatum as comparable to mammalian neocortex, although this certainly should not imply that it is homologous to the entire neocortex. Despite the tremendous proliferation of anatomical information, it must still be recognized that only a small proportion of telencephalic neurons has been characterized in the bird; the same must also be said of the majority of neocortical neurons in mammalian brain.

The accumulation of recent evidence concerning the specific afferentation of definable neuronal populations within the avian thalamus and telencephalon now provides a comprehensible and rational means of understanding the organization of the avian brain. If the organization of the avian forebrain is viewed as assemblages of neurons with specific afferent–efferent relationships in both linear and parallel arrays and less with regard to whether or not the telencephalic neuronal assemblies are laminated, then many of the previously conceived notions based on the gross morphology of the avian forebrain may be discarded. Thus, Karten (1969) has proposed that several of the discrete cell populations within the avian thalamus directly correspond to particular cell groups of the mammalian thalamus. A similar approach may be applied to the interpretation of the telencephalon; that is, structures such as the core of the ectostriatum, Field L and others may be conceived as equivalent to populations of neocortical neurons lying within a single lamina of a designated neocortical area. However, to account for the diversity of neuronal types found in a given cortical area we must postulate that it contains neurons comparable to those of the periectostriatal area, parts of the lateral neostriatum, and portions of the anterior archistriatum. Thus, in a sense, the seemingly specific constellation of neurons constituting a single cortical area may actually be the result of the assembling of neurons of several different ontogenetic origins into a laminated smorgasbord. The implications of this hypothesis, if correct, provide a novel, but far more rational, explanation of the seemingly unusual nature of the reptilian and avian forebrains. Perhaps even more exciting are the implications for the consequent understanding of the possible mode of evolution and the ontogeny of the individual components of the mammalian neocortex.

VII. Concluding Statement

During the past decade, comparative neuroanatomy has experienced a notable rejuvenation, largely as a consequence of advances in experimental methodology and changes in attitudes regarding the organization of nonmammalian vertebrate nervous systems. In particular, recent investigations of the sensory and motor systems of the avian brain have provided a far clearer picture of the overall organization of the avian nervous system and have pointed out numerous similarities to the more widely studied mammalian nervous systems.

Although the present chapter has been limited in scope, we feel it does provide an indication of the present state of knowledge of central neural organization in birds, and it is hoped that the summary diagrams serve to outline representative major systems. Based on such considerations, it seems reasonable to conclude that the main features of the longitudinally organized sensory and motor lemniscal channels are only minimally different from those of mammalia. Such differences as may exist appear primarily in regard to quantitative features, relative shifts in the positions of particular nuclear groups, degree of development of specific structures and morphological details of complex multineuronal fields.

Considerably more research will be necessary before our knowledge of the anatomy of the avian brain even approaches that currently enjoyed by students of mammalian brains. Yet, the discoveries within the past decade have provided good cause to state optimistically that the avian brain is different in detail but not in total nature from the mammalian brain. In fact, in several instances, findings in the avian nervous system have anticipated subsequent discoveries in mammalian forms, as well as clarifying the roles of particular structures in mediating certain important behaviors. Perhaps the most dramatic contribution of avian neuroanatomy, however, pertains to the insights it has provided in our understanding of one of the most seemingly enigmatic problems of neural evolution, namely the origins of the mammalian neocortex (Karten, 1969; Nauta & Karten, 1970).

References

Adamo, N. J. Connections of efferent fibers from hyperstriatal areas in chicken, raven and African love-bird. *Journal of Comparative Neurology*, 1967, **131**, 337–356.

Altman, J., & Carpenter, M. B. Fiber projections of the superior colliculus. *Journal of Comparative Neurology*, 1961, **116**, 157–178.

Baker-Cohen, K. F. Comparative enzyme histochemical observations on sub-mammalian brains. I. Striatal structures in the reptiles and birds. II. Basal structures of the brainstem in reptiles and birds. *Ergebnisse der Anatomie und Entwicklungsgeschichte*, 1968, **40**, 1–70.

Barnikol, A. Zur Morphologie des Nervus Trigeminus der Vögel unter besonderer Berucksichtigung der Accipitres, Cathartidae, Striges, und Passeriformes. *Zeitschrift für wissenshaftliche Zoologie*., 1954, **157**, 285–232.

Belekhova, M. G. Subcortico-cortical relationships in birds. *Neuroscience Translations*, 1968, **2**, 195–203.

Bertler, A., Falck, B., Gottfries, C. G., Ljunggren, L., & Rosengren, E. Some observations of adrenergic connections between mesencephalon and cerebral hemispheres. *Acta Pharmacologica*, 1964, **21**, 283–289.

Biederman-Thorson, M. Auditory responses of neurones in the lateral mesencephalic nucleus (inferior colliculus) of the Barbary dove. *Journal of Physiology*, 1967, **193**, 695–705.

Biederman-Thorson, M. Auditory evoked responses in the cerebrum (Field L) and ovoid nucleus of the ring dove. *Brain Research*, 1970, **24**, 235–246. (a)

Biederman-Thorson, M. Auditory responses of units in the ovoid nucleus and cerebrum (Field L) of the ring dove. *Brain Research*, 1970, **24**, 247–256. (b)

Boord, R. L. Ascending projections of the primary cochlear nuclei and nucleus laminaris in the pigeon. *Journal of Comparative Neurology*, 1968, **133**, 523–541.

Boord, R. L. The anatomy of the avian auditory system. *Annals of the New York Academy of Sciences,* 1969, **167**, 186–198.

Boord, R. L., & Rasmussen, G. L. Projection of the cochlear and lagenar nerves on the cochlear nuclei of the pigeon. *Journal of Comparative Neurology*, 1963, **120**, 463–475.

Botezat, E. Die Nervenendapparate in den Mundteilen der Vögel. *Zeitschrift für wissenshaftliche Zoologie*, 1906, **84**, 205–360.

Brodal, A., Kristiansen, K., & Jansen, J. Experimental demonstration of a pontine homologue in birds. *Journal of Comparative Neurology*, 1950, **92**, 23–69.

Cajal, S. R. Estructura del lóbulo óptico de las aves y origen de los nervios ópticos. *Revista Trimestral de Histologie*, 1889, **3/4**.

Cajal, S. R. *Histologie du système nerveux de l'homme et des vertébrés.* Madrid- Consejo Superior de Investigaciones Cientificas, 1952.

Cobb, S. A comparison of the size of an auditory nucleus (n. mesencephalicus lateralis, pars dorsalis) with the size of the optic lobe in twenty-seven species of birds. *Journal of Comparative Neurology*, 1964, **122**, 271–279.

Cohen, D. H. The neural pathways and informational flow mediating a conditioned autonomic response. In L. DiCara (Ed.), *The limbic and autonomic nervous systems: advances in research.* New York: Plenum Press, in press.

Cohen, D. H. The effects of lesions of the archistriatal complex and its outflow on heart rate conditioning in the pigeon. In preparation.

Cohen, D. H. & Dooley, S. An electrophysiological analysis of the responsiveness of the avian thalamofugal and tectofugal visual pathways to papillary stimulation and diffuse retinal illumination. In preparation.

Cohen, D. H., & Schnall, A. M. Medullary cells of origin of vagal cardioinhibitory fibers in the pigeon. II. Electrical stimulation of the dorsal motor nucleus. *Journal of Comparative Neurology*, 1970, **140**, 321–342.

Cohen, D. H., Schnall, A. M., Macdonald, R. L., & Pitts, L. H. Medullary cells of origin of vagal cardioinhibitory fibers in the pigeon. I. Anatomical studies of peripheral vagus nerve and the dorsal motor nucleus. *Journal of Comparative Neurology*, 1970, **140**, 299–320.

Cowan, W. M. Centrifugal fibers to the avian retina. *British Medical Bulletin*, 1970, **23**, 112–118.

Cowan, W. M., Adamson, L., & Powell, T. P. S. An experimental study of the avian visual system. *Journal of Anatomy*, 1961, **95**, 545–563.

Crosby, E. C., & Woodburne, R. T. The comparative anatomy of the preoptic area and the hypothalamus. *Research Publications of the Association for Research in Nervous and Mental Disorders*, 1940, **20**, 52–169.

Delius, J. D., & Bennetto, K. Cutaneous sensory projections to the avian forebrain. *Brain Research,* 1972, **37**, 205–222.

Diamond, I. T., & Hall, W. C. Evolution of neocortex. *Science*, 1969, **164**, 251–262.

Dorward, P. L. Response patterns of cutaneous mechanoreceptors in the domestic duck. *Journal of Comparative Biochemistry and Physiology*, 1970, **35**, 729–735. (a)

Dorward, P. L. Response characteristics of muscle afferents in the domestic duck. *Journal of Physiology*, 1970, **211**, 1–17. (b)

Dorward, P. L., & McIntyre, A. K. Responses of vibration-sensitive receptors in the interosseus region of the duck's hind limb. *Journal of Physiology*, 1971, **219**, 77–87.

Dow, R. S., & Moruzzi, G. *The physiology and pathology of the cerebellum.* Minneapolis, Minnesota. The Univ. of Minnesota Press, 1957.

Dubbeldam, J., & Karten, H. J. The trigeminal system in the pigeon (*Columba livia*). I. Projections of the Gasserian ganglion. In preparation.

Ebbesson, S. O. E. On the organization of central visual pathways in vertebrates. *Brain, Behavior and Evolution*, 1970, **3**, 178–194.

Edinger, L. *Vorlesungen über den Bau der Nervosen Centralorgane des Menschen und der Thiere.* Leipzig: F. C. W. Vogel, 1908.

Edinger, L., Wallenberg, A., & Holmes, G. Untersuchungen über des Gehirn der Tauben. *Anatomischer Anzeiger*, 1903, **15**, 245–271.

Erulkar, S. D. Tactile and auditory areas in the brain of the pigeon. *Journal of Comparative Neurology*, 1955, **103**, 421–457.

Erulkar, S. D. Comparative aspects of spatial localization of sound. *Physiological Reviews*, 1972, **52**, 237–360.

Flourens, P. *Recherches expérimentales sur les propriétés et les fonctions du système nerveux dans les animaux vertébrés.* Paris: Baillière, 1824.

Fox, C. A., Hillman, D. E., Siegesmund, K. A., & Sether, L. A. The primate globus pallidus and its feline and avian homologues: A Golgi and electron microscopic study. In R. Hassler & H. Stephan (Eds.), *Evolution of the forebrain.* New York: Plenum Press, 1967.

Fuxe, K. Evidence for the existence of monoamine neurons in the central nervous system. IV. Distribution of monoamine nerve terminals in the central nervous system. *Acta Physiologica Scandinavica*, 1965, **64**, Supplement 247.

Haartsen, A. B., & Verhaart, W. J. C. Cortical projections to brainstem and spinal cord in the goat by way of pyramidal tract and bundle of Bagley. *Journal of Comparative Neurology*, 1967, **129**, 189–202.

Haelfelfinger, H. R. *Beiträge zur vergleichenden ontogenese des Vörderhirns bei vogeln.* Basel: Verlag Helbing und Lichtenhahn, 1958.

Harman, A. L., & Phillips, R. E. Responses in the avian midbrain, thalamus and forebrain evoked by click stimuli. *Experimental Neurology*, 1967, **18**, 276–286.

Herrmann, A. Beiträge zur Anatomie des Vogelhirns. *Zeitschrift für Anatomie und Entwicklungsgeschichte*, 1922, **63**, 415–418.

Hodos, W. Color discrimination deficits after lesions of nucleus rotundus in pigeons. *Brain, Behavior and Evolution*, 1969, **2**, 185–200.

Hodos, W., & Karten, H. J. Brightness and pattern discrimination deficits in the pigeon after lesions of nucleus rotundus. *Experimental Brain Research*, 1966, **2**, 151–167.

Hodos, W., & Karten, H. J. Visual intensity and pattern discrimination deficits after lesions of ectostriatum in pigeons. *Journal of Comparative Neurology*, 1970, **140**, 53–68.

Huber, J. F. Nerve roots and nuclear groups in the spinal cord in the pigeon. *Journal of Comparative Neurology*, 1936, **65**, 43–91.

Huber, G. C., & Crosby, E. C. The nuclei and fiber paths of the avian diencephalon, with consideration of telencephalic and certain mesencephalic centers and connections. *Journal of Comparative Neurology*, 1929, **48**, 1–225.

Ilyichev, V. D. External part of auditory analyzer in birds. I. General morphology and functional peculiarities. *Zoologisches Zentralblatt*, 1960, **39**, 1871–1878.

Jungherr, E. L. The neuroanatomy of the domestic fowl (*Gallus domesticus*). *Avian Diseases*, Special Issue, 1969.

Juorio, A. V., & Vogt, M. Monoamines and their metabolites in the avian brain. *Journal of Physiology*, 1967, **189**, 489–518.

Kaiser, L. L'innervation segmentale de la peau chez le pigeon (*Columba livia var. domestica*). *Archives néerlandaises de physiologie de l'homme et des animaux*, 1924, **9**, 299–379.

Kalischer. O. Das Grosshirn der Papageien in anatomischer und physiologischer Beziehung. *Abhandlungen der Preussischen Akademieder Wissenschaften*, 1905, 1–105.

Kappers, C. U. A., Huber, G. C., & Crosby, E. C. *The comparative anatomy of the nervous systems of vertebrates, including man.* New York: Macmillan Co., 1936.

Karten, H. J. Ascending pathways from the spinal cord in the pigeon. *Proceedings of the sixteenth International Congress of Zoology*, 1963, **2**, 23.

Karten, H. J. Projections of the cerebellar nuclei of the pigeon (*Columba livia*). *Anatomical Record*. 1964, **148**, 297–298.

Karten, H. J. Projections of the optic tectum of the pigeon (*Columba livia*). *Anatomical Record*, 1965, **151**, 369.

Karten, H. J. The organization of the ascending auditory pathway in the pigeon (*Columba livia*). I. Diencephalic projections of the inferior colliculus (nucleus mesencephalicus lateralis, pars dorsalis). *Brain Research*, 1967, **6**, 409–427.

Karten, H. J. The ascending auditory pathway in the pigeon (*Columba livia*). II. Telencephalic projections of the nucleus ovoidalis thalami. *Brain Research*, 1968, **11**, 134–153.

Karten, H. J. The organization of the avian telencephalon and some speculations on the phylogeny of the amniote telencephalon. *Annals of the New York Academy of Sciences*, 1969, **167**, 164–179.

Karten, H. J. Efferent projections of the wulst of the owl. *Anatomical Record*, 1971, **169**, 353.

Karten, H. J. The trigeminal system in the pigeon. (*Columba livia*). II. Projections of the principal nucleus of the trigeminus. In preparation.

Karten, H. J., & Dubbeldam, J. L. The organization and projections of the paleostriatal complex in the pigeon (*Columba livia*). *Journal of Comparative Neurology*, 1973, **148**, 61–90.

Karten, H. J., & Hodos, W. *A stereotaxic atlas of the brain of the pigeon* (*Columba livia*). Baltimore, Maryland: Johns Hopkins Press, 1967.

Karten, H. J., & Hodos, W. Telencephalic projections of the nucleus rotundus in the pigeon (*Columba livia*). *Journal of Comparative Neurology*, 1970, **140**, 35–52.

Karten, H. J., Hodos, W., Nauta, W. J. H., & Revzin, A. M. Neural connections of the "Visual Wulst" of the avian telencephalon. Experimental studies in the pigeon (*Columba livia*) and owl (*Speotyto conicularis*). *Journal of Comparative Neurology*, 1973, **150**, 253–277.

Karten, H. J., & Nauta, W. J. H. Organization of retinothalamic projections in the pigeon and owl. *Anatomical Record*, 1968, **160**, 373.

Karten, H. J., & Revzin, A. M. The afferent connections of the nucleus rotundus in the pigeon. *Brain Research*, 1966, **2**, 368–377.

Kimberly, R. P., Holden, A. L., & Bamborough, P. Response characteristics of pigeon forebrain cells to visual stimulation. *Vision Research*, 1971, **11**, 475–478.

Koelle, G. B. The histochemical localization of cholinesterases in the central nervous system of the rat. *Journal of Comparative Neurology*, 1954, **100**, 222–235.

Langley, J. N. On the sympathetic system of bird, and on the muscles which move the feathers. *Journal of Physiology*, 1904, **30**, 221–252.

Larsell, O. The development and subdivisions of the cerebellum of birds. *Journal of Comparative Neurology*, 1948, **89**, 123–190.

Larsell, O. *The comparative anatomy and histology of the cerebellum from myxinoids through birds.* Minneapolis, Minnesota: The Univ. of Minnesota Press, 1967.

Lashley, K. S. Persistent problems in the evolution of mind. *Quarterly Review of Biology*, 1949, **24**, 28–42.

Lavail, J. H., & Cowan, W. M. The development of the chick optic tectum. I. Normal morphology and cytoarchitectonic development. *Brain Research*, 1971, **28**, 391–420.

Leonard, R. B., & Cohen, D. H. The spinal distribution of lumbar, thoracic and cervical dorsal roots in the pigeon (*Columba livia*). In preparation.

Macdonald, R. L., & Cohen, D. H. Cells of origin of sympathetic pre- and postaganglionic cardioacceleratory fibers in the pigeon. *Journal of Comparative Neurology*, 1970, **140**, 343–358.

Macdonald, R. L., & Cohen, D. H. Heart rate and blood pressure responses to electrical stimulation of the central nervous system in the pigeon (*Columba livia*). *Journal of Comparative Neurology*, 1973, **150**, 109–136.

Malinovsky, L. Contributions to the anatomy of the vegetative nervous system in the neck and thorax of the domestic pigeon. *Acta Anatomica* 1962, **50**, 326–347.

Medway, Lord. Echo-Location among Collocalia. *Nature*, 1959, **184**, 1352–1353.

Münzer, E., & Wiener, H. Beiträge zur Anatomie und Physiologie des Centralnervensystems der Taube. *Monotsschrift für Psychiatrie und Neurologie*, 1898, **3–4**, 379–406.

Nauta, W. J. H., & Karten, H. J. A general profile of the vertebrate brain, with sidelights on the ancestry of cerebral cortex. In G. C. Quarton, T. Melnechuk, & F. O. Schmitt (Eds.), *The neurosciences: Second study program.* New York: The Rockefeller Univ. Press, 1970.

Nieuwenhuys, R. Comparative anatomy of the spinal cord. In J. C. Eccles, & J. P. Schadé (Eds.), *Progress in brain research, Vol. 11: Organization of the spinal cord.* Amsterdam: Elsevier, 1964.

Novick, A. Acoustic orientation in the cave swiftlet. *Biological Bulletin*, 1959, **117**, 497–503.

Oscarsson, O., Rosen, I., & Uddenberg, N. Organization of ascending tracts in the spinal cord of the duck. *Acta Physiologica Scandinavica*, 1963, **59**, 143–153.

Papez, J. W. *Comparative neurology.* New York: Thomas Y. Crowell Co., 1929.

Parker, D. M., & Delius, J. D. Visual evoked potentials in the forebrain of the pigeon. *Experimental Brain Research*, 1972, **14**, 198–209.

Payne, R. S. How the barn owl locates prey by hearing. *In the living bird: First annual of the Cornell laboratory of ornithology.* Ithaca, New York: Cornell Univ. Press, 1962.

Pearson, R. *The avian brain.* London: Academic Press, 1972.

Portmann, A., & Stingelin, W. The central nervous system. In A. J. Marshall (Ed.), *Biology and comparative physiology of birds (Vol. 11).* New York: Academic Press, 1961.

Powell, T. P. S., & Cowan, W. M. The thalamic projection upon the telencephalon in the pigeon (*Columba livia*). *Journal of Anatomy*, 1961, **95**, 78–109.

Powell, T. P. S., & Kruger, L. The thalamic projection upon the telencephalon in Lacerta viridis. *Journal of Anatomy*, 1960, **94**, 528–542.

Quilliam, T. A., & Armstrong, J. Mechanoreceptors. *Endeavour*, 1963, **22**, 55–60.

Revzin, A. M. A specific visual projection area in the hyperstriatum of the pigeon (*Columba livia*). *Brain Research*, 1969, **15**, 246–249.

Revzin, A. M., & Karten, H. J. Rostral projections of the optic tectum and the nucleus rotundus in the pigeon. *Brain Research*, 1966, **3**, 264–276.

Rogers, F. T. Studies on the brain stem. VI. An experimental study of the corpus striatum of the pigeon as related to various instinctive types of behavior. *Journal of Comparative Neurology*, 1922, **35**, 21–60.

Rose, M. Uber die Cytoarchitektonische Gliederung des Vorderhirns der Vögel. *Zeitschrift für Psychologie und Neurologie*, 1914, **21**, 278–352.

Sanders, E. B. A consideration of certain bulbar, midbrain, and cerebellar centers and fiber tracts in birds. *Journal of Comparative Neurology*, 1929, **49**, 155–222.

Schneider, G. E. Two visual systems. *Science*, 1969, **163**, 895–902.

Schwartzkopf, J. Auditory, proprioception, equilibrium and cutaneous receptors. In D. S. Farner & J. R. King (Eds.), *Avian biology.* New York: Academic Press, in press.

Sillman, A. Avian vision. In D. S. Farner & J. R. King (Eds.), *Avian biology.* New York: Academic Press, in press.

Smith, C. A., & Takasaka, T. Auditory receptor organs of reptiles, birds, and mammals. In W. D. Neff (Ed.), *Contributions to sensory physiology, Vol. 5.* New York: Academic Press, 1971.

Ssinelnikov, R. Das intramurale nervensystem des vogelherzens. *Zeitschrift für Anatomie und Entwicklungsgeschichte*, 1928, **86**, 563–568.

Stingelin, W. Qualitative und quantitative Untersuchungen an Kerngebieten der Medulla oblongata bei Vögeln. *Br.*, 1965, **6**, 1–116.

Stopp, P. E., & Whitfield, I. C. Unit responses from brain stem nuclei in the pigeon. *Journal of Physiology*, 1961, **158**, 165–177.

Streeter, G. L. The structure of the spinal cord of the ostrich. *American Journal of Anatomy*, 1904, **3**, 1–27.

Takasaka, T., & Smith, C. A. The structure and innervation of the pigeon's basilar papilla. *Journal of Ultrastructure Research*, 1971, **35**, 20–65.

Terni, T. Ricerche anatomiche sul sistema nervoso autonomo degli uccelli. *Archivio italiano di anatomia e di embriologia*, 1923, **21**, 55–86.

Van den Akker, L. M. *An anatomical outline of the spinal cord of the pigeon*. The Netherlands: Royal VanGorcum Ltd., 1970.

Wallenberg, A. Eine zentrifugal leitende direkte Verbindung der frontalen Vorderhirnbasis mit der Oblongata (+ Rückenmark?) bei der Ente. *Anatomischer Anzeiger*, 1902, **22**, 289–292.

Wallenberg, A. Der Ursprung des Tractus isthmo-striatus (odor bulbostriatus) der Taube. *Neurologisches Zentralblatt*, 1903, **22**, 98–101.

Wallenberg, A. Neue untersuchungen über den Hirnstamm der Taube. *Anatomischer Anzeiger*, 1904, **24**, 357–369.

Watanabe, T., Isomura, G., & Yasuda, M. Comparative and topographical anatomy of the fowl. XXX. Distribution of nerves in the oculomotor and ciliary muscles. *Japanese Journal of Veterinary Science*, 1967, **29**, 151–158.

Winkelmann, R. K., & Myers, T. T. The histochemistry and morphology of the cutaneous sensory end-organs of the chicken. *Journal of Comparative Neurology*, 1961, **117**, 27–36.

Winter, P. Vergleichende Qualitative und Quantitative untersuchungen an der Hörbahn von Vogeln. *Zeitschrift für Morphologie und Ökologie der Tiere*, 1963, **52**, 365–400.

Yasuda, M. Comparative and topographical anatomy of the fowl. XXXIV. Distribution of cutaneous nerves of the fowl. *Japanese Journal of Veterinary Science*, 1964, **26**, 241–248.

Zecha, A. The "pyramidal tract" and other telencephalic efferents in birds. *Acta morphologica neerlando-scandinavica*, 1962, **5**, 194–195.

Zeier, H., & Karten, H. J. The archistriatum of the pigeon: Organization of afferent and efferent connections. *Brain Research*, 1971, **31**, 313–326.

Zeigler, H. P., & Karten, H. J. Brain mechanisms and feeding behavior in the pigeon. (*Columba livia*). Quinto-frontal structures. *Journal of Comparative Neurology*, 1973, **152**, 59–81.

Zeigler, H. P., Karten, H. J., & Green, H. L. Neural control of feeding in the pigeon. *Psychonomic Science*, 1969, **15**, 156–157.

Zeigler, H. P., & Witovsky, P. The main sensory trigeminal nucleus in the pigeon: A single-unit analysis. *Journal of Comparative Neurology*, 1968, **134**, 255–264.

3 NEUROBEHAVIORAL FINDINGS

Hearing and Vocalization in Songbirds

Masakazu Konishi[1]
Princeton University

Bird songs and calls have played a significant role in studies of animal communication. Field and laboratory investigations have yielded ample evidence to support the contention that bird songs and calls have communicatory functions. Some preliminary studies were initiated to discover how birds use different physical parameters of sound to code their vocal signals (Brémond, 1967; Falls, 1963). The most obvious physical feature of bird vocalizations pertains to the frequency of sound. Some birds produce low-pitched sounds, some high-pitched ones. However, we know little about what songbirds can hear; are birds capable of hearing the entire frequency range of their own voices? Can different species hear one another? Is the songbird's ear more sensitive than the human ear? Partial answers to some of these questions are now available (Konishi, 1969, 1970).

In this chapter an attempt is made to correlate anatomical, neurophysiological and behavioral data on the auditory system with its function in relation to vocalizations.

I. Auditory Mechanisms

In contrast to the coiled mammalian inner ear, the avian cochlea is almost straight. The elasticity of the avian basilar membrane varies system-

[1] The work reported here was supported in part by Grant No. GB12729 from the National Science Foundation.

atically along its length (Békésy, 1944). The basal part of the membrane (near the oval window) is stiffest, the middle part more elastic, and the apical part is the most flexible. This gradient in elasticity is responsible for the differential displacement of the membrane in response to sounds of different frequencies.

The primary auditory fibers preserve the orderly sequence of innervation on the basilar membrane as they travel from the inner ear to the medulla oblongata. There, they terminate in the cochlear nucleus in a systematic manner, suggesting a point-to-point projection of the basilar membrane on the hind brain (Fig. 1).

A primary or secondary auditory neuron reacts to a restricted range of frequencies within which its response is a simple function of frequency and intensity of sound. The frequency to which a neuron has the lowest threshold is called the characteristic frequency of that cell. Different neurons have different characteristic frequencies. In the avian cochlear nucleus neurons are systematically arranged according to their characteristic frequencies (Fig. 2). This arrangement is also known in mammals and called the tonotopic organization (Rose, Galambos, & Hughes, 1960). The avian cochlear nucleus consists of two parts, the nucleus magnocellularis and the nucleus angularis.

In the magnocellular nucleus the characteristic frequency increases in the rostromedial direction. The rostromedial third of the nucleus contains higher characteristic frequencies, the middle third includes intermediate frequencies, and the caudal third lower frequencies. The nucleus angularis exhibits a much clearer tonotopic organization than does the nucleus magnocellularis. In the angular nucleus the highest characteristic frequency occurs in the most caudal and lateral end. The characteristic frequency decreases gradually as more medial or rostral areas are probed. A conspicuous feature of the angular nucleus is the presence of a tonotopic organization in the vertical direction. The most superficial layer, which corresponds to the floor of the fourth ventricle in the most caudal part of the nucleus angularis, contains the highest characteristic frequency. As the electrode descends deeper in the nucleus it encounters cells with lower characteristic frequencies.

The same tonotopic organization has been found in all the species studied so far. Interspecific variation occurs only in the range of characteristic frequencies and in their relative abundance. The tonotopic organization in the avian cochlear nucleus is obviously correlated with the orderly pattern of fiber projection mentioned before (compare Figs. 1 and 2). Comparison of these two sets of data indicates that the areas of the cochlear nucleus where the fibers from the basal part of the basilar membrane terminate yield high characteristic frequencies, the zones receiving the fiber innervating the medial part contain intermediate characteristic frequencies and the regions connected to the apical fibers show low characteristic frequencies.

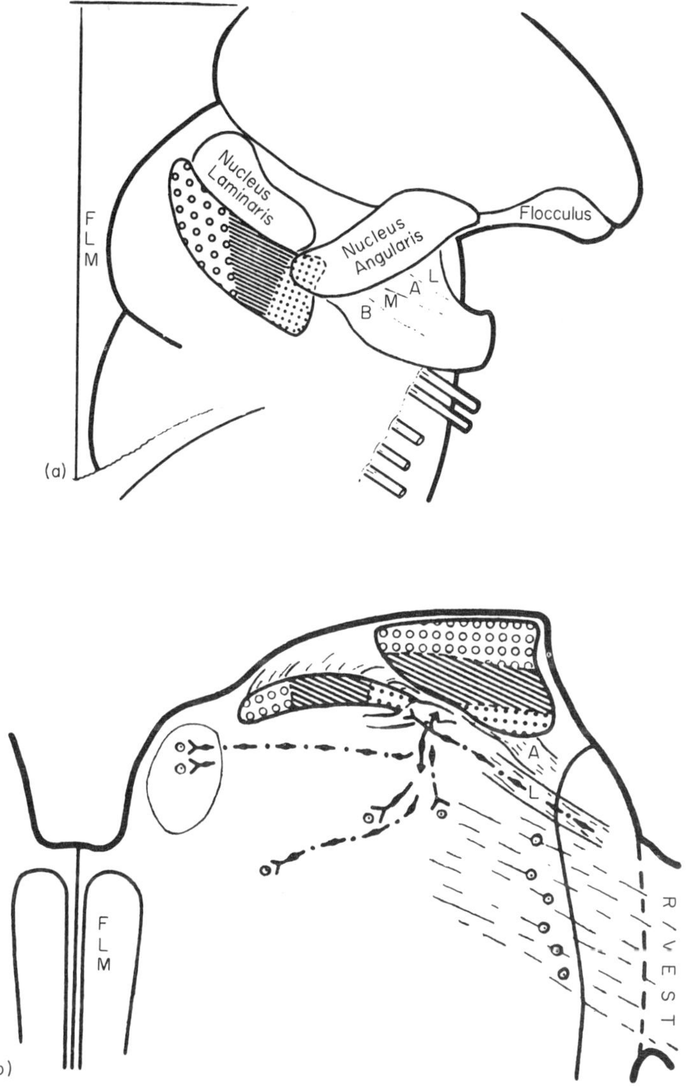

Fig. 1. (a). Diagram of the dorsal aspect of the pigeon cochlear nucleus. Cochlear fibers preserve the spatial order of innervation on the basilar membrane as they travel from the inner ear to the hind brain. B, M, A indicate fiber bundles innervating the basal, middle and apical parts of the basilar membrane respectively. L is for lagena fibers. Cochlear fibers innervating the basal one-third of the basilar membrane project onto the rostromedial third of the *nucleus magnocellularis*, as indicated by circles. The middle and apical thirds of the membrane project onto the middle (oblique lines) and caudal (dots) thirds of the same nucleus. (b). Diagram showing a cross section of the *nucleus angularis*. Basal, middle, and apical cochlear fibers are systematically arranged in that sequence from the most superficial to bottom layers. [From Boord & Rasmussen (1963).]

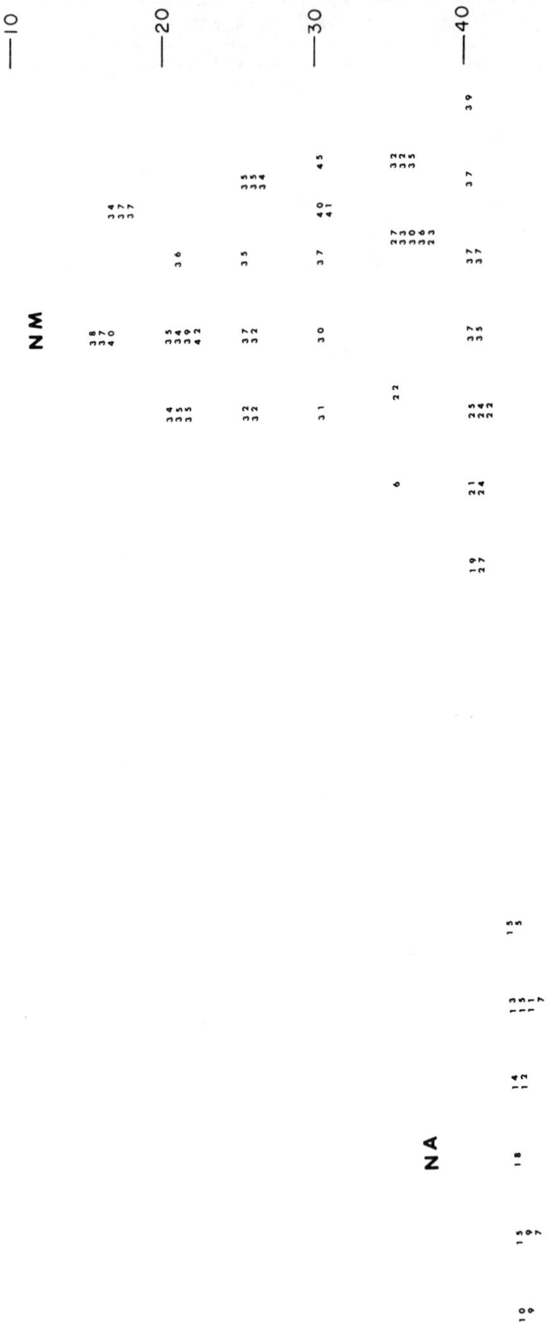

—10

NM

38 37 40 · 34 37 37

—20

34 35 35 · 35 34 39 42 · 36

32 32 · 37 32 · 35 · 35 35 34

—30

31 · 30 · 37 · 40 41 · 45

6 · 22 · 27 33 30 36 23 · 32 32 35

—40

19 27 · 21 24 · 25 24 22 · 37 35 · 37 37 · 37 · 39

NA

10 9 · 15 9 7 · 18 · 14 12 · 13 15 11 7 · 15 5

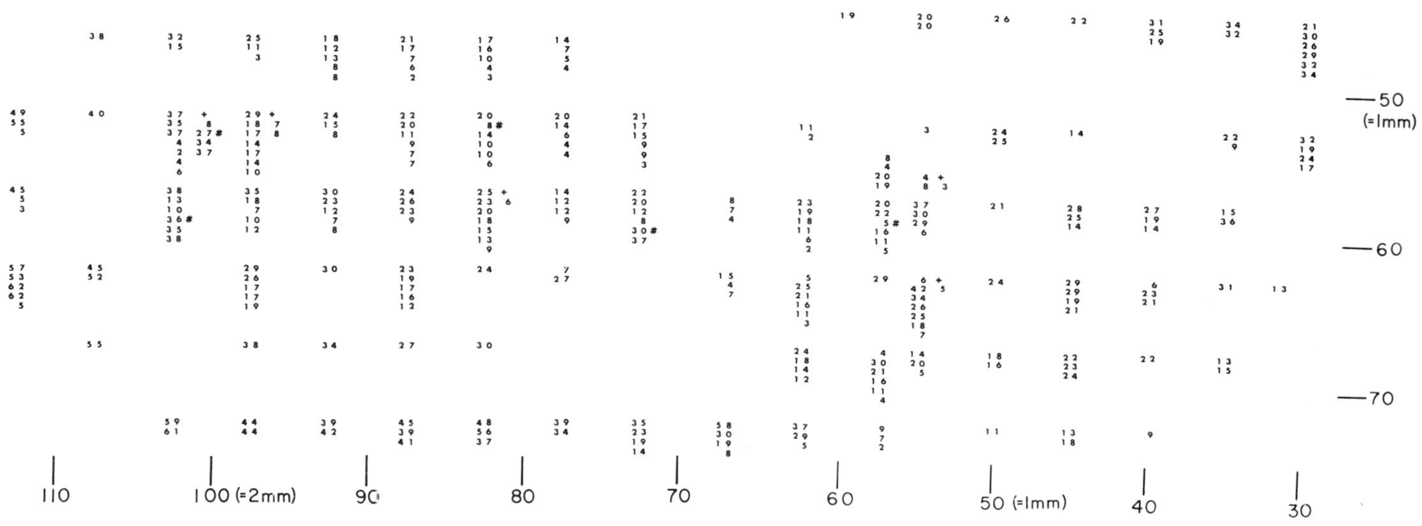

—50 (=1mm)

—60

—70

110 | 100 (=2mm) | 90 | 80 | 70 | 60 | 50 (=1mm) | 40 | 30

Fig. 2. Tonotopic organization in the cochlear nucleus of the house sparrow. Numbers indicate characteristic frequencies (CF) in hundreds of hertz. Each column of numbers indicates the vertical sequence in which the units were encountered during one unrepeated penetration at the coordinate point occupied by the topmost number. Characteristic frequencies increase toward the medial margin and the anterior tip in the *nucleus magnocellularis* (NM). In the *nucleus angularis* (NA) CFs increase caudad, laterad and dorsad. + denotes the continuation of the columns from left to right for the lack of space. # points out irregularities in tonotopic arrangement. Abscissa; distance from the midline of the medulla. Ordinate; distance from the ventral lip of the aqueduct of Sylvius. Unit of scale 1/50 mm.

The tonotopic organization, the pattern of fiber projection, and the site-dependent frequency response of the membrane indicate that the site of its innervation on the membrane determines the frequency response of an auditory fiber. Thus, birds use the same principle of mechanical wave analysis employed by mammals. Unless compelling reasons arise, there is no need to postulate separate mechanisms of frequency analysis for birds and mammals.

There is another phenomenon that deserves a close examination. Each point on the basilar membrane can be thought to have a characteristic frequency, even though the membrane is continuous. In mammals the length of the membrane and the interval between characteristic frequencies are not linearly correlated. For example, in the cat's basilar membrane the points tuned to 250 and 500 Hz are separated by 1.7 mm, whereas a distance of 2.6 mm corresponds to the interval between 8000 and 16,000 Hz (Schuknecht, 1960). This nonlinear distribution is inevitable, if a wide range of frequencies is to be accommodated in a limited length. The bird cochlea is generally much shorter than the mammalian cochlea. Pumphrey (1961), assuming the same nonlinear relationship in the shorter avian basilar membrane, correctly predicted a lower range of maximum audible frequencies for birds. Whether the avian basilar membrane is similar to the mammalian one in this respect is unknown. However, the relationship between characteristic frequencies and their spatial separation in the cochlear nucleus is nonlinear (see Fig. 2). If the basilar membrane projects onto the cochlear nucleus linearly, then the distribution of characteristic frequencies in the membrane will be nonlinear as in mammals.

The orderly arrangement of auditory neurons enables us to predict and sample the entire range of characteristic frequencies quickly and systematically. It is not always possible or easy to correlate single-unit and behavioral data. The study of auditory neurons would have little use if it could not be related to hearing. In birds the frequency–intensity characteristic of hearing, i.e., the audibility curve, can be constructed from single-unit data. If we plot the thresholds of auditory neurons for their characteristic frequencies, the resultant distribution approximates the audibility curve (Fig. 3). This relationship has been established at least for two bird species, the canary (*Serinus canarius*) and the starling (*Sturnus vulgaris*), and the cat (Kiang, 1966) in which both behavioral and neurophysiological data are available.

Another predictable relationship is in the number of neurons in each interval of frequencies. When neurons are classified by their characteristic frequencies, more of them fall in the most sensitive range of hearing than in either end of the audible frequency range. How this skewed distribution of neurons is related to auditory perception is not clear in birds. Perhaps more parallel lines in the best frequency range are useful for simultaneous analysis

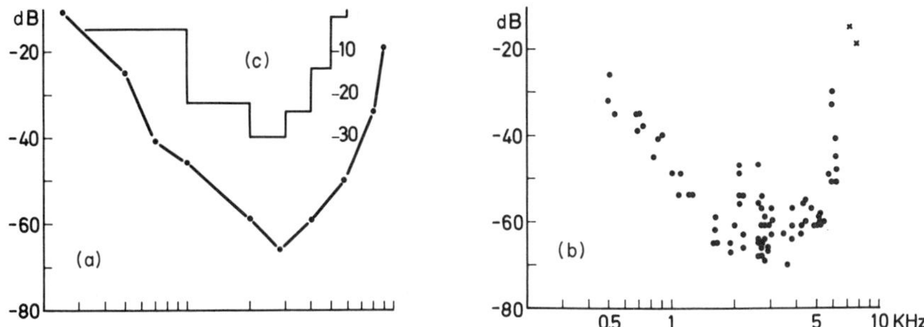

Fig. 3. Behavioral audibility curve, single unit thresholds and numbers in the canary. (a) Behavioral audiometric curve, average of four birds. [From Dooling (1969).] (b) Thresholds of single units at their CFs in one individual. × indicates that units with the CF of 6.2 KHz (near the highest limit) respond to still higher frequencies 7.2–7.7 KHz with a corresponding rise in their thresholds. (c) Numbers of units classified according to their CF's from another bird.

of different types of information. This leads us to the consideration of vocal signals in relation to hearing.

II. Vocalization and Audition

Although it is natural to predict a good match between hearing and vocalization in each species, there is no a priori reason to suppose such an agreement. Birds may be able to hear a restricted range of frequencies in their vocal signals or they may register a much wider range of frequencies than that in their own voice. Dooling, Mulligan, and Miller (1971), based on their study of the canary, concluded that the dominant frequency range of the species' song coincided with the best audible frequency range, whereas the range of dominant frequencies in the canary calls extended into less sensitive ranges of hearing. Whether the observation made on a domestic species applies to wild birds remains to be tested. Available information on wild songbirds is inadequate for comparison with their results, mainly because of the difficulty in recording all existing vocalizations of each species in the field.

However, there is one clear correlation between voice and hearing. The maximum audible frequency of a species tends to be correlated with the average range of its vocal frequencies. A similar relationship does not hold for the minimum audible frequency; all birds studied so far are sensitive to frequencies far below the lowest frequency in their vocalizations. For

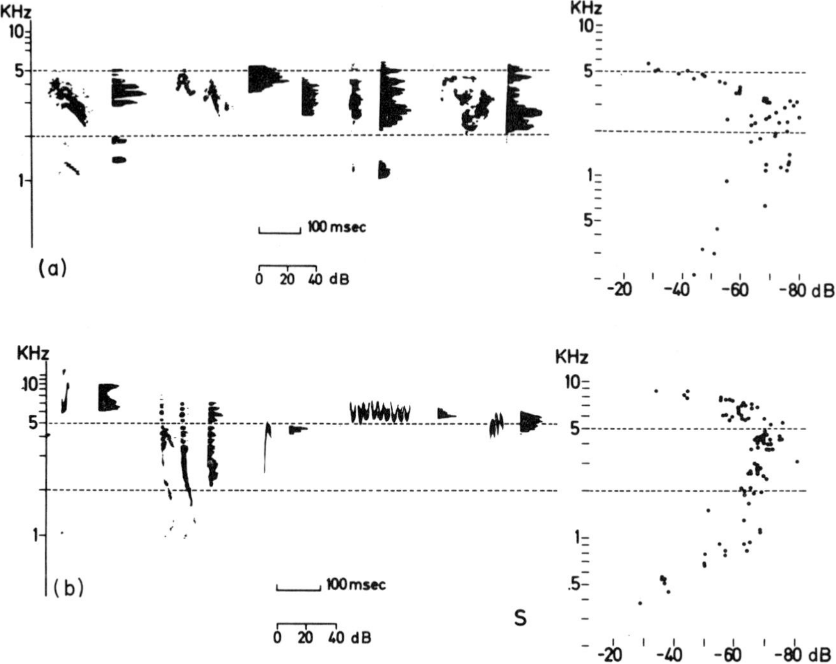

Fig. 4. Single-unit thresholds and vocalizations for (a) house sparrow and (b) slate-colored junco. Narrow-band time–frequency sound-spectrograms of vocalizations and their frequency-amplitude profiles are compared with single-unit thresholds obtained from one individual of each species; S denotes the typical sound component in the species song. Sound frequencies are shown in logarithmic scales. All decibel values with a negative sign refer to 1 dyne cm^{-2}. The time scale is for time-frequency spectrograms, and the bar scale in decibels is for frequency-amplitude profiles.

example, the house sparrow (*Passer domesticus*) and the slate-colored junco (*Junco hyemalis*) differ from each other in their vocal frequencies (Fig. 4). Most of the sparrow's vocalizations are below 5000 Hz, whereas those of the junco are above that frequency. The two species clearly differ from each other in the range of characteristic frequencies and the distribution of single-unit thresholds; the highest characteristic frequency in the sparrow is about 6000 Hz, whereas that of the junco reaches 9000 Hz. Whereas the sparrow's auditory sensitivity for 6000 Hz is about -25 dB (re 1 dyne cm^{-2}), that of the junco for the same frequency is greater by as much as 50 dB. In the low-frequency end of the audibility curve, the two species do not differ from each other.

Comparison of voice and hearing among several species shows that the songbirds can hear most of the frequencies in their vocalizations. Since similar ranges of vocal frequencies occur among different species, including those which live in the same habitat, it is rather unlikely that the range of frequency alone is used in species recognition. Recent field studies show that songbirds can use more than one type of auditory cue for species recognition (Falls, 1963).

Despite the type of correlation just mentioned, the songbird's ear does not show the kind and degree of specialization for communication seen in orthopterans and frogs. The tympanum of the cricket (*Teleogryllus commodus*) is mechanically tuned to the frequency range of its song (Johnstone, Saunders, & Johnstone, 1970). The distribution of characteristic frequencies in the auditory nerve of the bullfrog (*Rana catesbeiana*) matches that of sound spectral energy in its territorial call (Frishkopf & Goldstein, 1963). The lack of peripheral specialization in the avian ear suggests that the higher auditory centers process biologically relevant information. Biederman-Thorson (1970) reported that the forebrain auditory center, field L of the Ringdove (*Streptopelia risoria*) contained neurons which responded better or exclusively to complex sounds such as hisses and squeaks than to tone bursts.

With the knowledge of the avian auditory pathway that has been most clearly mapped out (Boord, 1969; Karten, 1967, 1968), songbirds can offer unique opportunities for various types of studies of general applicability, such as the neural mechanisms of signal processing, of learning and of behavioral development.

References

Békésy, G. V. Über die mechanische Frequenzanalyse in der Schnecke verschiedener Tiere. *Akustische Zeitschrift*, 1944, **8**, 3–11.

Biederman-Thorson, M. Auditory responses of units in the ovoid nucleus and cerebrum (Field L) of the ring dove. *Brain Research*, 1970, **24**, 247–256.

Boord, R. L. The anatomy of the avian auditory system. *Annals of the New York Academy of Science*, 1969, **167**, 186–198

Boord, R. L. & Rasmussen, G. L. Projection of the cochlear and lagenar nerves on the cochlear nuclei of the pigeon. *Journal of Comparative Neurology*, 1963, **120**, 463–475.

Brémond, J. C. Reconnaissance de schemas reactogenes lies a l'information contenue dans le chant territorial du rouge-gorge. *Proceedings of the 14th International Ornithological Congress*, 1967. Pp. 217–229.

Dooling, R. J. An audibility curve for the common canary as determined by instrumental avoidance conditioning. Master's Thesis, St. Louis Univ., 1969.

Dooling, R. J., Mulligan, J. A., & Miller, J. D. Auditory sensitivity and song spectrum of the common canary (*Serinus canarius*). *Journal of the Acoustical Society of America*, 1971, **50**, 702–709.

Falls, J. B. Properties of bird song eliciting responses from territorial males. *Proceedings of the 13th International Ornithological Congress*, 1963. Pp. 259–271.

Frishkopf, L. S. & Goldstein, M. M., Jr. Responses to acoustic stimuli from single units in the eighth nerve of the bullfrog. *Journal of the Acoustical Society of America*, 1963, **35**, 1219–1228.

Johnstone, B. M., Saunders, J. C., & Johnstone, J. R. Tympanic membrane response in the cricket. *Nature*, 1970, **227**, 625–626.

Karten, H. J. The organization of the ascending pathway in the pigeon (*Columba livia*). I. Dienephalic projections of the inferior colliculus (*nucleus nesencephalicus lateralis pars dorsalis*). *Brain Research*, 1967, **6**, 409–427.

Karten, H. J. The ascending auditory pathway in the pigeon (*Columba livia*). II. Telencephalic projections of the nucleus ovoidalis thalami. *Brain Research*, 1968, **11**, 134–153.

Kiang, N. Y. S. *Discharge patterns of single fibers in the cat's auditory nerve*. Cambridge, Massachusetts Press, 1966.

Konishi, M. Hearing, single-unit analysis, and vocalizations in songbirds. *Science*, 1969, **166**, 1178–1181.

Konishi, M. Comparative neurophysiological studies of hearing and vocalizations in songbirds. *Zeitschrift für Vergleichende Physiologie*, 1970, **66**, 257–272.

Pumphrey, R. J. Sensory organs: Hearing. In A. J. Marshall (Ed.), *Biology and comparative physiology of birds*. Vol. 2, New York: Academic Press, 1961.

Rose, J. E., Galambos, R., & Hughes, R. J. Organization of frequency sensitive neurons in the cochlear nuclear complex of the cat. In G. L. Rasmussen & W. F. Windle (Eds.), *Neural mechanisms of the auditory and vestibular systems*. Springfield, Illinois: Thomas, 1960. Chapter 9.

Schuknecht, H. F. Neuroanatomical correlates of auditory sensitivity and pitch discrimination in the cat. In G. L. Rasmussen & W. F. Windle (Eds.), *Neural mechanisms of the auditory and vestibular systems*. Springfield, Illinois: Thomas, 1960. Chapter 6.

Brain Stimulation Parameters Affecting Vocalization in Birds[1]

Jerram L. Brown[1]
University of Rochester

I. Anatomical Considerations

The neural basis of vocalization was reviewed in 1969 (Brown, 1969). Since then, considerable progress has been made in mapping brain areas from which vocalizations and related behaviors can be evoked by electrical stimulation of the brain (ESB) in a variety of species, and in studying the functional organization of the vocalization system (mallard duck: Maley, 1969; *Coturnix* quail: Potash, 1970a; domestic chicken: Andrew, 1969; Murphey & Phillips, 1967; Phillips & Youngren 1971; Phillips, Youngren, & Peek, 1972; Peek & Phillips, 1971; passerine birds: Newman, 1970; Nottebohm, 1971; Brown, 1971, 1973; gulls and pigeons: Delius, 1971; ring doves: Harwood & Vowles, 1967). In addition, work on the auditory system (Newman, 1970; Potash 1970b) has partially clarified the apparently close anatomical relationship between auditory and vocalization systems in the avian brain.

A. The Vocalization System

For convenience, those parts of the brain from which vocalization can be evoked by weak to moderate electrical stimulation will be referred to as

[1] This research was supported by Grants MH 07700 from the National Institute of Mental Health and NB 08450 from the National Institute of Neurological Diseases and Stroke, United States Public Health Service.

the vocalization system. Certainly, other areas are at least indirectly involved in vocalization, but it is with those areas which yield positive results that we are primarily concerned in these early stages of investigation. The vocalization system in birds has basic anatomical and functional similarities with corresponding parts of the mammalian brain, especially those concerned with threat behavior. Points of similarity are the existence of responsive zones in the archistriatum of the forebrain, the hypothalamus, the midbrain, and the medulla in birds and mammals. Evidence that these areas are linked in a descending series is adequate in mammals but incomplete in birds. Perhaps most convincing is the finding that bilateral lesions in the midbrain vocalization area in and near the torus semicircularis of blackbirds (Brown, 1965) and chickens (R. E. Phillips, Univ. of Minnesota, personal communication) causes apparent muteness.

The anatomical connections between the archistriatum (in the forebrain), the midbrain, and the medulla are only vaguely known. The tractus occipitomesencephalicus has long been considered to be possibly important in this respect, since it runs from the archistriatum in the cerebral hemispheres through the hypothalamus and passes near the midbrain vocalization area, but less conspicuous tracts paralleling this tract might be more important. Decisive experiments to discriminate among the possibilities have yet to be done. Anatomical connections in this system could be fruitfully investigated using studies of degeneration from small lesions in the various vocalization areas.

Agreement on the locations of responsive areas in the midbrain by persons working on various species of birds is widespread. The area of lowest threshold is in the torus semicircularis, in the part known as torus externus (Brown, 1971) and in the tegmental area just beneath it known as the area (or nucleus) mesencephalicus lateralis ventralis. It is in this area that bilateral lesions cause mutism. In the hypothalamus, the medial region has been found to have responsive areas by several workers. Responsive areas in the paleostriatum and neostriatum were most conspicuous in Åkerman's work on pigeons, but the existence of such areas in these structures in pigeons and in other bird species remains to be confirmed. However, most workers agree that responsive areas can be found in the archistriatum.

Basically, there is agreement on the existence of a posterior system descending from anterior parts of the brain that are involved more in information processing, decision making, and motivation to brain parts more concerned with coordination and production of relatively fixed patterns of action. Although there is little firm agreement on the question of homologies between birds, mammals, fishes, reptiles, and amphibians, the similarities that do exist are suggestive of a basic vertebrate pattern.

B. An Auditory Communication System?

Recent work on the avian nervous system has focused on its role in auditory communication. Since birds both produce stereotyped vocalizations and respond selectively to them, there must be functional connections between auditory and vocalization systems. The connections need not be of a direct, oligo-synaptic, reflex nature. In view of the necessity of some sort of auditory–vocal connections, it is a curious, but still unexplained, fact that the known auditory and vocalization structures in midbrain and diencephalon are contiguous. In the hypothalamus responsive areas for vocalization border the nucleus ovoidalis, an auditory structure. In the midbrain the torus externus borders the nucleus mesencephalicus lateralis dorsalis (MLD), also an auditory structure. Except for border regions, these auditory areas appear to be mainly unresponsive to electrical stimulation so far as evoked vocalization is concerned. Work on the problem of unraveling the neural mechanisms of auditory–vocal communication in birds is still in the crudest stages.

C. Anatomical Representation of Different Calls

In mammals, it has been possible to evoke a wide variety of vocalizations with ESB, and to map responsive parts of the brain for different calls into different neural systems (e.g., Jurgens & Ploog, 1970). This approach has yet to reach fruition in birds. In the most thorough analyses so far, those on domestic chickens and on passerine birds, there is little evidence of more than one vocalization system. Perhaps two systems can be recognized, as suggested by Åkerman (1966), for reproductive and agonistic behavior in pigeons with each system characterized by different types of calling. But such systems would be motivational in nature and the production of separate types of calls would require separate patterns of motor control. Exactly where the motivational properties begin anatomically and functionally is far from clear. We expect the motivational properties to lie anteriorly and the pattern forming properties to lie posteriorly, closer to the motor outflow. Yet in the experience of several workers using ESB, stimulation more posteriorly in the system yields fewer types of calls and less variability than stimulation anteriorly. Are anterior areas, such as the preoptic area and striatal areas required for complex patterns, such as song? Or are there cell assemblies in the lower brainstem which require only activation to produce a fixed pattern of vocalization?

Two mechanisms of pattern production may be envisioned. In the first, the "switchboard model," spatially separate assemblies of neurons on the motor side would exist. There would be an assembly for each type of call, and it might or might not involve peripheral or central feedback mechanisms. In the second, the "computer model," complex assemblies capable of producing various types of calls (depending on how they are activated) are postulated. In *Coturnix*, separate localization of calls evoked from the midbrain has been reported; this finding would appear to favor the switchboard model. In other studies in passerines and chickens, separate localization of calls in the midbrain is notably lacking. The question will not be settled until more is known about the control of motor neurons in vocalization of birds.

D. Motivation and Vocalization

In mammals, analysis of brain regions in terms of their roles in positive and negative reinforcement has reached a relatively advanced stage in which the anatomical and neurochemical bases of reinforcement mechanisms are becoming known. In birds, the basic circuits mediating these functions are virtually unknown, although a few studies have been done in hopes of locating them. Probably the first impetus for using brain stimulation to study bird behavior came from ethological analyses of the motivation of displays and vocalization. These studies relied on a set of hypotheses that were based on the concept of unitary drives.

The analysis of neurobehavioral mechanisms responsible for such concepts as drives for attack, escape, threat, eating, pair-formation, incubation, nest-building, and drinking in birds remains virtually a virgin field. Are the neural mechanisms of these behaviors fundamentally alike in mammals, birds, and other vertebrates?

Vocalization plays a supporting role in most of these behaviors, and should be useful as an aid in motivational analysis. Some preliminary attempts have been made to correlate evoked vocalization and other behavior with the reinforcement value of self-stimulation (e.g., Goodman & Brown, 1966; Goodman, 1970; Andrew, 1969) but the surface has only been scratched.

II. Methodological Considerations

I will now turn to a consideration of some of the parameters of electrical stimulation as they affect the behavioral responses observed in ESB work. An understanding of the relationship between stimulus parameters and evoked behaviors is crucial for determining the characteristics of the vocalization system.

A. Stimulus Spread

Stimulus spread is an ever-present problem in studies using ESB. Bipolar electrodes are used to reduce the spread in the brain of the electric current delivered at the electrode tips. Yet in practice it appears that an adequate degree of precision can be obtained with monopolar electrodes, especially if low current levels are used. This is illustrated in Fig. 1, which shows thresholds for calls evoked by ESB in red-winged blackbirds (*Agelaius phoeniceus*) with electrodes that were lowered in small steps after each threshold determination (Brown, 1971). A change in electrode depth of as little as .7 mm resulted in a change in threshold from negative ($> 500\ \mu A$) to 50 μA or less. A change in depth of .3 mm lowered the threshold from over 300 μA to 50 μA, and changes of .1 mm often produced a noticeable change in response. Thus it is possible to stimulate monopolarly with a strong current as close as .7 or .3 mm to the part of the brain with the lowest threshold for vocalization and still not

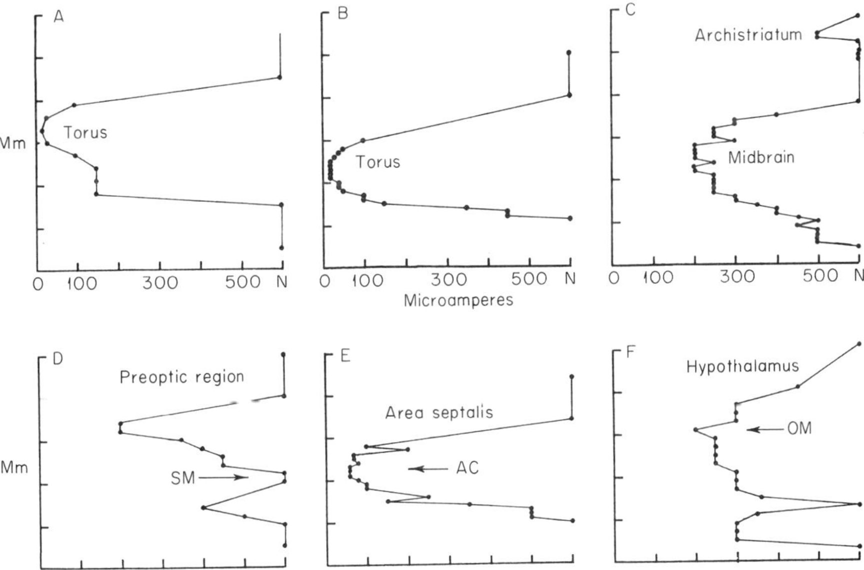

Fig. 1. Thresholds for vocalization evoked by ESB in male red-winged blackbirds as a function of electrode depth. Subjects were restrained under a local anesthetic while the brain was explored with movable monopolar electrodes. A Wyss-type stimulator was used to deliver asymmetrically biphasic pulse pairs at 30 Hz. The rounded cathodal phase had a duration of 10.5 msec while the anodal phase was longer. The standard test stimulus was at 500 μA for 30 sec. N: no vocalization to standard test stimulus. SM: tractus septo-mesencephalicus; AC: anterior commissure; OM: tractus occipito-mesencephalicus. [From Brown (1971).]

evoke a vocalization. With lower test currents the precision is greater. Figure 1 shows that the responsive areas are sharply demarcated vertically in terms of threshold current.

Probably the main source of error in studies using ESB to plot the location of responsive parts of the brain is not stimulus spread but the process of plotting the electrode sites on atlas sections and preparing them for publication. Where many points must be plotted on the same section, most must be somewhat anterior or posterior to the section illustrated. Some brains are inevitably sectioned in planes not parallel to the plane of illustration and skill is required to plot these electrode sites correctly. Moreover, there is a tendency to use as few sections as possible for illustration of the anatomical findings, and this tends to increase the number and severity of plotting errors.

B. Pulse Duration

Threshold for calling evoked by ESB in birds is a function of pulse duration (Kramer, St. Paul, & Heinecke, 1964 and later authors). As shown in Fig. 2 for seven sites in red-winged blackbirds and Steller's jays (*Cyanocitta stelleri*), pulse duration was nearly interchangeable with pulse height (current) over a wide range (from .015 to 10 msec). Similar curves were obtained in chickens by Phillips and Youngren (1971). Pulse durations greater than 1.0 msec seemed to be less efficient than shorter pulses, in the sense that at threshold the product of pulse height and length was greater than for shorter pulses (more coulombs required to reach threshold). There was no evidence of a species difference or of a difference between midbrain and hypothalamus in this relationship.

The type of call uttered was not a function of pulse duration; it remained about the same at all pulse durations tested, with minor variations. The differences in calling and associated behavior at long and at short pulse durations are illustrated in Fig. 2, curve C, where an electrode was implanted in the area mesencephalicus lateralis ventralis of a red-winged blackbird. At 10.0 and 1.0 msec, calling near threshold came at regular intervals and the bird's posture was normal but slightly crouched; at .1 msec, the crouch was much lower and toward the ipsilateral side, the wings were held slightly away from the body (especially the ipsilateral wing), and the bird looked to the contralateral side; at .014 msec the crouch was more severe, the bill was kept wide open in silent periods, and both wings were raised. The calls given at .014 and .1 msec were irregular in rhythm and strained in utterance. The general effect of shorter pulse durations was to exaggerate the accompanying posture and reduce the conspicuousness of the calling. Responses at 1.0 and 10.0 msec appeared to be relatively normal, while those at .014 and .1 msec

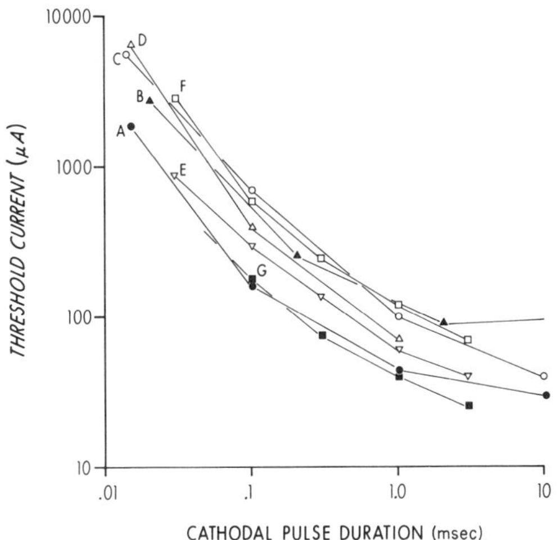

Fig. 2. Thresholds for calling evoked by ESB as a function of pulse duration in red-winged blackbirds and Steller's jays with implanted electrodes. All stimulations were performed with a Nuclear–Chicago constant-current stimulator with symmetrically biphasic rectangular pulses at 30 Hz. Duration and height of the cathodal pulses were determined on a Hewlett-Packard dual trace 1200 A oscilloscope. Stimulations were timed for 10 sec of constant current by an electronic timer and were given every 2 or 3 min. The shortest pulse durations (highest thresholds) were tested last. Thresholds were determined to within at least 10 per cent of the given value. A. Red-winged blackbird (RW) 135, border torus externus (TX) and nucleus Mesencephalicus lateralis pars dorsalis (MLD). B. RW 156, TX medial to MLD. C. RW 157, area mesencephalicus lateralis pars ventralis (MLV). D. Steller's jay (SJ) 6, TX anterior, E. SJ 8, MLV. F. SJ 8, hypothalamus just above dorsal supraoptic decussation. G. SJ 9, TX-MLD border.

seemed rather abnormal. Calling could be evoked at the shortest pulse durations only by also inducing a severe defensive posture, but at the longer pulse durations regular calling was easily evoked with only a trace of the defensive posture. Extremely short pulses did not have the effects on evoked behavior that high frequencies did; several calls could be given and latencies could be rather long (8–9 sec). Once started, the reaction continued until termination of the stimulus (10 sec).

C. Pulse Frequency

It is well known that behavior evoked by ESB varies with stimulus frequency, but systematic studies of the relationship are uncommon. When frequency is raised, the threshold current for the behavior being evoked may generally be expected to drop. This is because with more pulses per second

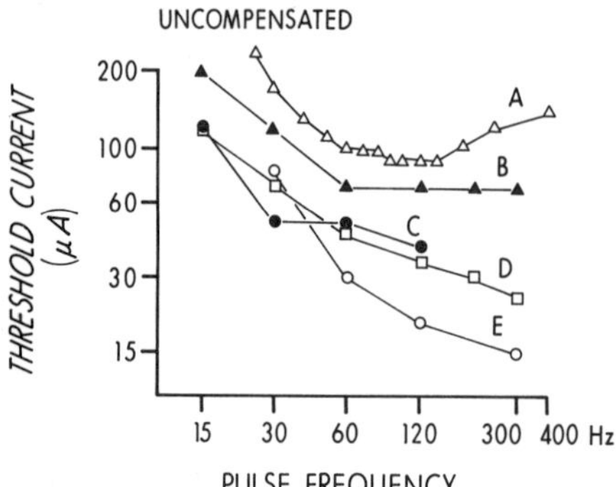

Fig. 3. Thresholds for calling evoked by ESB as a function of pulse frequency in red-winged blackbirds and Steller's jays with implanted electrodes. Procedure as in Fig. 2 except that frequency was varied and pulse duration was constant at 1.0 msec. Abbreviations as in Fig. 2. A. RW 154, anterior hypothalamus. B. SJ 8, anterior hypothalamus. C. SJ 6, TX anterior. D. SJ 8, TX. E. RW 157, MLV.

more coulombs are delivered. Some examples of thresholds for calling as a function of pulse frequency with a constant pulse duration of 1 msec are shown in Fig. 3. At the lower frequencies (15–60 Hz) the expected relationship was found. At the higher frequencies a drop in threshold was found in the midbrain sites but not in the anterior hypothalamus. One of the hypothalamic sites even showed a rise in threshold with frequencies above 120 Hz. Further testing is needed to determine whether or not these findings represent real functional differences between the hypothalamus and the midbrain with respect to the system concerned with ESB-evoked calling.

It is impossible to vary pulse frequency without also varying some other parameter of the stimulus. In the preceding examples, the coulombs delivered to the brain are increased by raising pulse frequency even if the current level (pulse height) remains the same. This effect of raising pulse frequency can be avoided by compensating for the increased coulombs by decreasing pulse duration. Figure 4 shows threshold curves obtained by varying pulse frequency but keeping the product of frequency and pulse duration (in milli-

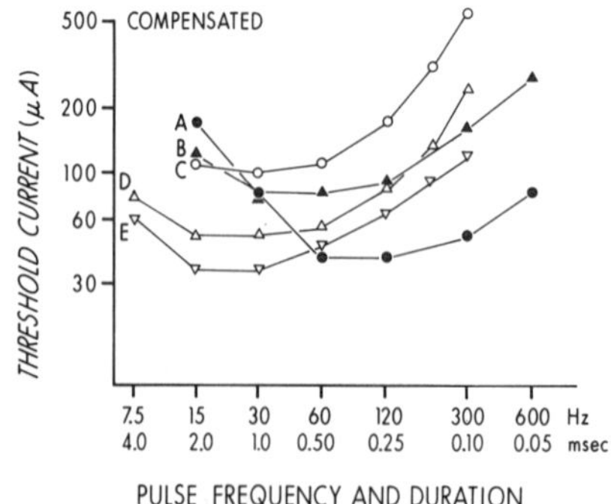

Fig. 4. Thresholds for calling evoked by ESB as a function of pulse frequency and duration in red-winged blackbirds and Steller's jays with implanted electrodes. Procedure as in Fig. 3. A: RW 157, MLV. B: RW 156, TX-MLD border. C: SJ 8, anterior hypothalamus. D: SJ 9, anterior commissure. E: SJ 9, TX-MLD border.

seconds) constant at 30. Similar curves result with other constants, but they are displaced up or down. If thresholds for calling were simple functions of the coulombs delivered and independent of the particular combination of pulse duration and number, then the curves in Fig. 4 should be horizontal straight lines (diagonal straight lines in Fig. 3). Clearly they are not. The lowest part of the curve in Fig. 4 indicates that combination of pulse duration and frequency which is most economical in terms of coulombs for the particular product of frequency and pulse duration chosen. The magnitude of the deviation from the most economical frequency is indicated by the fact that the thresholds at 300 Hz and .1 msec range from 1.4 to 5.5 times the most efficient threshold, which usually included 30 Hz and 1.0 msec. For six electrodes in the midbrain of blackbirds and jays (torus externus or area mesencephalicus lateralis ventralis) the most economical frequencies were 15–30, 25–30, 30, 30–60, 30–60, and 60–120 Hz, respectively. For two electrodes in the hypothalamus the most economical frequencies were 15–30 and 30. This evidence is at least consistent with the possibility that vocal response evoked by stimulation of the hypothalamus is more adversely affected by high frequencies than that of the midbrain.

Since in Fig. 4 both frequency and pulse duration were varied simultaneously, it may be asked which variable had the greatest effect. A suggestive answer may be obtained by examining the curves in Fig. 2. These show that a change in pulse duration from 1.0 to .1 msec, when not accompanied by a change in frequency, *increases* the efficiency of the stimulus in terms of coulombs. Consequently, the *decrease* in the efficiency of the stimuli in Fig. 4 in going from 1.0 msec at 30 Hz to .1 msec at 300 Hz must be due mainly to the increase in frequency.

Behaviorally the elements of the response at high frequencies (200–600 Hz) tend to be compressed into a shorter period and to adapt or end sooner than at low frequencies (15–60 Hz). Thus we see both "telescoping" and "abbreviation" effects of high frequencies. At lower frequencies (15–60 Hz) the various components of a behavioral response to ESB tend to appear in a normal sequence and to be repeated at intervals; calling evoked by ESB may slow down but it persists for many minutes if stimulation is continued at suitable intensities. In contrast, at 300 Hz calling and certain associated behaviors tend to end abruptly a few seconds after onset; the bird does not call again until the stimulus is shut off and a new stimulus initiated.

The telescoping phenomenon affects evoked vocalizations by causing the calls to be run together and the duration of calling abbreviated. For example in a red-winged blackbird (Fig. 4, curve B) stimulated in the torus of the midbrain, calling evoked by ESB at 30 Hz continued as long as the stimulus was on (30 sec) at the rate of one call every 1–5 sec at threshold; at 60 Hz, at threshold calling was more rapid (a call every .2–1.0) and ceased sooner (after 15 sec); at 100–300 Hz, at threshold there was a burst of 2–6 calls in the first second followed by silence for the duration of stimulation; and at 600 Hz there were 1–3 calls in the first second, then silence. Similarly, another red-winged blackbird that produced repeated normal chek calls at 30, 50, 100, and 150 Hz when stimulated in the torus, gave a blurred-together series of cheks that sounded more like a trill at 200 Hz and a single garbled high-pitched sound at 500 Hz. Still another blackbird (Fig. 3, curve E, Fig. 4, curve A) at threshold gave long wide-vertical calls repetitively for over 30 sec with slight adaptation at 30 Hz, shorter calls at 120 Hz with adaptation after several seconds, and a single long scream followed by a burst of pips with rapid adaptation at 300 Hz.

In a Steller's jay (Fig. 3, curve B), calling came at medium and long latencies (several to many seconds) at 15 and 30 Hz and continued while the stimulus was on: At 200 and 300 Hz latencies were less than 1 sec and only one or two calls were given per stimulation. The abbreviation effect began to appear at 120 Hz. Curiously, the abbreviation in this case affected calling but not aggressive pecking evoked in the same stimulations. The frequency response of pecking differed from that of calling. Pecking was inconsistent and weak at 15 and 30 Hz while calling was regular and strong. From 120–300 Hz, calling showed the telescoping and abbreviation effects but pecking continued to be regular and strong without noticeable adaptation. In the same jay stimulated with a different electrode in the midbrain (Fig. 3, curve D), the calls and all other parts of the response were telescoped in time at 300 Hz; calling ceased immediately and after 8–10 sec the jay appeared unstimulated although the current was still on. To summarize, the telescoping and abbreviation effects are present in blackbirds and jays with a variety of electrode sites.

D. Stimulation Test Environment

Calling evoked by ESB is notable for its predictability with proper stimuli and lack of dependence on features of the environment. Unlike certain types of behavior evoked by ESB in mammals that require adequate environmental stimuli (such as eating, drinking, gnawing), vocalization is seemingly possible in any environment and indeed can easily be evoked while the bird is under anesthesia. Nevertheless, calling evoked by ESB in passerine birds is susceptible to environmental influence. One example illustrates the effect of presence or absence of bright illumination on the occurrence and strength of calling evoked by ESB. A red-winged blackbird was stimulated in the torus at two-minute intervals in a small box with its own internal fluorescent illumination alternately turned on and off 1 min before stimulation. When its stimulation-box lights were off, the bird was in dim light but not in darkness since the room lights remained on. The experimenter observed the bird from a separate room through one-way vision glass. Stimulation consisted of a train of pulses at 90 μ and 30 Hz for 30 sec with cathodal pulse length 1.0 msec; the symmetrically biphasic rectangular waves were produced by a Nuclear–Chicago stimulator. In each of ten successive pairs of stimulations, the blackbird called more with the lights on than off ($p < .01$, two tailed randomization test for matched pairs), with an average of 5.9 calls per stimulation with the light on and .6 with it off. The bird called in every trial with the lights on and in only two with the lights off. The calls were mainly whistles ($N = 56$) but included normal ($N = 9$) and subdued cheks ($N = 9$).

Temperature may also influence the behavioral response to ESB. When Steller's jays are warm, their plumage is ruffled out and the crest moderately elevated. Plumage is not ruffled and crests are not elevated in jays resting at moderate temperatures. When a jay was induced to raise its crest by warming the box in which it was being tested, stimulation at 100 μA in the anterior hypothalamus caused the jay to lower its crest in twelve successive trials. When tested with the same electrode at cooler temperatures with the crest down between stimulations, the crest was raised slightly at 100 μA but was raised moderately to greatly by stimulation at 120–150 μA. Thus the same electrode had opposite effects on crest position depending on the environmental temperature and the stimulus intensity.

The significance of these observations is that they demonstrate that even relatively subtle environmental factors, such as changes in environmental temperature or in intensity of illumination, can have definite and conspicuous effects on the nature of behavior evoked by ESB.

E. Stimulus Parameters

In practice the neglect of stimulation parameters in birds seems not to have been critical since in most cases the behaviors can be easily evoked with a variety of stimulus parameters. The most critical parameter is pulse frequency, and most workers have rather arbitrarily selected a frequency in the range from 30–100 Hz, which is quite suitable. This amount of interinvestigator variability in stimulus frequency may be expected to cause variation in thresholds by a factor of two but not to alter greatly the overall character of the evoked behavior.

Investigators of stimulation parameters in birds have tended to emphasize their interchangcability within limits (von Holst & von St. Paul, 1960; Kramer *et al.*, 1964; Phillips, Youngren, & Peek, 1972). Only Potash (1970a) has claimed an important role for stimulus parameters: "In some instances only relatively narrow ranges of stimulating frequencies will drive CNS activity so as to produce vocalizations that resemble natural calls [p. 164]." He described an example in *Coturnix* where this seemed to be true. I suggest, therefore, that the role of pulse frequency deserves more systematic testing and documentation, since differential effects from small variations have not been found even though looked for in chickens, jays, and blackbirds, and frequency effects have not been mentioned in reports on ducks, gulls, pigeons, and ring doves (Phillips *et al.*, 1972; Brown, 1971, 1973; Maley, 1969; Delius, 1971; Goodman & Brown, 1966; Harwood & Vowles, 1967; Åkerman, 1966; Putkonen, 1967). In the squirrel monkey Jürgens and Ploog (1970) were impressed with the importance of frequency as a factor determining the type of call evoked by ESB of certain brain structures, but details were omitted.

The greater susceptibility of hypothalamic than mesencephalic sites to threshold elevation at high frequency, as already suggested, may have a parallel in some findings of Phillips *et al.* (1972). They found that the rate of calling evoked by ESB at anterior and medial sites tended to fade at high pulse frequencies, while in the lowest-threshold mesencephalic sites this effect was absent or reduced at the same frequencies.

References

Åkerman, B. Behavioural effects of electrical stimulation in the forebrain of the pigeon. I. Reproductive behaviour. II. Protective behaviour. *Behaviour*, 1966, **26**, 323–350.

Andrew, R. J. Intracranial self-stimulation in the chick and the causation of emotional behavior. *Annals of the New York Academy of Sciences*, 1969, **159**, 625–639.

Brown, J. L. Loss of vocalization caused by lesions in the *Nucleus mesencephalicus lateralis* of the Redwinged Blackbird. *American Zoologist*, 1965, **5**, 693 (abstract).

Brown, J. L. The control of avian vocalization by the central nervous system. In R. A. Hinde (Ed.), *Bird vocalizations*. New York and London: Cambridge Univ. Press, 1969. Pp. 79–96.

Brown, J. L. An exploratory study of vocalization areas in the brain of the Red-winged Blackbird *(Agelaius phoeniceus)*. *Behaviour*, 1971, **39**, 91–127.

Brown, J. L. Behavior elicited by electrical stimulation of the brain of the Steller's jay. *Condor*, 1973, **75**, 1–16.

Delius, J. D. Neural substrates of vocalization in gulls and pigeons. *Experimental Brain Research*, 1971, **12**, 64–80.

Goodman, I. J. Approach and avoidance effects of central stimulation: An exploration of the pigeon fore- and midbrain. *Psychonomic Science*, 1970, **19**, 39–41.

Goodman, I. J. & Brown, J. L. Stimulation of positively and negatively reinforcing sites in the avian brain. *Life Sciences*, 1966, **5**, 693–704.

Harwood, D. and Vowles, D. M. Defensive behaviour and the after-effects of brain stimulation in the ring dove (*Streptopelia risoria*). *Neuropsychologia*, 1967, **5**, 345–366.

Jurgens, U. & Ploog, D. Cerebral representation of vocalization in the squirrel monkey. *Experimental Brain Research*, 1970, **10**, 532–554.

Kramer, E., Saint Paul, U. v. & Heinecke, P. Die Bedeutung der Reizparameter bei elektrischer Reizung des Stammhirnes. *Biologische Jahresheft*, 1964, **4**, 119–134.

Maley, M. J. Electrical stimulation of agonistic behavior in the mallard. *Behaviour*, 1969, **34**, 138–160.

Murphey, R. K. & Phillips, R. E. Central patterning of a vocalization in fowl. *Nature*, 1967, **216**, 1125–1126.

Newman, J. D. Midbrain regions relevant to auditory communication in songbirds. *Brain Research*, 1970, **22**, 259–261.

Nottebohm, F. Neural lateralization of vocal control in a passerine bird. *Journal of Experimental Zoology*, 1971, **177**, 229–261.

Peek, F. W. & Phillips, R. E. Repetitive vocalizations evoked by local electrical stimulation of avian brains: II Anesthetized chickens *(Gallus gallus)*. *Brain, Behavior and Evolution*, 1971, **4**, 417–438.

Phillips, R. E. & Youngren, O. M. Brain stimulation and species-typical behavior: Activities evoked by electrical stimulation of the brains of chickens *(Gallus gallus)*. *Animal Behaviour*, 1971, **19**, 757–779.

Phillips, R. E., Youngren, O. M. & Peek, F. W. Repetitive vocalizations evoked by electrical stimulation of avian brains: I. Awake chickens *(Gallus gallus)*. *Animal Behaviour*, 1972, **20**, 689–705.

Potash, L. M. Vocalization elicited by electrical brain stimulation in *Coturnix coturnix japonica*. *Behaviour*, 1970, **36**, 149–167. (a)

Potash, L. M. Neuroanatomical regions relevant to production and analysis of vocalization within avian Torus semicircularis. *Experientia*, 1970, **26**, 1104–1105.(b)

Putkonen, P. T. Electrical stimulation of the avian brain. *Annales Academinae Scientiarum Fennicae*. Series A. V. Medica, 1967, **130**, 1–95.

von Holst, E. & von St. Paul, U. Vom Wirkungsgefüge der Triebe. *Naturwissenschaften*, 1960, **18**, 409–422.

Feeding Behavior in the Pigeon: A Neurobehavioral Analysis

H. Philip Zeigler[1]
Hunter College and American Museum of Natural History, New York

In his classic paper on "The Experimental Analysis of Instinctive Behavior, "Lashley (1938) pointed out that species-typical behavior patterns such as eating or drinking confront the investigator with two sets of problems. There is, first, the problem of what Lashley called "the sensory-motor mechanism" underlying the behavior. Feeding behavior in the pigeon, for example, may be subdivided into a series of response sequences: pecking, mandibulation (the process by which food is moved from the beak tip to the back of the mouth), and swallowing. Elucidation of the sensory–motor mechanism underlying feeding thus involves an analysis of the morphology, sensory control, and developmental history of these three movement patterns. A second problem is raised by the fact that most animals do not eat constantly, even when the stimulus which elicits eating (i.e., food) is constantly present. Instead, their responsiveness to food varies continuously over time. Because feeding is not completely under sensory control we are confronted by what Lashley termed the problem of "variations in the excitability of the sensory–motor mechanism." Such variation in responsiveness to a constant stimulus is one of the hallmarks of "motivated" behavior (Hinde, 1970) and, in the case of feeding, is usually described by the term "hunger."

Although the analysis of feeding behavior has a long history in psychology, it has been focused largely on a few mammalian species—most

[1] This work was supported by Research Grant MH-08366 and Research Career Development Award No. K-2-6391, both from The National Institute of Mental Health.

notably the rat—and few systematic data are available for other vertebrate classes. This is an unfortunate situation because one aim of comparative psychology is to clarify the similarities and differences in the behavior of animals representing many levels of evolutionary history. To achieve this aim requires comparisons among a variety of species whose morphology, behavior, and ecology are sufficiently diverse to illustrate something of the range of solutions which have evolved to meet the adaptive requirements of the environment.

From the standpoint of a comparative analysis of feeding behavior, the pigeon (*Columba livia*) is an excellent representative of the class Aves. By contrast with some other avian groups (e.g., hummingbirds, pelicans, ducks, or parrots) the pigeon's feeding apparatus is relatively unspecialized structurally, but its eating and drinking response patterns are behaviorally distinct from each other (Farner, 1960; Wolin, 1968). Like others of class Aves, however, its alimentary and digestive systems differ in many respects from those of mammals, and include a number of distinctive features such as the absence of teeth and the subdivision of its gastric apparatus into a glandular and a muscular stomach. The pigeon is also one of a number of birds possessing a crop sufficiently well developed to serve as a storage organ (Farner, 1960). In contrast with mammals such as the rat, the morphology and innervation of the pigeon's oral region make it an extremely useful preparation for deafferentation studies (Zeigler, 1973) and for the analysis of sensory control of feeding behavior (Zeigler & Witkovsky, 1968). Furthermore, the anatomical studies by Karten and his associates (Karten & Hodos, 1967; Karten, 1969) have made it possible to identify a group of structures at several levels of the pigeon brain which may be defined, on both structural and functional grounds, as constituting a feeding behavior "system" (Zeigler, Karten, & Green, 1969; Zeigler & Karten, 1973).

Over the past few years we have been carrying on a program of research on the feeding behavior of the pigeon. In view of the paucity of data on avian feeding behavior, one of our earliest aims was to provide for the pigeon a body of normative data on its feeding behavior patterns under *ad lib* conditions and following varying degrees of food deprivation, and to define some of the relationships among eating, drinking, and body weight regulation. More recently, we have begun an analysis of the morphology and neurosensory control of the movement patterns involved in feeding. A third group of studies, carried out in collaboration with Karten, has been devoted to a neurobehavioral analysis of the brain mechanisms underlying hunger in the pigeon. The present occasion provides a welcome opportunity to systematically review some of our findings, to relate them to those of other investigators and to indicate possible directions for future research.

I. Normative Studies of Feeding Behavior in the Pigeon

A. Analysis of the Movement Patterns Constituting Feeding

For purposes of analysis, feeding may be divided into a series of distinct response sequences forming a cycle of movements (pecking, mandibulation, swallowing) whose repetition constitutes a bout of feeding. Analyses of film records taken at 64 frames/sec indicate that these movement patterns are both complex and highly stereotyped. Figure 1 is derived from a series of enlarged single frames and illustrates the morphology of the movement patterns and the temporal relationships between the various components of the feeding response. *Pecking*, with which the response cycle begins, consists of a downward movement of the head with the mouth initially closed but gradually opening as the bird's head approaches the grain. Contact with the grain terminates the downward movement and initiates *mandibulation*, which involves an upward movement of the head synchronized with series of tongue

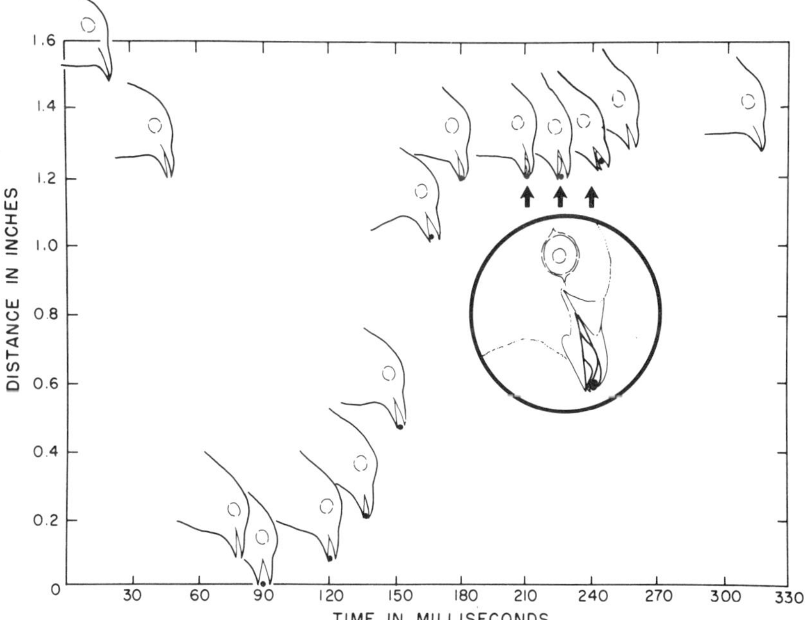

Fig. 1. Feeding behavior movement patterns in the pigeon: The spatial and temporal organization of the response sequences constituting feeding. The circular inset diagrams the relation of the tongue to the grain at the start of the period indicated by the arrows. Drawings are based upon individual motion picture frames taken at 64 frames/sec.

movements which propel the grain from the beak tip to the rear of the buccal cavity, at which point *swallowing* is initiated.

This analysis indicates something of the complexity of the feeding response, and suggests that it involves the integration of visual, tactile and proprioceptive information. The extreme brevity and stereotypy of the feeding response are apparent from the data in Table I which indicate that the entire process takes about 300 msec, that the temporal relationships among the three components are quite constant for an individual bird, and that differences between birds are minimal. Furthermore, the temporal characteristics of the response do not appear to vary with the motivational state of the bird. Whereas the feeding behavior of the deprived pigeon is characterized by an increase in the number of feeding responses per minute this increase is produced by reducing the interresponse interval rather than by shortening the total duration of the feeding response cycle or any of its components. The stereotypy of these response sequences is similar in many respects to that of the action patterns described by students of species-typical behavior (Hinde, 1970) and raises many of the same problems of peripheral versus central control of movement patterns.

B. Patterns of Feeding Behavior under ad lib Conditions

An important prerequisite for the study of feeding behavior mechanisms in any species is an analysis of the way in which its individual feeding responses are distributed across time to produce the organized feeding behavior patterns typical of the species. For this purpose, we developed a "pigeon feedometer" that enabled us to monitor feeding automatically and to obtain independent measures of the number of feeding responses made during a given period of time and the total food intake for that period (Fig. 2). Pecks into the food magazine interrupt a light beam across the food, triggering a photocell circuit whose output may be connected to recorders or response

Table I. *Temporal Organization of the Components of the Consummatory Response (Duration in Milliseconds)*

Bird no.	Observ. no.	Peck *M*	Peck *SD*	Observ. no.	Mandibulation *M*	Mandibulation *SD*	Observ. no.	Swallow *M*	Swallow *SD*	Total
45	13	73.2	18.4	10	140.4	34.3	10	78.0	18.7	291.6
95	12	84.4	10.5	10	152.3	40.2	10	74.3	21.8	311.0
97	9	79.7	22.6	10	156.0	33.7	10	81.1	28.1	316.8
150	12	74.1	9.7	10	106.8	16.1	10	82.6	36.0	263.5
152	13	75.5	10.7	10	163.8	30.4	10	79.5	22.6	318.8

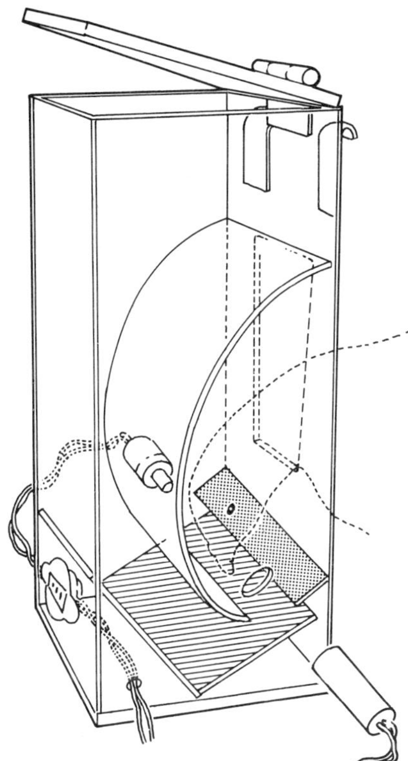

Fig. 2. Schematic diagram of a photocell feedometer designed for use with the pigeon. [From Zeigler & Feldstein (1971).]

counters. A modification of the device may be used in a free-operant conditioning situation, and permits the independent measurement of key-pecking and feeding responses (Zeigler & Feldstein, 1971).

By connecting a feedometer to an event recorder it is possible to obtain data on the daily distribution of feeding responses for individual birds. Analysis of these data indicates that, in contrast to the rat, which eats a few large meals at widely spaced intervals and does the bulk of its feeding at night (Le Magnen, 1969), the feeding behavior of pigeons maintained under ad lib conditions consists of a large number of relatively brief feeding bouts (Fig. 3). However, feeding bouts are not distributed normally throughout the day, and the pigeon ingests the bulk of its food during prolonged feeding periods in the morning and afternoon (Fig. 4). Analysis of the data from individual birds over several weeks of testing indicates that whereas the number of bouts and the total time spent feeding may vary from bird to bird, the parameters for a given bird are stable and quite characteristic of that bird (Zeigler, Green, & Lehrer, 1971).

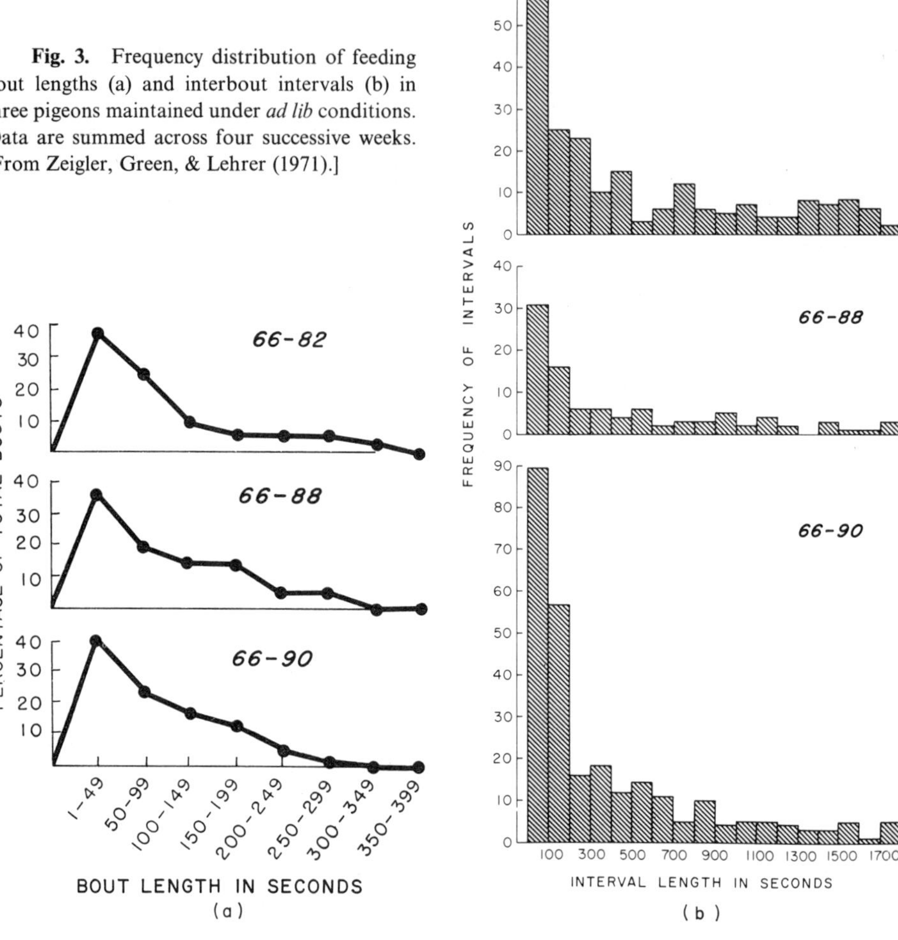

Fig. 3. Frequency distribution of feeding bout lengths (a) and interbout intervals (b) in three pigeons maintained under *ad lib* conditions. Data are summed across four successive weeks. [From Zeigler, Green, & Lehrer (1971).]

C. Food and Water Intake and Body Weight Regulation

Another aim of our early studies was to provide for the pigeon a body of normative data on intake and weight regulation comparable to that which now exists for a number of mammalian species. In addition to its comparative interest, the acquisition of such data is a first step in the study of regulatory processes in any species. Furthermore, it provides a base line against which to evaluate the significance of changes in feeding behavior or weight regulation

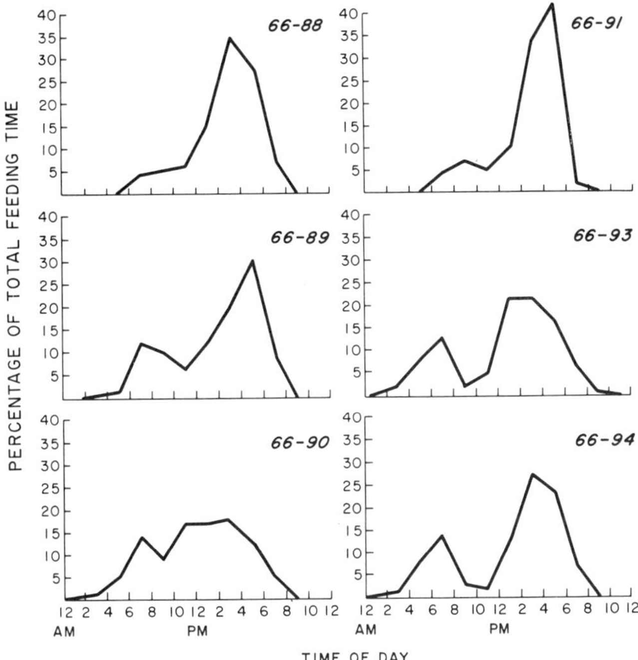

Fig. 4. Individual differences in the temporal distribution of feeding bouts over 24-hr periods under *ad lib* conditions for a group of six pigeons. Data are summed across four successive weeks. [From Zeigler, Green, & Lehrer (1971).]

produced in neurobehavioral studies of hunger or thirst. Two groups of studies proved to be of particular significance in this respect, and their findings will be summarized briefly (Zeigler, Green, & Siegel, 1972).

1. The Interaction of Eating and Drinking in the Pigeon

Our experiments have confirmed the existence of a complex interaction between eating and drinking in pigeons, such as has been previously demonstrated for a number of mammals and some other avian species (Cizek, 1959; Strominger, 1947; McFarland, 1964). Under ad lib conditions, the average daily water intake of the pigeon is about 150% of its food intake; and there is a significant correlation between daily food and water intake. The interdependence of eating and drinking is even more apparent during periods of food or water deprivation. Total water deprivation is followed by a drastic reduction in food intake, whereas food deprivation produces a reduction in water intake proportional to the size of the available food ration, reaching a limiting value under conditions of total food deprivation (Fig. 5).

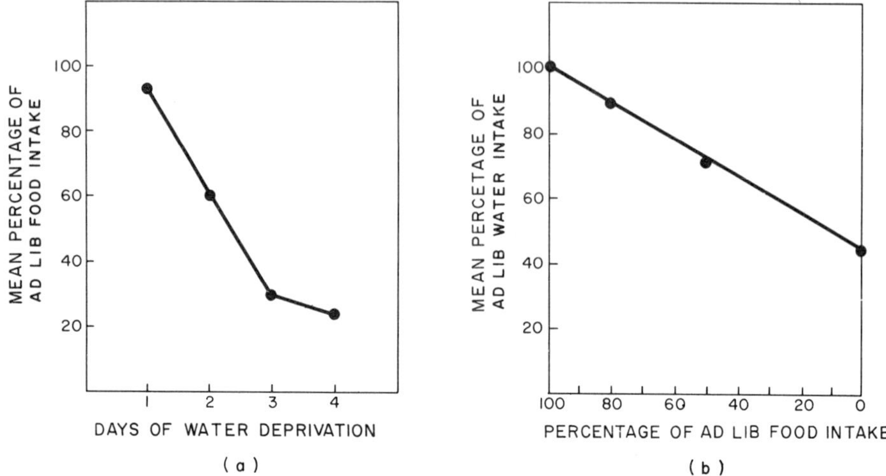

Fig. 5. The interaction of eating and drinking in the pigeon. (a). Effects of four days of total water deprivation upon food intake ($N = 10$). (b). The relation between food ration size and water intake in pigeons given varying proportions of their *ad lib* food ration. [Adapted from Zeigler, Green, & Siegel (1972).]

2. *Food Intake and Body Weight Regulation in the Pigeon*

Under ad lib conditions in a controlled and familiar environment, the pigeon's daily food (and water) intake varies within a range of values which is characteristic for a given bird (Fig. 6, bottom). The short-term variations may be considerable but over a more extended period the rate of eating and drinking is maintained at a fairly constant value (Fig. 6, middle). Thus, although daily variations in a bird's food (or water) intake may be on the order of 20–30%, daily variations in its body weight rarely exceed 2 or 3%. Similar data have been reported for a number of mammalian species, suggesting the operation of mechanisms controlling intake over extended periods and regulating body weight within fairly narrow limits. The most striking evidence for the existence of such regulatory processes in the pigeon comes from studies of the relation between food intake and body weight.

There is, in the pigeon, as in many mammalian species (Collier, 1969), a linear relation between the log intake of food and the log body weight. This relation holds not only under ad lib conditions but also under food deprivation. Thus, during a period of total food deprivation, log body weight declines linearly with time, whereas under partial food deprivation, body weight declines to an asymptotic value which is a linear function of log food intake. After a period of total food deprivation, the pigeons' food intake during a 1-hr test period is directly proportional to its body weight loss (Megibow &

Zeigler, 1968), and a similar relationship was found to hold when intake was measured over a 24-hr period (Zeigler, Green, & Siegel, 1972). Figure 7 illustrates the relation between food intake and body weight in three groups of birds during a recovery from varying levels of body weight loss over a still more extended period. Intake levels during recovery are significantly above normal declining gradually as body weight approaches its ad lib value. Furthermore, analysis of similar data reveals that the linear relation between log intake and log body weight is also present during recovery from depriva-

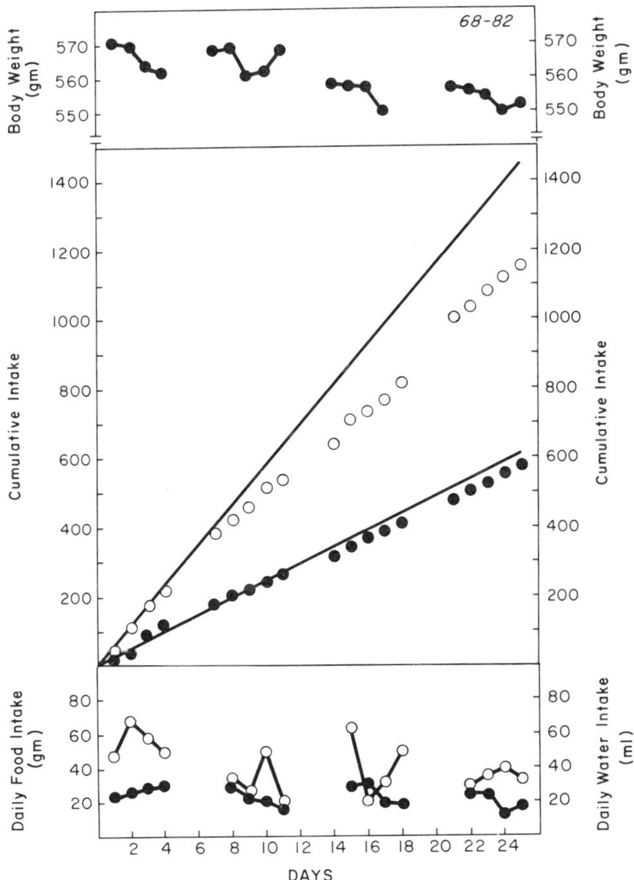

Fig. 6. Food and water intake and body weight regulation in the pigeon. Data are for a single representative bird over four successive weeks. Weekend data, excluded from the bottom portion of the figure, are included in calculating the rate of intake plotted in the middle portion. The top portion indicates variability in body weight over the 4-week period. [From Zeigler, Green, & Siegel (1972).]

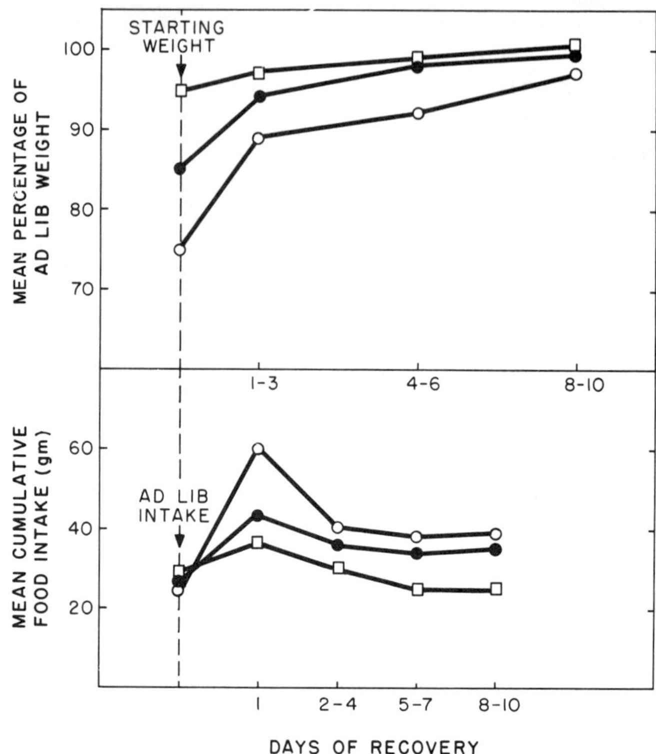

Fig. 7. Food intake and body weight during recovery from varying degrees of body weight loss in the pigeon. Percentages of body weight lost during deprivation: 5%, $N = 6$, 15%, $N = 6$; 25%, $N = 6$. [From Zeigler & Karten (1973)]

tion since log body weight *gain* during recovery is a linear function of log food intake (Zeigler, Green, & Siegel, 1972). Such data suggest that the pigeon can detect the extent of its body weight loss and respond to it by means of both short- and long-term adjustments in its food intake. Since the food intake of any species is a product of the frequency and duration of individual feeding bouts, our next task was to examine the temporal organization of feeding responses in the hungry pigeon.

D. Hunger: Methodological Considerations and Experimental Findings

Prolonged periods of food deprivation are typically followed by an increased responsiveness to food, which may be measured in a variety of ways. Such measures do not always agree, and it has become increasingly clear that comparisons among several measures are necessary to dissociate the processes underlying hunger (Miller, 1967). Furthermore, the work of Bolles (1967) and Collier (1969), as well as our own studies of body weight regulation, have shown the utility of treating body weight loss rather than hours of food deprivation as the operative variable producing changes in an animal's motivational state. Considerations such as these have led us to develop a group of measures that may be used to assess the effects of varying degrees of weight loss upon the "appetitive" and "consummatory" behaviors involved in feeding.

Perhaps the simplest measure of an animal's appetitive behavior with respect to food is its readiness to eat; that is, the latency with which it will initiate feeding, either in its home cage or in a novel test situation (Bolles, 1965). As Fig. 8 shows, readiness to eat in both situations varies directly with body weight loss. Figure 9 provides similar data for the consummatory response; it presents a series of cumulative records of the feeding response obtained from a single bird at different weight levels, in 1-hr test periods during which a feedometer was attached to the home cage. Effects of body weight loss upon food intake and the patterning of feeding responses are shown, for a group of 12 birds, in Fig. 10. As noted earlier, food intake is directly proportional to body weight loss over a wide range of deprivation levels. Analysis of the temporal distribution of feeding responses within the test period shows that this increase in food intake is the result of increases in

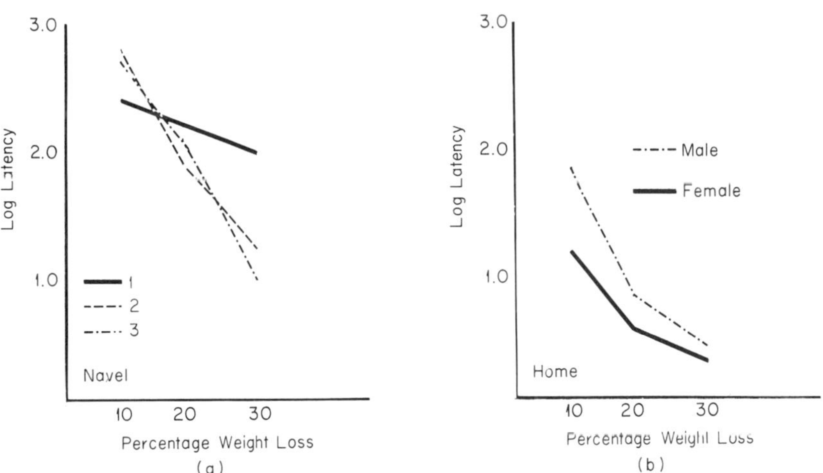

Fig. 8. Readiness to eat as a function of body weight loss in the pigeon. Both the novel cage data (a) and the home cage data (b) are shown. The parameter in the "novel" graph is replications, whereas that in the "home" graph is sex. [From Megibow & Zeigler (1968).]

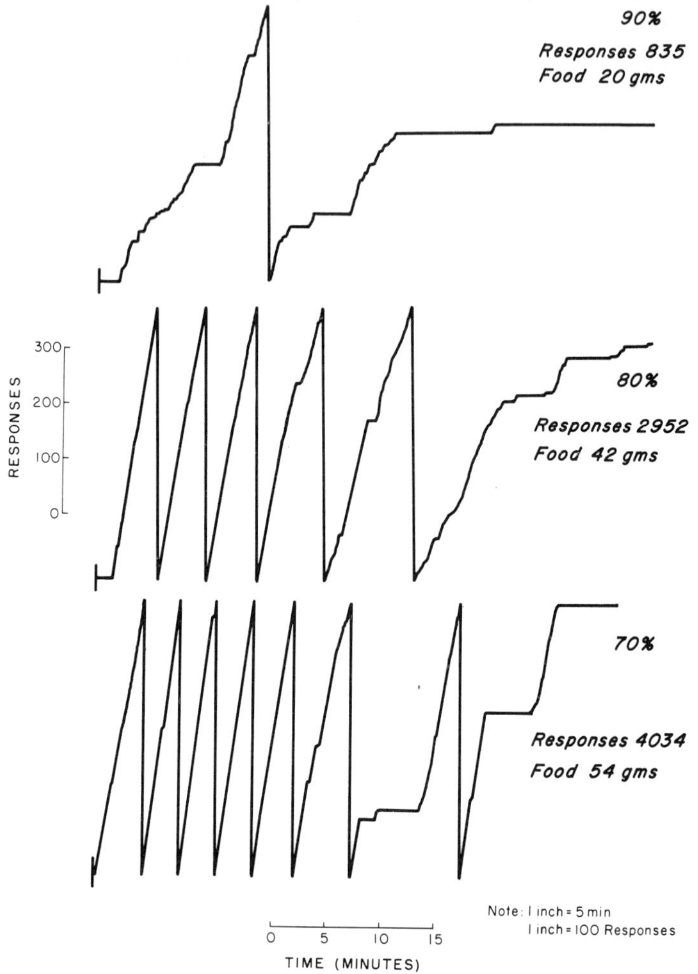

Fig. 9. Cumulative feeding responses and food intake in one hour test sessions for a single bird tested at 90%, 80%, and 70% of its ad lib body weight. [From Zeigler & Feldstein (1971).]

the initial rate of feeding, the length of the first feeding bout, and the duration of subsequent bouts but does not involve an increase in the number of feeding bouts. Similar findings have been reported for the licking response of rats in studies of intake patterns following periods of deprivation (Stellar & Hill, 1952).

Studies such as those outlined in the first part of this chapter have provided a body of normative data on the movement patterns constituting feed-

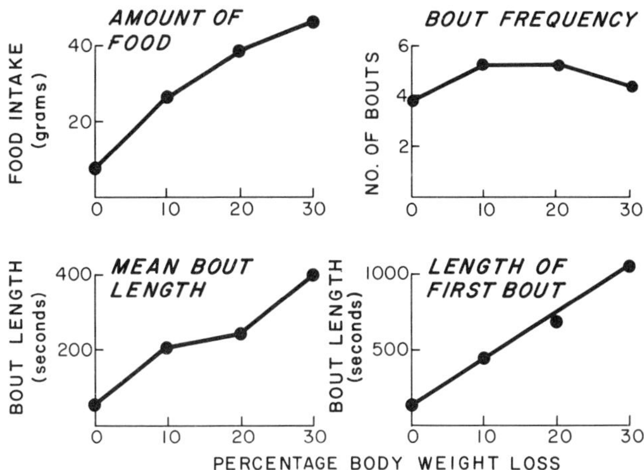

Fig. 10. Functions relating body weight loss and several measures of the consummatory response during 1-hr test period ($N = 10$). [From Zeigler, Green & Lehrer (1971).]

ing behavior, their temporal distribution under ad lib conditions and after periods of food deprivation, and their role in the control of food intake. They have also provided a body of techniques and measures for the neurobehavioral analysis of feeding in the pigeon, the problem to which the remainder of this paper is devoted.

II. A Feeding Behavior "System" in the Pigeon

It has long been known that removal of the cerebral hemispheres in the pigeon produces an apparently permanent aphagia. The decerebrate pigeon walks, grooms, and swallows food placed in its mouth but does not feed (Flourens, 1824; Åkerman, Fabricius, Larsson, & Steen, 1962). Moreover, although electrical stimulation of hypothalamic regions is reported to elicit feeding in the intact pigeon it has no such effect in the decerebrate bird (Åkerman, Andersson, Fabricius, & Svensson, 1960). These data indicate that endbrain structures are involved in the neural control of feeding in the pigeon; but the location of these structures has remained uncertain. Massive ablations of the dorsal endbrain do not produce feeding behavior deficits (Zeigler, 1963), but aphagia has been reported after lesions of the basal telencephalon in the pigeon (Rogers, 1922) and electrical stimulation of basal endbrain regions elicits a variety of "feeding movements" in chickens, ducks, and doves (Putkonen, 1967; Phillips, 1964; Harwood & Vowles, 1966).

Within the past few years our collaborative studies with Karten have made possible a more precise delimitation, within the avian telencephalon, of the regions whose destruction is responsible for the aphagia of the decerebrate pigeon (Zeigler, Karten, & Green, 1969; Zeigler & Karten, 1973). Furthermore, analysis of the afferent and efferent connections of these regions indicates that they are part of a network of structures at several levels of the pigeon brain and lesion studies have implicated all these structures in the neural control of feeding behavior. Figure 11 illustrates the location of these structures in a mid saggital section through the pigeon brain.

The *afferent* limb of this network includes three central components of the avian trigeminal system; the Main Sensory Trigeminal Nucleus (PrV), the quinto-frontal tract (QFT), and the nucleus basalis (NB). PrV is innervated by afferents of the trigeminal nerve and is the origin of an ascending projection to the endbrain, the quinto-frontal tract, which projects bilaterally (and apparently without intervening synapses) to the basal portion of the hemisphere (Wallenberg, 1903; Woodburne, 1936). At the level of the diencephalon, QFT is visible in normal material as a discrete bundle ventral to the ansa lenticularis and lateral to the hypothalamus. It enters the telencephalon in the lateral forebrain bundle, passes through the paleostriatum primitivum and the lobus parolfactorius and terminates in the nucleus basalis in the basolateral portion of the anterior telencephalon. On the basis of comparative anatomical data from a number of avian species Stingelin (1961) noted that the relative size of NB is directly proportional to the magnitude of PrV which, in turn, reflects the relative extent of beak development in the species.

Control lesions placed in regions adjacent to PrV, QFT, and NB, *including lesions of the lateral hypothalamic area*, have no significant effect

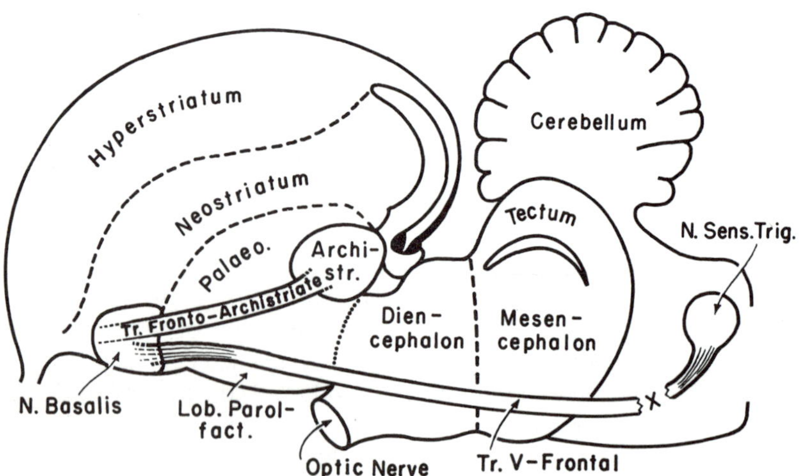

Fig. 11. A mid-sagittal section through the pigeon brain indicating schematically the location of some structures involved in the neural control of feeding in the pigeon.

upon food or water intake in the pigeon. By contrast, bilateral lesions of quinto-frontal structures are followed by periods of aphagia lasting from several days to several weeks. The rate of body weight loss in aphagic birds is comparable with that of normal birds deprived of food for equivalent periods and they may be maintained satisfactority by intubation of a liquid diet. In many birds, the postoperative period of aphagia is followed by an extended period of hypophagia which may persist for several months. Moreover, the compensatory overeating characteristic of normal, food-deprived birds (see Fig. 7) may be absent or considerably delayed in lesioned birds, so that body weight may be reduced significantly below its ad lib value and maintained at these levels for many months. Several lines of evidence suggest that these deficits are not attributable to digestive or metabolic dysfunction, but reflect an interference with neural mechanisms underlying the pigeons' responsiveness to food. Figure 12 illustrates the relation between food and water intake and body weight in birds with lesions of quinto-frontal structures. In these figures, body weight is plotted as a percentage of its ad lib value while food and water intake are cumulated over successive days and the cumulative total for a given day is divided by the number of elapsed days to give a mean cumulative intake value. In computing the values for postoperative intake, elapsed time is calculated from the day on which the resumption of eating (or drinking) occurs.

It is evident from these illustrative cases that the drinking behavior of lesioned birds is affected far less drastically than their feeding behavior. Birds with quinto-frontal lesions typically resume drinking within 24–48 hr postoperatively and do not differ in this respect from birds with control lesions. Whereas the reduction of water intake in many of the birds may be interpreted as a lesion-produced hypodipsia, paralleling their hypophagia, the available evidence suggests strongly that the reduction in water intake is an indirect effect of the decreased food intake of lesioned birds.

In addition to their effect upon the pigeon's food intake, lesions of quinto-frontal structures also affect the neural control of movement patterns involved in eating. Birds with such deficits can peck normally and will swallow food placed manually at the back of the mouth but have difficulty grasping and mandibulating kernels of grain. The existence of such deficits raises the possibility that the aphagia of lesioned birds is attributable to a disruption of sensory–motor mechanisms rather than to a reduction in the birds' responsiveness to food. Fortunately it is possible to dissociate these two effects experimentally by using a feedometer to record feeding responses postoperatively (Fig. 13). It is clear from the feedometer data that during the postoperative period of aphagia (as defined by the total absence of food intake) feeding responses are almost completely nonexistent and their resumption tends to parallel the recovery of food intake. For this reason, we have con-

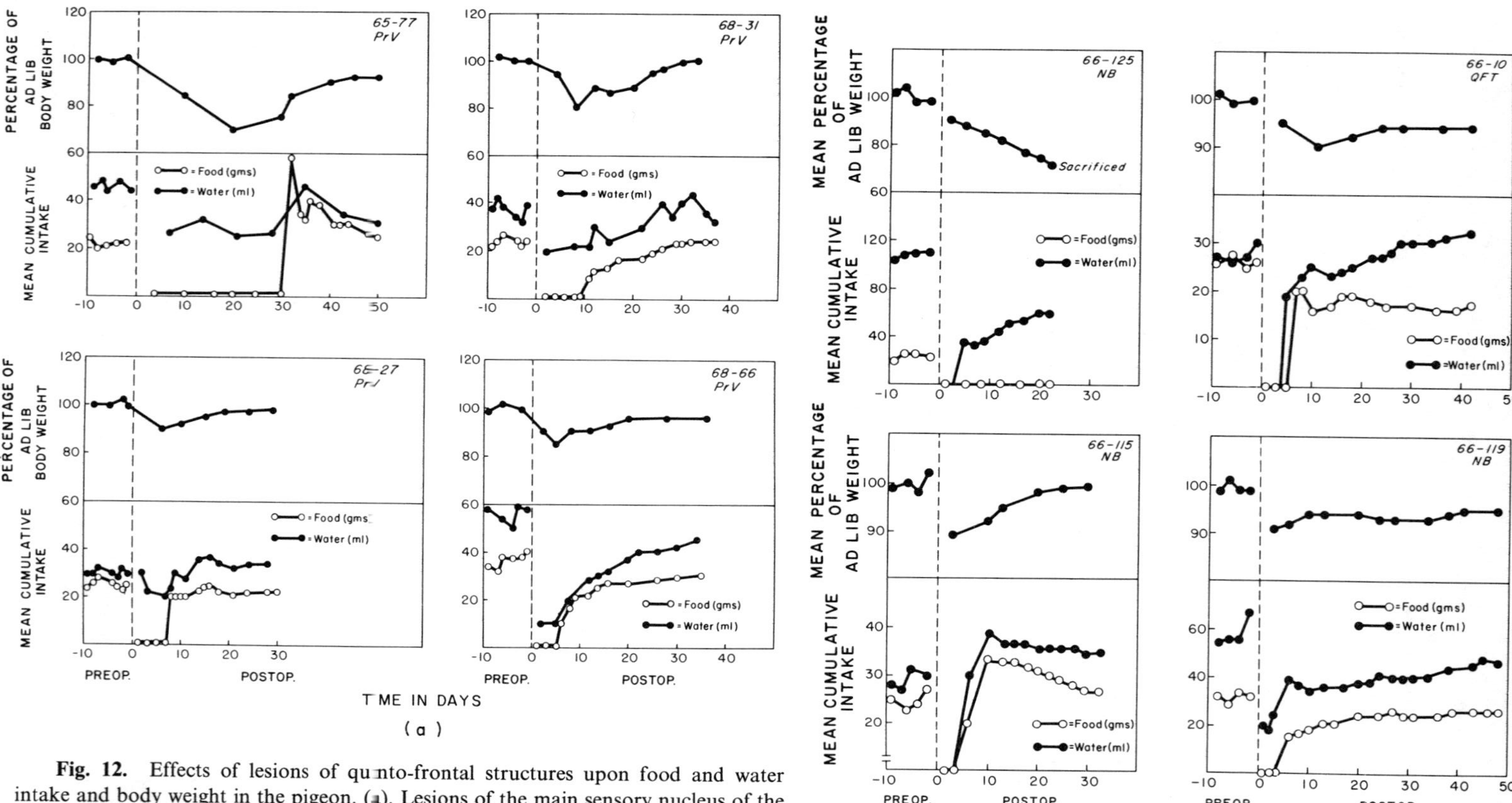

Fig. 12. Effects of lesions of quinto-frontal structures upon food and water intake and body weight in the pigeon. (a). Lesions of the main sensory nucleus of the trigeminus (PrV); (b). Lesions of the quinto-frontal tract (QFT) and the nucleus basalis (NB). [From Zeigler & Karten (1973)].

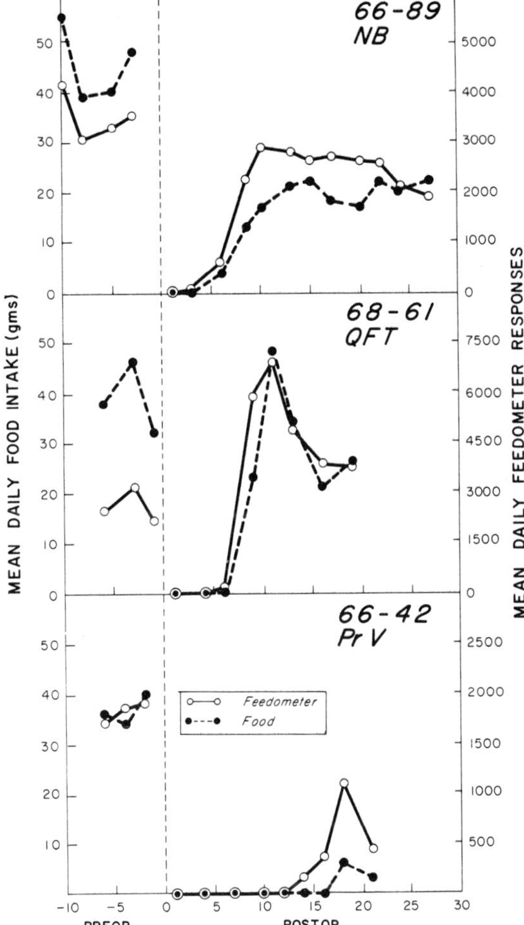

Fig. 13. The relation between food intake and feeding responses, as measured independently by means of a feedometer, in three birds with lesions of quinto-frontal structures. [From Zeigler & Karten (1973)]

cluded that lesions of quinto-frontal structures affect both the movement patterns involved in feeding and the pigeons' responsiveness to food (Zeigler & Karten, 1973).

Several additional structures involved in the neural control of feeding have been identified in recent lesion studies based upon anatomical findings from Karten's laboratory. The first of these studies was based upon data indicating that the fronto-archistriate tract, an efferent projection of the nucleus basalis, terminates in the dorsolateral nucleus of the archistriatum and in the overlying region of the caudal neostriatum (Zeier & Karten, 1971).

Accordingly, electrolytic lesions were placed stereotaxically in these regions and their effects upon food and water intake and body weight were studied. These effects are similar in some respects to those of quinto-frontal lesions, producing periods of aphagia varying from several days to several weeks, but having no significant effect upon water intake (Zeigler, Silver, & Karten, 1969).

Data from a recently completed lesion study have made it possible to tentatively identify an *efferent* component of the pigeons' feeding behavior system. Zeier and Karten (1971) have shown that the avian archistriatum is the origin of a descending tract, the occipito-mesencephalic tract (OMT), one of whose components has been traced into the brainstem down to the level of the pontine and spinal trigeminal nuclei. On the basis of its origin, distribution, and termination, Karten has suggested that the OMT may be an avian homologue of the mammalian pyramidal tract; our lesion data provide some support for this hypothesis. Lesions of the OMT, placed stereotaxically at the mesencephalic level of the tract, produce periods of aphagia as prolonged as those found after lesions of quinto-frontal structures. Furthermore, in addition to their effects upon food intake, preliminary studies indicate that OMT lesions also produce deficits in the pigeon's feeding behavior movement patterns. The pecking movements of these birds are hesitant and frequently incomplete and, in many cases, even when the peck is completed the oral aperture fails to open sufficiently to permit grasping and mandibulation of the grain kernels. Such observations suggest that by contrast with the disruption of neurosensory processes produced by quinto-frontal lesions, the deficits of OMT birds reflect an interference with the neuromotor control of feeding behavior movement patterns.

III. Feeding Behavior Mechanisms in the Pigeon: Neurobehavioral Analysis

Having delimited a group of brain structures which, on both anatomical and functional grounds, may be considered to be components of a feeding behavior "system," our next step is to determine the contribution of each of these structures to the neural control of feeding in the pigeon. Our approach to this problem has been to attempt to relate the physiological characteristics of a given structure to the nature of the behavior deficits produced by lesions of that structure. Such an approach involves both electrophysiological studies and analysis of lesion-produced deficits in feeding behavior. This combined approach has helped to clarify certain of the behavioral effects of lesions of quinto-frontal structures.

A. Electrophysiological Studies of Quinto-Frontal Structures

Analysis of single-unit activity in PrV (Zeigler & Witkovsky, 1968) and NB (Witkovsky, Zeigler, & Silver, 1973) indicates that both these nuclei contain cells responsive to mechanical stimulation of the beak and mouth region but unresponsive to mechanical stimulation of the tongue. The majority of neurons in PrV are responsive to light touch and pressure stimuli, but the nucleus also contains cells responsive to jaw displacement produced by opening or closing the mouth (Fig. 14). The receptive fields of PrV tactile units are relatively small, covering only a few square millimeters, and the majority of these units are slow adapting to maintained stimulation. By contrast, the NB does not contain units signaling beak displacements, and the majority of its units have large receptive fields, are fast adapting and are driven most effectively by the movement of a tactile stimulus relative to the surface of the

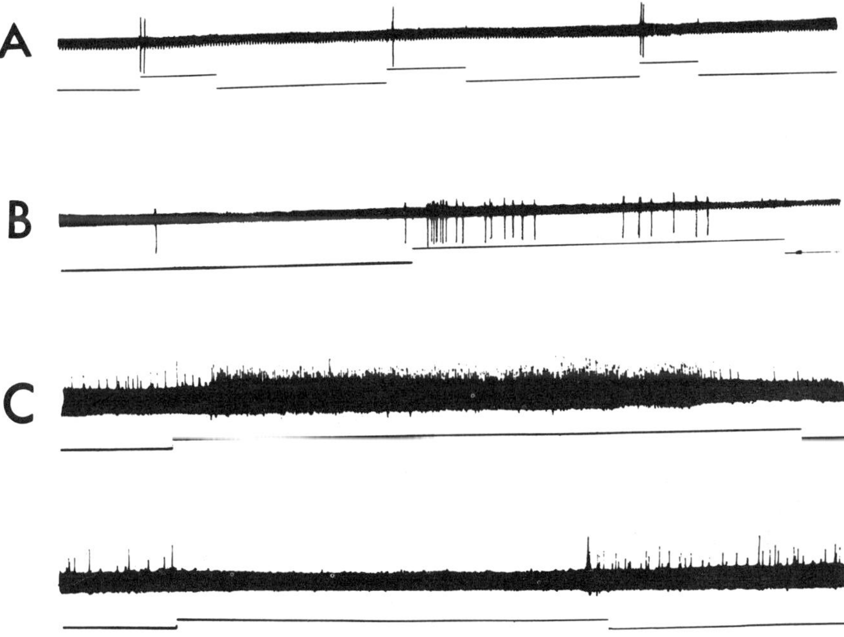

Fig. 14. Samples of unit discharges recorded from PrV cells in the pigeon. A. a fast adapting cell responding to light touch of the outer ventral beak surface. B. a more slowly adapting unit responsive to tactile stimulation of the edge of the dorsal beak. C. unit responses to beak displacement. Spontaneous firing is increased by opening of the beak and inhibited by closing. [From Zeigler & Witkovsky (1968).]

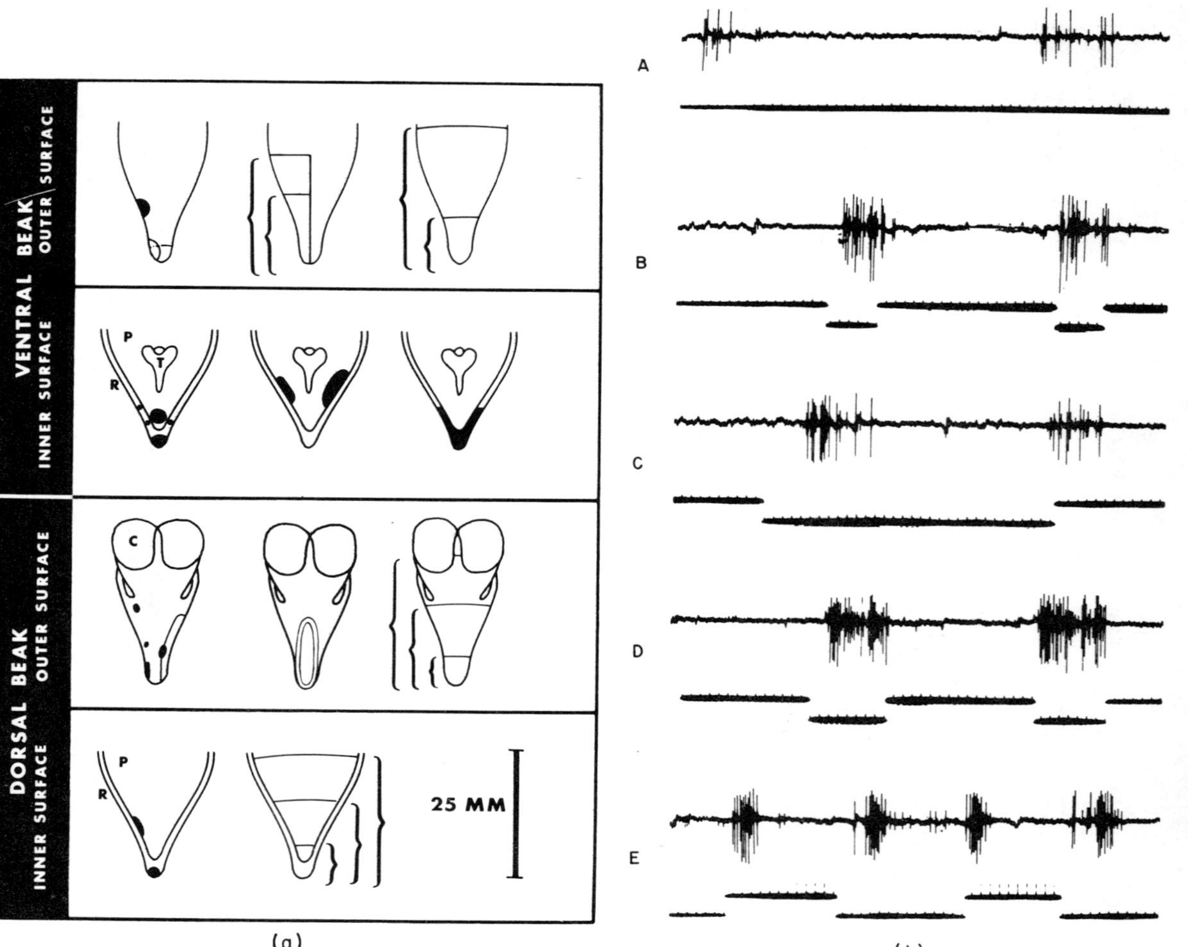

Fig. 15. (a). Examples of size and location of receptive fields of units recorded in PrV and NB of the pigeon. Abbreviations: C-cere: P-palate; T-tongue; R beak rim. (b). Response properties of an NB unit with a small receptive field located inside the mouth; A. spontaneous firing; B. brief, repeated contacts with the sensitive region; C. a prolonged contact; D. the probe is moved through the receptive area in one direction only; E. the probe is moved back and forth through the receptive area. A downward deflection of the lower trace signals the onset of movement in one direction and upward deflection signals movement of the probe in the opposite direction. Time marker: 50 msec. [From Witkovsky, Zeigler, & Silver (1973)]

beak or mouth (Fig. 15). Measurement of unit response latencies to electrical stimulation support the anatomical finding that PrV projects directly to NB without intervening synapses.

B. Analysis of Deficits after Lesions of Quinto-Frontal Structures

In attempting to assess lesion effects upon feeding behavior movement patterns we have used as one measure the ratio of feeding responses to food intake, on the assumption that this ratio should increase significantly after any type of interference with the sensory or motor control of the consummatory response. Such ratios may be obtained either by direct observation of the pigeons' feeding behavior for a brief (10 min) period in a test chamber, or calculated indirectly for longer periods by using a feedometer attached to the home cage to obtain independent measures of feeding responses and food intake.

Table II presents data on total food intake and total number of feeding responses for 7-day periods immediately preceding and following bilateral lesions of the nucleus basalis in five pigeons. In addition to a decline in food, intake there is a postoperative increase in the number of feeding responses required to obtain a unit quantity of food. Comparable data have been obtained following lesions of PrV and QFT. Direct observation of lesioned birds indicates that the reduction in their feeding efficiency is due to impairment of the sensory control of mandibulation, rather than of pecking or swallowing. (The identification of this deficit as neurosensory rather than neuromotor is supported by the finding that a similar deficit is produced by section of trigeminal sensory nerves while damage to the trigeminal motor

Table II. *Effects of Nucleus Basalis Lesions upon Food Intake and Feeding Efficiency in the Pigeon.*

	Preoperative			Postoperative		
Bird no.	Food intake (gm)	Feeding responses	Responses per gram	Food intake (gm)	Feeding responses	Responses per gram
66 82	184	10,790	58.1	95	9,535	100.3
66–87	155	11,912	76.8	26	3,416	131.4
66–89	293	22,818	77.8	20	3,936	196.8
66–90	157	20,980	133.6	80	22,959	286.9
66–93	295	17,790	60.3	148	21,219	143.4

nucleus produces flaccidity of the lower beak and severely disrupts the movement patterns involved in *both* eating and drinking.) Furthermore, the electrophysiological data suggest that this impairment is due to the deprivation of tactile and proprioceptive information normally involved in the sensory control of mandibulation. Thus, analysis of the temporal organization of the pigeon's feeding response (Fig. 1) indicates that mandibulation takes 100–150 msec and is initiated and maintained by contact of the grain with the beak tip. A comparison of the sensory control and time course of mandibulation with the receptive field characteristics and adaptive properties of PrV and NB units suggests that these nuclei contain neurons that signal the presence of a kernel of grain at the beak tip, provide complementary information about its static position and movement within the mouth, and monitor the extent of beak displacement. In conjunction with the spinal trigeminal nucleus (Silver & Witkovsky, 1973) quinto-frontal structures may thus act as "grain-detectors" for the pigeon, functioning within the avian trigeminal system in a manner analogous to that of the "bug detectors" described for the visual system of the frog (Lettvin, Maturana, McCulloch, & Pitts, 1959).

In addition to their role in the sensory control of feeding, the fact that lesions of quinto-frontal structures produce prolonged periods of aphagia and hypophagia suggest that they are also involved in the neural control of hunger, i.e., in the pigeons' responsiveness to food. Birds with basalis lesions show abnormally long latencies to feed in a novel test situation and they eat less and satiate faster than normal pigeons when tested after varying periods of food deprivation (Zeigler, unpublished data). In a recent study a group of birds was trained on an operant key-pecking response reinforced with food and then subjected to basalis lesions. Key-pecking was completely abolished for periods of up to a month postoperatively, although the birds would eat in their home cage or feed from a dish placed in the Skinner box at the end of a test session. By contrast, basalis lesions had no effect upon a key-pecking response in birds reinforced with water (Zeigler & Karten, 1973).

The presence of such deficits may indicate that the avian trigeminal system mediates "motivational" as well as "sensory–motor" processes. Alternatively, the "motivational" deficits seen after lesions of quinto-frontal structures may be an indirect effect of the deafferentation of other brain structures in the pigeon's feeding behavior "system" (e.g., the archistriatal "somato–motor" region or the OMT). Although such an interpretation is supported by the results of a recent study of trigeminal deafferentation (to be discussed), a wide variety of physiological and behavioral investigations will be required to distinguish among these alternatives and to clarify the contribution of each of these structures to the neural control of feeding.

C. Trigeminal Deafferentation: Sensory and Motivational Effects

A large and growing body of evidence indicates that orophyaryngeal factors play an essential role in the control of food intake in mammals. However, for a variety of reasons, much of the mammalian data have been obtained indirectly in experiments that involve a combination of intragastric feeding techniques and operant conditioning procedures rather than the direct observation of feeding behavior in surgically deafferented animals (Teitelbaum & Epstein, 1963). On the basis of such studies, many investigators have concluded that the rat can regulate its body weight in the absence of sensory feedback from the oral region—a conclusion that has recently been challenged on methodological grounds (Holman, 1969). The finding that lesions of central trigeminal sensory nuclei produce both motivational and sensory deficits suggested to us that the pigeon might be a useful preparation for an analysis of the interaction of sensory and neural mechanisms in the control of feeding behavior and body weight regulation. The location and peripheral distribution of the trigeminal sensory nerves in the pigeon are such as to permit trigeminal deafferenation of the oral region without disrupting motor function. Because the pigeon is a granivorous bird, we hypothesized that somatosensory stimuli would be essential for the control of feeding while olfaction and taste would play relatively minor roles. In a recently completed series of experiments we have studied the effects of trigeminal deafferentation upon a variety of measures of sensory–motor and motivational processes.

A combination of gross dissection (Zeigler & Witkovsky, 1968) and electrical stimulation of the trigeminal nerve in the pigeon indicates that its sensory branches innervate the region of the mouth and beak, exclusive of the tongue. There is limited innervation of the orbital region, both the ophthalmic and maxillary branch distributing primarily to the upper beak, while the mandibular sensory branch distributes to the lower beak. The absence of trigeminal innervation of the tongue is in accord with previous electro physiological studies of the avian gustatory system (Kitchell, Strom, & Zotterman, 1959) and with our own studies of the response properties of units in PrV, NB, and the spinal trigeminal complex. Unilateral or bilateral section of one or more sensory branches was carried out under direct observation using a Zeiss dissection microscope.

By contrast with the absence of effects upon drinking, trigeminal deafferentation produces disruptions of feeding behavior as severe and persistent as any we have encountered after lesions of central structures in the pigeon's feeding behavior "system." Indeed, bilateral section of all three sensory branches has produced periods of aphagia ranging from 1 to 6 weeks

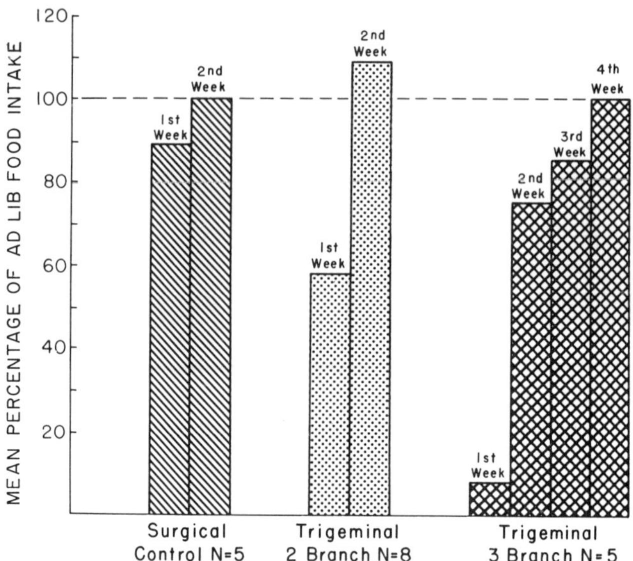

Fig. 16. Effects of trigeminal deafferentation upon food intake in the pigeon. Comparison of the effects of control procedures, bilateral denervation of two branches and bilateral denervation of three branches.

and followed by equally prolonged periods of hypophagia. As Fig. 16 indicates, surgical control procedures produce only minor and transient effects, whereas bilateral section of trigeminal sensory nerves is followed by a disruption of food intake (and a consequent loss of body weight) whose persistence and severity are approximately proportional to the number of nerves sectioned bilaterally.

In addition to its effects upon food intake, trigeminal deafferentation disrupts the sensory control of feeding and produces a striking reduction in the efficiency of the consummatory response. An independent measure of such deficits was obtained, for a second group of pigeons, by direct observation of their feeding behavior in a test chamber containing a dish in which 20 kernels of grain were available. Birds were maintained at 80 % of their ad lib weight to ensure a prompt response to the food, and were tested two to four times weekly. They were introduced individually into the test chamber and their latency to feed was recorded. The number of pecks made by each bird was counted and observations of feeding continued either until all the grains were consumed or until a 10-min test period had elapsed. The number of kernels consumed was then counted and a ratio of pecks per grain was calculated for each bird. Under these testing conditions a normal pigeon usually requires one or two pecks per grain to obtain the 20 kernels (mean

1.2). Figure 17 illustrates the effects of trigeminal deafferentation upon the ratio of feeding responses to kernels of grain obtained in such a test. Control surgery (#126) has a minor and transient effect but the ratio increases dramatically with progressive increases in the extent of sensory denervation. In the most successful cases of bilateral three-branch denervation (e.g., # 222) the deficit was so severe that birds were observed to make 200 or more pecks within a 10-min period without obtaining a single kernel of grain. Over time, there is a gradual improvement in the bird's feeding efficiency but performance eventually reaches an asymptotic level which has been observed to remain stable over several months of testing. Because pecking and swallowing appear to take place normally we have concluded that trigeminal deafferentation disrupts the sensory control of mandibulation.

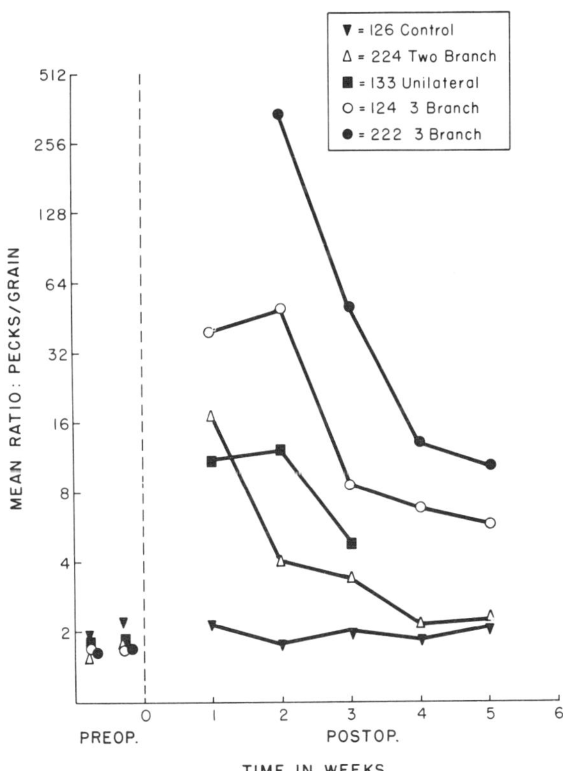

Fig. 17. Effects of varying degrees of trigeminal deafferentation upon the efficiency of the pigeon's feeding behavior in a 10 min "pick-up" test.

Thus, the effects of trigeminal deafferentation upon both sensory and motivational processes are strikingly similar to those seen after lesions of central trigeminal (quinto-frontal) structures in the pigeon. Indeed, by automatically monitoring feeding responses independently of the measurement of food intake it is possible to experimentally dissociate the two effects in a manner comparable to the dissociation shown for quinto-frontal birds (Fig. 13).

Illustrative data for a single pigeon are presented in Fig. 18 and indicate that, in this bird, trigeminal deafferentation is followed by a period of aphagia lasting for about 6 weeks. (In order to prevent inanition and death, intubation of a liquid diet was begun during the fourth postoperative week and continued daily until the resumption of spontaneous feeding.) Throughout the entire 40-day period of aphagia the bird made very few feeding responses, although its drinking behavior was unaffected and there were no disturbances of posture or locomotion which might account for the prolonged failure to eat. Feeding resumed during the sixth postoperative week but intake remained below its preoperative level for several weeks. Moveover, despite its considerable weight loss the bird showed no significant degree of compensatory overeating. Recovery of body weight appears to reach its maximum at between Weeks 9 and 12 postoperatively, remaining at 80–85 % of its preoperative value until the bird's sacrifice 18 weeks post-

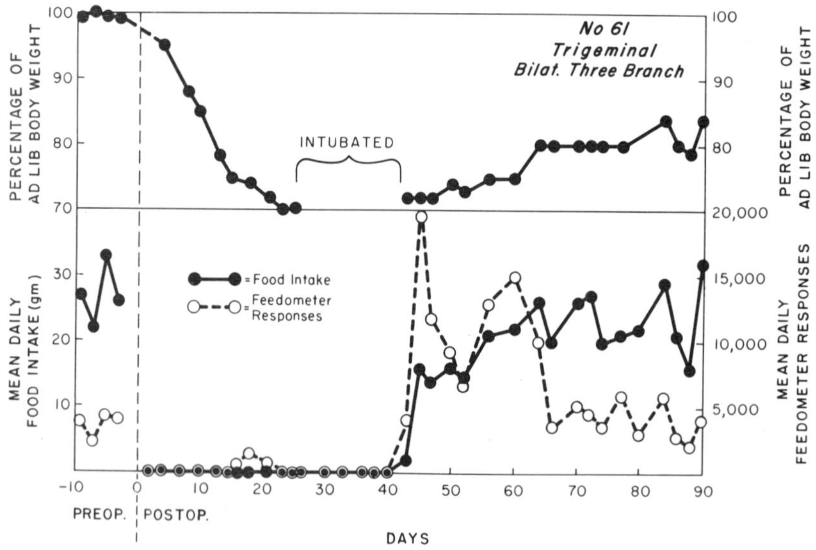

Fig. 18. Effects of trigeminal deafferentation upon food intake, feeding responses and body weight in a single bird.

operatively. The stability of body weight during this period suggests that it was being regulated, although at a level about 20% below its ad lib value.

In the first few weeks after the resumption of feeding, there is a striking reduction in the bird's feeding efficiency as measured by the ratio of feedometer responses to food intake. During this period the ratio rises from its preoperative mean level of about 130 to a high of 1000–2000 feeding responses per gram of food in the first few days of this period. From this point on feeding efficiency improves gradually with ratios dropping over the next few weeks until, by the twelfth postoperative week, the ratio has declined to about 160 responses per gram of food.

The feedometer data thus indicate that trigeminal deafferentation produces a severe but gradually declining deficit in the somato–sensory control of feeding, a finding confirmed independently by the observational data presented in Fig. 17. However, the virtual absence of feeding responses during the prolonged period of aphagia and the persistence of hypophagia and reduced body weight despite the marked improvement in feeding efficiency suggest that, in addition to its effect upon the "sensory–motor mechanism" for feeding, trigeminal deafferentation also produces a deficit in the bird's responsiveness to food.

This hypothesis is supported by the latency data obtained during the feeding behavior observation tests, which provide a measure of responsiveness to food that is distinct from either food intake or the consummatory response. In view of the prolonged periods during which deafferented birds are unresponsive to food in the home cage it is not surprising that their latencies to eat in the novel test situation are significantly greater than those of normal pigeons tested at equivalent weight levels (Zeigler, in preparation).

In an attempt to measure appetitive behavior in a test situation where it could be still further dissociated from the consummatory response, we have studied the effects of trigeminal deafferentation upon a learned, key-pecking response reinforced with food in an operant-conditioning situation. A modified feedometer was used as a food magazine, and it was thus possible to obtain, for each test session, independent measures of the number of feeding responses, the number of key-pecking responses and the amount of food eaten, and to calculate an intake-per-feeding-response ratio which might be used as a measure of the bird's feeding efficiency.

Figure 19 presents data for a representative bird, reinforced on a variable interval 1-min schedule and subjected to bilateral section of the ophthalmic and mandibular (sensory) nerves, carried out in two stages. Key-pecking is not markedly affected by unilateral section, although there is a decline in food intake. The slight rise in the feedometer-per-food-intake ratios suggests that this decline is caused by a deficit in mandibulation such

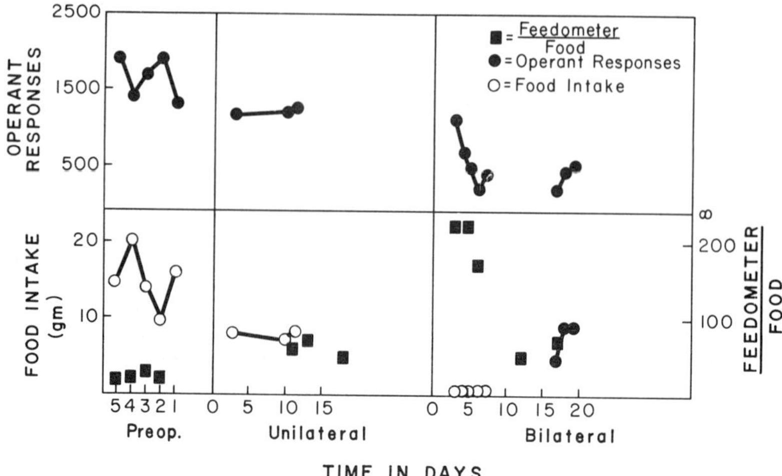

Fig. 19. The effects of unilateral and bilateral section of two trigeminal sensory nerves upon food intake, operant key-pecking and feeding efficiency in a single bird. [From Zeigler & Feldstein (1971).]

that the bird obtains less food in each 6-sec reinforcement interval. Following the contralateral section, key-pecking is maintained at close to its previous level but declines gradually over the next few days. The marked elevation of the feedometer-per-food-intake ratios indicates that the bird is making many feeding responses but getting little or no food. The decline in key-pecking is thus an extinction phenomenon due to an inability to obtain reinforcement, rather than a direct effect of denervation upon key pecking. In a group of birds subjected to bilateral section of all three branches, key-pecking is also intact postoperatively but *the bird is not responsive to food during the reinforcement interval,* so that extinction of the response gradually takes place. With the eventual return of responsiveness to food, reconditioning of the operant takes place even though the existence of elevated feedometer-per-food-intake ratios indicates the persistence of a somato sensory deficit in mandibulation. It is of interest that birds trained on a *water-reinforced* key-pecking response, were unaffected by trigeminal deafferentation.

IV. Conclusion: Implications for the Study of Feeding Behavior

With the exception of Dethier's (1969) studies of hunger in the blowfly the analysis of feeding behavior has focused *largely* upon one vertebrate species, the rat, and upon one brain structure, the hypothalamus. As has been pointed out, "virtually every effect on feeding behavior obtained by manipulating the lateral hypothalamus can be produced equally well by manipulating the activity of other structures [Albert, Storlien, Wood, & Ehman, 1970]." Because the hypothalamic region is richly endowed with fibers of passage to and from these structures, the relation between hypothalamic and extrahypothalamic mechanisms constitutes a persistent problem in the neurobehavioral analysis of feeding (Morgane, 1969). Moreover, as Grossman (1969) notes, "the excitatory hypothalamic feeding center overlaps so extensively with the hypothalamic drinking center in all common laboratory animals that lesions or electrical stimulation almost invariably induce concurrent effects on food and water intake." Furthermore, the role of sensory factors in the neural control of hunger also requires clarification. Although the contribution of peripheral sensory factors (particularly oropharyngeal sensations) is well documented in the rat (Teitelbaum & Epstein, 1963; Epstein, 1967) and some data on sensory projections to hypothalamic and extrahypothalamic "feeding sites" are available (Wyrwicka, 1969; Wyrwicka & Chase, 1970) there is little direct evidence on the relation between central sensory structures and feeding behavior. Finally, the problem of lesion effects upon sensory–motor processes and the relation between sensory–motor and motivational mechanisms has received only cursory attention in the mammalian literature (Rodgers, Epstein, & Teitelbaum, 1965).[2]

Viewed against this background, the experiments just reviewed suggest that the pigeon has many advantages as a neurobehavioral preparation for the analysis of feeding behavior mechanisms. First, the temporal organization of its feeding behavior movement patterns involves a time scale (milliseconds) that approximates that of neural activity patterns (e.g., Zeigler & Witkovsky, 1968; Witkovsky, Zeigler, & Silver, 1973). Second, by dissociating feeding responses from food intake it seems possible to clarify the relation between sensory–motor and motivational mechanisms. Third, the fact that lesions of quinto-frontal structures affect eating without affecting drinking suggest that the neural substrates of hunger and thirst may be more easily dissociated in the pigeon than in the rat. Finally, the feeding-behavior deficits produced by trigeminal deafferentation, and lesions of central trigeminal structures and the absence of such deficits after hypothalamic lesions, suggest that the pigeon may be an ideal preparation for the dissociation of hypothalamic and extrahypothalamic mechanisms as well as for an analysis of the interaction of sensory and neural processes in the control of feeding behavior.

[2] A recent report from Teitelbaum's laboratory indicates that in rats made aphagic by lateral hypothalamic lesions the transition from stage 1 (complete aphagia) to stage 2 (accepting only highly palatable food) was preceded by or coincident with the recovery of head orientation responses to olfactory and trigeminal stimuli. [Marshall, J. F., Turner, B. H. and Teitelbaum, P. Sensory neglect produced by lateral hypothalamic damage. *Science*, 1971, 174, 523–525.]

Acknowledgment

Toward the end of his paper, "The Experimental Analysis of Instinctive Behavior," Lashley (1938) refers to Woodworth's suggestion that "habits might acquire dynamic functions, that a mechanism might become a drive" and then concludes as follows: "I should carry this notion a step further and suggest that physiologically all drives are no more than expressions of the activity of specific mechanisms." This suggestion, like so many of Lashley's insights and the papers in which they are embedded, constitutes an intellectual endowment to psychology which continues to pay us heuristic dividends. The work reviewed in this paper is intended as a partial repayment of a long-standing intellectual debt.

References

Åkerman, B., Andersson, B., Fabricius, E. & Svensson, L. Observations on central regulation of body remperature and of food and water intake in the pigeon *(Columba livia). Acta Physiologica Scandinavica*, 1960, **50**, 328–336.

Åkerman, B., Fabricius, E., Larsson, B. & Steen, L. Observations on pigeons with prethalamic radiolesions in the nervous pathways from the telencephalon. *Acta Physiologica Scandinavica*, 1962, **56**, 286–298.

Albert, D. J., Storlein, L. H., Wood, D. J. & Ehman, G. K. Further evidence for a complex system controlling feeding behavior. *Physiology and Behavior*, 1970, **5**, 1075–1082.

Bolles, R. C. Readiness to eat: effects of age, sex and weight loss. *Journal of Comparative and Physiological Psychology*, 1965, **60**, 88–92.

Bolles, R. C. *Theory of motivation.* New York: Harper, 1967.

Cizek, L. J. Long-term observations on relationship between food and water ingestion in the dog. *American Journal of Physiology*, 1959, **197**, 242–246.

Collier, G. Body weight loss as a measure of motivation in hunger and thirst. In P. J. Morgane (Ed.), *Neural regulation of food and water intake. Annals of the New York Academy of Sciences*, 1969, **157**, 594–609.

Dethier, V. G. Feeding behavior in the blowfly. In D. S. Lehrman, R. A. Hinde, and E. Shaw (Eds.), *Advances in the study of behavior*, Vol. 2, New York: Academic Press, 1969.

Epstein, A. N. Oropharyngeal factors in feeding and drinking. In C. E. Code (Ed.), *Handbook of physiology.* Section 6. *Alimentary canal.* Vol. 1. *Control of food and water intake.* Washington, D.C.: American Physiological Society, 1967.

Farner, D. S. Digestion and the digestive system. In A. J. Marshall (Ed.), *Biology and comparative physiology of birds*, Vol. 1. New York: Academic Press, 1960.

Flourens, P. *Recherches experimentales sur les propriétiés et les functions du systéme nerveux dans les animaux vertébrés.* Paris: Baillière, 1824.

Grossman, S. P. A neuropharmacological analysis of hypothalamic and extrahypothalamic mechanisms concerned with the regulation of food and water intake. In P. J. Morgane (Ed.), *Neural regulation of food and water intake. Annals of the New York Academy of Sciences*, 1969, **157**, 902–917.

Harwood, D. & Vowles, D. M. Forebrain stimulation and feeding behavior in the ring dove *(Streptopelia resoria). Journal of Comparative and Physiological Psychology,* 1966, **62**, 388–396.

Hinde, R. A. *Animal behaviour.* New York: McGraw-Hill, 1970.

Holman, G. L. Intragastric reinforcement effect. *Journal of Comparative and Physiological Psychology*, 1969, **69**, 432–441.

Karten, H. J. The organization of the avian telencephalon and some speculations on the phylogeny of the amniote telencephalon. *Annals of the New York Academy of Sciences*. 1969, **167**, 164–179.

Karten, H. J. & Hodos, W. *A stereotaxic atlas of the brain of the pigeon (Columba livia)*. Baltimore: John Hopkins Univ. Press, 1967.

Kitchell, R. L., Ström, L. & Zotterman, Y. Electrophysiological studies of thermal and taste reception in chickens and pigeons. *Acta Physiologica Scandinavica*, 1959, **46**, 133–151.

Lashley, K. S. Experimental analysis of instinctive behavior. *Psychological Review*, 1938, **45**, 445–471.

Le Magnen, J. Peripheral and systemic actions of food in the caloric regulation of intake. In P. J. Morgane (Ed.), *Neural regulation of food and water intake*. Annals of the New York Academy of Sciences, 1969, **157**, 1126–1157.

Lettvin, J. Y., Maturana, H. R., McCulloch, W. & Pitts, W. H. What the frog's eye tells the frog's brain. *Proceedings of the Institute of Radio Engineers*, 1959, **47**, 1940–1951.

McFarland, D. Interaction of hunger and thirst in the Barbary dove. *Journal of Comparative and Physiological Psychology*, 1964, **58**, 174–179.

Megibow, M. & Zeigler, H. P. Readiness to eat in the pigeon. *Psychonomic Science*, 1968, **12**, 17–18.

Miller, N. E. Behavioral and physiological techniques. In C. E. Code (Ed.), *The handbook of physiology*. Section 6. *The alimentary canal*. Vol. 1. *Control of food and water intake*. Washington, D.C.: American Physiological Society, 1967.

Morgane, P. J. The function of the limbic and rhinic forebrain-limbic midbrain systems and reticular formation in the regulation of food and water intake. In P. J. Morgane (Ed.), *Neural regulation of food and water intake. Annals of the New York Academy of Sciences*, 1969, **157**, 806–848.

Phillips, R. E. Wildness in the Mallard duck: Effects of brain lesions and stimulation on "escape behavior" and reproduction. *Journal of Comparative Neurology*, 1964, **116**, 139–155.

Putkonen, P. T. S. Electrical stimulation of the avian brain. *Annales Academiae Scieniarum Fennicae*, 1967, Series A. V. Medica No. 130.

Rodgers, W. L., Epstein, A. N. & Teitelbaum, P. Lateral hypothalamic aphagia: motor failure or motivational deficit? *American Journal of Physiology*, 1965, **208**, 334–342.

Rogers, F. T. Studies on the brain stem. VI. An experimental study of the corpus striatum of the pigeon as related to various instinctive types of behavior. *Journal of Comparative Neurology*, 1922, **35**, 21–60.

Silver, R. & Witkovsky, P. The spinal trigeminal nucleus of the pigeon: a single-unit analysis. *Brain, Behavior and Evolution*, 1973, **8**, 287–303.

Stellar, E., & Hill, H. H. The rat's rate of drinking as a function of water deprivation. *Journal of Comparative and Physiological Psychology*, 1952, **45**, 96–102.

Stingelin, W. Grössenunterschiede des sensiblen Trigeminuskerns bein verschiedenen Vögeln. *Revue Suisse de Zoologie*. 1961. **68**, 247–251.

Strominger, J. L. The relation between water intake and food intake in normal rats and rats with hypothalamic hyperphagia. *Yale Journal of Biology and Medicine*, 1947, **19**, 279–288.

Teitelbaum, P. & Epstein, A. N. The role of taste and smell in the regulation of food and water intake. In Y. Zotterman (Ed.), *Olfaction and taste*, Vol. 1. London: Pergamon Press, 1963, Pp. 347–360.

Wallenberg, A. Der Ursprung des Tractus isthmo-striatus (oder bulbostriatus) der Taube. *Neurologische Zentralblatt*, 1903, **22**, 98–101.

Witkovsky, P., Zeigler, H. P. & Silver, R. A single-unit analysis of the nucleus basalis in the pigeon. *Journal of Comparative Neurology*, 1973, **147**, 119–128.

Wolin, B. R. Difference in manner of pecking a key between pigeons reinforced with food and with water. In A. C. Catania (Ed.) *Contemporary research in operant behavior*, Glenview, Illinois: Scott, Foresman, 1968.

Woodburne, R. T. A phylogenetic consideration of the primary and secondary centers and connections of the trigeminal complex in a series of vertebrates. *Journal of Comparative Neurology*, 1936, **65**, 403–501.

Wyrwicka, W. Sensory regulation of food intake. *Physiology and Behavior*, 1969, **4**, 853–858.

Wyrwicka, W. & Chase, M. Projections from the buccal cavity to brain stem sites involved in feeding behavior. *Experimental Neurology*, 1970, **27**, 512–519.

Zeier, H. & Karten, H. J. The archistriatum of the pigeon: organization of afferent and efferent connections. *Brain Research*, 1971, 313–326.

Zeigler, H. P. Effects of forebrain lesions upon activity in pigeons. *Journal of Comparative Neurology*, 1963, **120**, 183–194.

Zeigler, H. P. Trigeminal deafferentation and feeding behavior in pigeons: sensory and motivational effects. *Science*, 1973, **182**, 1155–1158.

Zeigler, H. P. & Feldstein, S. A feedometer for the pigeon. *Journal of the Experimental Analysis of Behavior*, 1971, **16**, 181–187.

Zeigler, H. P. Green, H. L. & Lehrer, R. Patterns of feeding behavior in the pigeon. *Journal of Comparative and Physiological Psychology*, 1971, **76**, 468–477.

Zeigler, H. P., Green, H. L. & Siegel, J. Food and water intake and weight regulation in the pigeon. *Physiology and Behavior*. 1972, **8**, 127–134.

Zeigler, H. P. & Karten, H. J. Brain mechanisms and feeding behavior in the pigeon *(Columba livia)*: Quinto-frontal structures. *Journal of Comparative Neurology*, 1973, **152**, 59–101.

Zeigler, H. P., Karten, H. J. & Green, H. L. Neural control of feeding in the pigeon. *Psychonomic Science*, 1969, **15**, 156–157.

Zeigler, H. P., Silver, R. & Karten, H. J. Archistriate lesions and feeding behavior in the pigeon. Paper presented at Eastern Psychological Association, 1969.

Zeigler, H. P. & Witkovsky, P. The main sensory trigeminal nucleus in the pigeon: a single-unit analyses. *Journal of Comparative Neurology*, 1968, **134**, 255–264.

The Study of Sleep in Birds

Irving J. Goodman[1]
West Virginia University

Sleep, in the last few years particularly, has become a topic responsible for a proliferation of speculative and research literature. Excitement has been generated by a number of findings. Among them are electrographic correlates of sleep; different sleep stages; the initiation, increase, or decrease of sleep by local agents within the brain; biochemical and sleep correlations; and activity cycles.

Most of the correlational and experimental studies of sleep and wakefulness have been carried out in mammals; information in avians has been relatively sparse. However slowly, there is developing a small body of literature regarding the behavioral and physiological bases of sleep in birds which, perhaps, is worthy of attention.

This chapter, although not exhaustive, will attempt to describe some more recent findings from studies of normally occurring and experimentally manipulated sleep and wakefulness in birds.

I. Behavioral and Physiological Correlates of Sleep and Waking

Behavioral and physiological correlates of sleep have been investigated in several different avian forms: chicken (e.g., Hishikawa, Cramer, & Kuhlo, 1969; Ookawa & Gotoh, 1965; Peters, Vonderahe, & Schmid, 1965); pigeon (e.g., Klein, Michel, & Jouvet, 1964; LoPresti & Goodman, 1968; Tradardi,

[1] Financial support of this work was in part provided by a West Virginia University General Medical Research Grant.

1966; Van Twyver & Allison, 1972; Walker & Berger, 1972); burrowing owl (Berger & Walker, 1972); hawk and falcon (Rojas-Ramirez & Tauber, 1970).

Investigators have divided the sleep–awake continuum into categories that correspond somewhat to two or more of the following: wakefulness (W), drowsiness (D), non-rapid eye movement (NREM) sleep, and rapid eye movement (REM) sleep.

A. Wakefulness

A wide range of behaviors extending from highly excitable movements to relatively quiescent postures are observable during wakefulness; animals may show head and neck movements (e.g., during feeding or grooming), wing movements (e.g., during flight or attack), leg movements (e.g., during strutting or head scratching), vocalization, opened eyes, or may show fixed postures (e.g., standing erect, bowing or crouching), etc. The tendency to engage in one or another of these or other behavioral activities is influenced by a variety of environmental and physiological conditions. Such diverse conditions as enclosure size, lighting, temperature, extraneous noise, and tethering elements (e.g., recording cable) are among the experimental variables that must be considered as contributory to behavioral observations. From the standpoint of behavior, the waking period is potentially most variable.

Electroencephalographic (EEG) activity is recordable from locations over the entire surface of the brain (Berger, 1929). During wakefulness in adult pigeons Van Twyver and Allison (1972) found a predominance of high-frequency and low-amplitude EEG activity at various locations. They noted that the most prominent locus for low-frequency activity development was located over the posterior–medial area of the brain surface, overlying the hippocampal area. Despite some differences in EEG patterns at different electrode locations in the waking state, in all observed species, frequencies tend to be relatively high (above 10 Hz) and amplitudes low (below 75 μV) (see Fig. 1).

Electromyographic (EMG) activity, usually recorded from neck muscles, shows relatively high tonic levels during waking, accompanied by phasic elevations during behavioral excitability. Eye movements, reflected in electrooculographic (EOG) recordings, are almost continuously present; also reflected in the EOG record are eyelid movements (blinking) and clusters of sharp rhythmic waves, reportedly associated with eyeball oscillation (Paulson, 1964). There is one reported exception; the burrowing owl fails to show eye movements during waking (Berger & Walker, 1972). Heart rate is labile during wakefulness, varying markedly with behavioral

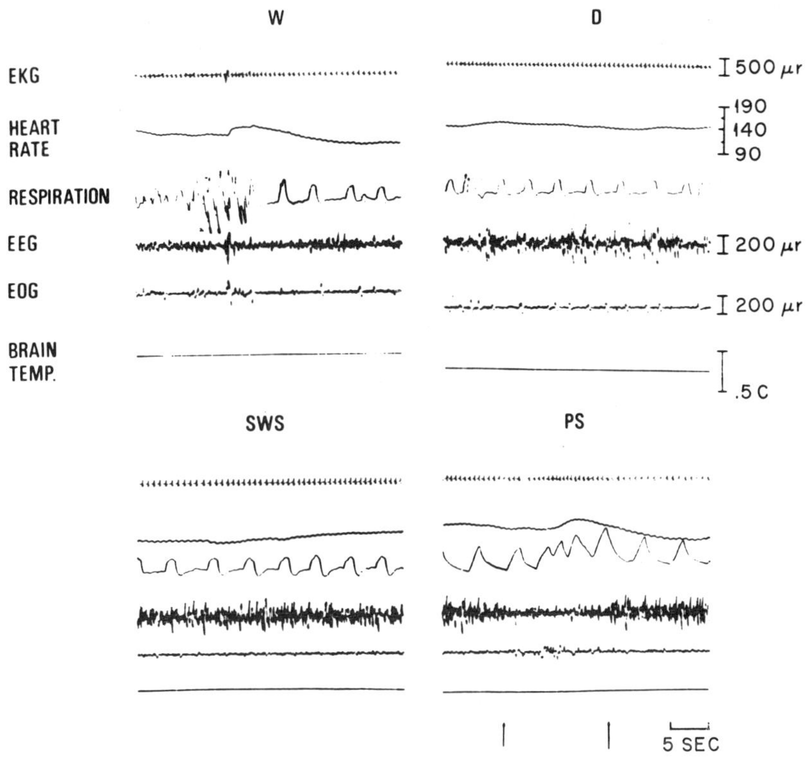

Fig. 1. Polygraphic record of waking and sleeping in the pigeon. W, awake; D, drowsy; SWS, slow-wave sleep; PS paradoxical sleep or REM sleep. [From Van Twyver & Allison (1972).]

activity but is, on the average, observed at higher rates than during sleep. Likewise, brain temperature and respiration are elevated during waking, with the latter also showing great irregularity (Van Twyver & Allison, 1972).

B. Drowsiness

Behaviorally, drowsiness is characterized by a posture of standing or perching quietly, a somewhat lowered head (drooping) and slow-paced eye blinking. The predominantly high-frequency, low-amplitude EEG activity is marked by short episodes of higher voltage, lower frequency waves in pigeons (Van Twyver & Allison, 1972). Ookawa and Gotoh (1965) observed a *predominance* of slow waves (6–12 and 3–4 Hz) at high voltages (200–300 μV) in the chicken. This discrepancy in slow-wave episode durations may reflect a species difference, a difference attributable to variability in electrical activity at different electrode placements or a difficulty in clearly defining the drowsy state. Eye movements and tonic neck muscle activity are reduced relative to that seen during wakefulness. Heart and respiration rates, while still variable, also show a tendency toward lower levels along with brain temperature (Van Twyver & Allison, 1972).

C. Non-Rapid Eye Movement Sleep (NREM Sleep)

Several behavioral changes are observed with sleep onset. Falconiformes (Rojas-Ramirez & Tauber, 1970) and chickens (Ookawa & Gotoh, 1965) are immobile, have closed eyes, and a lowered head, which becomes buried under a wing. In the burrowing owl (Berger & Walker, 1972), there is also an absence of head and body movements with sleep occurring while standing either one one or two legs and the head only slightly lowered. On the other hand, the eyes remain open throughout this period accompanied by periodic eye blinks and only maintain continuous closure just prior to the onset of rapid eye movement (REM) sleep. The pigeon (LoPresti & Goodman, 1968; Van Twyver & Allison, 1972; Walker & Berger, 1972) is characteristically in an immobile standing posture (on one or two legs) or sitting posture with its neck withdrawn, its head resting upon a puffed chest and its eyes either opened and blinking or continuously closed.

Electroencephalographic activity during NREM sleep shifts predominantly to slow waves (1–4 Hz) of high amplitude (100–300 μV) in all reported species except the hawk and falcon (Rojas-Ramirez & Tauber, 1970) in which no long trains of slow waves were observed; in these latter birds slow

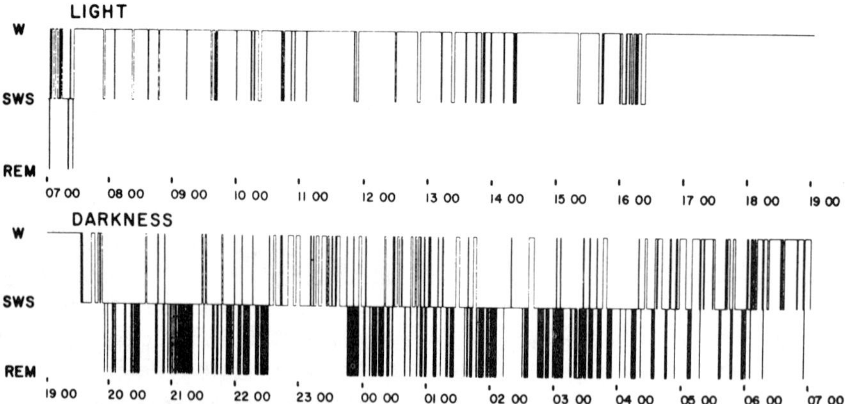

Fig. 2. States of wakefulness, slow-wave sleep and REM sleep observed in one pigeon during a 24-hr period. [From Walker & Berger (1972).]

waves 3–10 Hz reached amplitudes rarely greater than 45 μV and in trains of no greater than .3 sec durations. Electroencephalographic recordings in the hawk and falcon were reportedly made fairly lateral from the midline (10 mm) at locations indicated in the pigeon (Van Twyver & Allison, 1972) as somewhat less than optimal for recording high voltage, slow waves. Electroencephalographic sleep spindles commonly recorded in mammals were looked for but not observed in avian records during NREM sleep (e.g., Hishikawa *et al.*, 1969; Van Twyver & Allison, 1972; Walker & Berger, 1972).

Eye movements are minimal or absent but eye blink artifacts do occur during this period. Electromyographic records indicate a decrease in tonic neck muscle activity, relative to the waking and drowsy periods with heart rate, respiration rate, and brain temperature also declining.

D. Rapid Eye Movement Sleep (REM Sleep)

During REM sleep birds tend to maintain a general body immobility and maintain or initiate eye closure. The head descends to a position even lower than that seen during NREM sleep, sometimes dropping to the floor; this involves conspicuous jerking rotational or downward head movements. These episodes are only several seconds in duration and occur on a number of occasions throughout the sleep period. The eyes often open briefly following the head-fall sequence with a rapid upswing of the head to an upright position. A notable exception to this observation again is the burrowing owl which displays no jerking head movements during REM sleep (Berger & Walker, 1972).

The EEG record changes from a NREM sleep pattern to a desynchronized pattern of high-frequency, low-amplitude activity, similar to that seen during wakefulness. However, in the hawk and falcon (Rojas-Ramirez & Tauber, 1970) the contrast between REM and NREM sleep EEG was not as marked, primarily because of the reduced quantity of high-amplitude, low-frequency activity recordable during NREM sleep.

Eye movements, conjugate and disconjugate, evolve from the rather silent EOG record observed during NREM sleep in all reported species except the burrowing owl, which shows no eye movements during sleep or waking (Berger & Walker, 1972). Neck muscle activity is reduced from NREM sleep levels in pigeon (Church & Goodman, 1971; Van Twyver & Allison, 1972), chicken (Hishikawa *et al.*, 1969; Klein *et al.*, 1964), falcon and hawk (Rojas-Ramirez & Tauber, 1970); however, only the report on the latter two species indicated that a complete flattening of the EMG trace occurred during this stage. This latter finding is similar to that seen in mammals (Dement & Kleitman, 1957). In marked contrast to findings in mammals is the lack of a consistent change in tonic neck muscle activity during sleep characterized by EEG desynchronization in the burrowing owl.

Heart rate was slightly elevated above NREM sleep levels in pigeon (Van Twyver & Allison, 1972) but was decreased in chicks (Hishikawa *et al.*, 1969); respiration was increased and more variable; and brain temperature showed no reliable increase or decrease during REM sleep.

Birds in all stages of sleep and wakefulness tend to be easily aroused by various environmental stimuli including those in visual, auditory, and somatosensory modalities. When tested for arousal thresholds in the various stages, differences are found: The arousal threshold is lowest during waking with threshold increases found in NREM sleep and an even higher thresholds evident during REM sleep. This evidence supports the view that REM sleep is a deeper state of sleep.

E. Sleep and Waking Patterns

Prolonged and continuous observations extending over 24-hr periods have provided information regarding sleep and waking patterns in birds. In the burrowing owl, an average of 59.5% of the 24-hr period is spent in sleep with 5.1% of the total sleep time being spent in desynchronized sleep (Berger & Walker, 1972). In the pigeon, 40–50% of the 24-hr period is spent in sleep with 3–12% of that sleep being REM sleep (Church & Goodman, 1971; Van Twyver & Allison, 1972; Walker & Berger, 1972). In the hawk and falcon REM sleep constitutes 7–10% of the total sleep time (Rojas-Ramirez & Tauber, 1970); in the chick an average of 7.3% is spent in REM sleep (Hishikawa *et al.*, 1969). These percentages of REM sleep contrast sharply with earlier accounts of sleep in the bird showing less than 1% of total sleep time spent in REM sleep (Klein *et al.*, 1964; Tradardi, 1966). The larger percentages fit more closely with those reported in mammals (Ruckebusch, 1972; Sterman, Knauss, Lehmann, & Clemente, 1965).

Further analysis has been made of the frequencies, durations, and intervals of the various periods. Walker and Berger (1972), in describing the patterns in the pigeon, reported mean episode durations of waking and sleep as follows: Waking, 10.9 min; NREM sleep, 1.2 min; REM sleep, 11 sec (the latter occurring about once every 1.1 min or an average of 216 times in a 24-hr period). In almost all instances REM sleep was followed by NREM sleep. In contrast to the pigeon, waking and NREM sleep periods in the burrowing owl (Berger & Walker, 1972) are similar in length (each approximately 2 min in duration). The REM sleep episodes occur on an average of every 8.2 min or about 113 times in a 24-hr period. Also, REM sleep was not observed immediately following NREM sleep.

It appears that sleep in birds is comparable to that in mammals, showing discernable behavioral and physiological signs and a close correspondence regarding the presence of REM sleep. Perhaps the differences in avian and mammalian sleep (particularly head) postures, might account for difference in level of tonic neck muscle activity during REMS. Of all birds reported, the owl is the only exception to the reported presence of rapid eye movements during wakefulness but also during REM sleep. In this case it might be more appropriate to refer to this latter stage as desynchronized or paradoxical sleep as have Berger and Walker (1972).

II. Experimental Manipulation of Sleep

Attempts have been made to look beyond the behavioral and physiological correlates of sleeping and waking in birds by investigators who have used experimental intervention techniques that are able to induce or disrupt sleep. Generally, such strategies have been found useful in attempts to uncover the mechanisms involved in sleep and waking states. Typically, brain perturbation has been directed to restricted regions or structures. Occasionally, whole-brain disturbances have been employed (e.g., in some chemical studies).

A. Electrical Stimulation of the Brain

Reports have described the induction of sleep in mammals by circumscribed electrical stimulation applied to one of a number of different brain areas or structures: the cerebral cortex (Penaloza-Rojas, Elterman, & Olmos, 1964); basal forebrain and preoptic areas (Sterman & Clemente, 1962); amygdala, septal area (Kawakami *et al.*, 1966); caudate nucleus (Buchwald, Wyers, Okuma, & Heuser, 1961); hippocampus (Kawakami & Sawyer, 1959); thalamus (Hess, 1957); brainstem reticular formation (Rossi, 1963); etc. With the ability to induce sleep from such a widespread number of different brain areas, some investigators (e.g., Jouvet, 1967) have questioned the significance of such findings and the value of such a technique in uncovering the brain mechanisms of sleep. Perhaps, the finding that a number of neural access routes for sleep induction are simultaneously operating supports a view of a rather complex but integrated organization of sleep mechanisms rather than a simplistic view of isolated single centers responsible for arousal and the various sleep stages. Whereas any valid theory of neural control must eventually explain these complicated findings, investigators have continued to use electrical brain stimulation in their studies because of the promise it has, particularly when used in conjunction with other research techniques.

Information regarding the effects of electrical brain stimulation upon sleep in birds has, on some occasions, come from studies where sleep has been of primary interest and in other instances where it was of secondary interest.

Von Holst and von St. Paul (1963), for example, were primarily interested in the more general matter of explicating the functional organization of drives in the chicken with the use of brain stimulation. Among the varied evoked behaviors they observed, they reported that stimulation at some chronic electrode placements produced a sequence of events suggestive of sleep. A bird in the midst of feeding, when stimulated, suddenly stopped feeding, sat down, and eventually showed eye closure; following the termination of brain stimulation, the bird eventually awoke and engaged in other behaviors. Because of the methodological views and specific aims of these investigators, no attempt was made to identify the stimulated brain areas.

Harwood and Vowles (1966), in their study of forebrain stimulation and feeding behavior in the ring dove, noted that at least in some of their subjects a sleep reaction sometimes *followed* the termination of brain stimulation. Physiological measures of sleep and effective brain electrode locations also were not reported.

In a study in which the investigators (Goodman & Brown, 1966) were interested in comparing evoked behavioral effects and reinforcing effects of electrical brain stimulation in pigeons, sleeplike behavior (similar to that described by von Holst and von St. Paul, 1963) was reported at sites in the neostriatum intermediale. These subjects were apparently not immobilized as a result of an induced abnormal muscular hypertonus since a relatively low intensity of sound was sufficient to produce behavioral arousal, no matter whether such an arousal stimulus was presented during or after brain stimulation. Sleep was observed not only during brain stimulation but endured beyond stimulus termination (up to 15 min of observation) during periods of acoustic silence. Only behavioral evidence of sleep was reported.

In an exploration of the chicken telencephalon and diencephalon with electrical stimulation, Putkonen (1967) described a complex of evoked behavioral responses which he referred to as the "trophotrope" syndrome. The behaviors included locomotor inhibition, fluffing of feathers, a crouching or sitting posture, and a tendency to close the eyes, accompanied by miosis and cardiac slowing. Perhaps because of a lack of other electrographic evidence of sleep and the limited investigation of this evoked behavior, the author cautiously refrained from referring to this condition as sleep. The trophotrope reaction resulted from stimulation of loci near the oral part of, but not including, nucleus rotundus of the thalamus. Also implicated were nucleus reticularis superior, nucleus ventrolateralis of the thalamus, the lateral anterior thalamic nucleus, the lateral forebrain bundle, and tractus septomesencephalicus.

In a study designed explicitly for purposes of exploring the pigeon forebrain with electrical stimulation in an attempt to identify arousal and sleep sites, LoPresti and Goodman (1968) provided electrographic (electroencephalogram, electrooculogram, and electromyogram) and behavioral data to support their conclusions regarding sleep and/or arousal induction. Employing vertically repositionable stimulating electrodes (in freely moving subjects), they observed that of the 73 forebrain sites stimulated, 7 induced sleep, 15 produced drowsy behavior, and 8 evoked increased activity (arousal) A variety of specific evoked behavioral responses were also noted.

Sleep-inducing stimulation produced a characteristic progression of behavioral patterns: arrest of ongoing behavior; then a resting posture; and finally a sleep posture. Subjects remained in the sleep posture for up to 12 min after termination of brain stimulation (see Fig. 3). Accordingly, EEG activity changed from high-frequency and low-amplitude activity to low-frequency and high-amplitude activity, EOG was reduced in frequency and amplitude; and neck muscle EMG was reduced in amplitude (see Fig. 4). The high reliability of these evoked phenomena was supported by the reported observation of the behavioral and electrographic signs of sleep upon repeated (3–8) occasions at each sleep site. Careful attention was called toward the state of the organism prior to brain stimulation; facilitating and nonfacilitating environmental influences were discussed.

In this same study it also was reported that on several occasions REM-like episodes were observed during induced sleep; behaviorally this involved a "cogwheel" or jerking head descent plus the electrographic signs of high-frequency and low-amplitude EEG activity, rapid eye movements with an apparently reduced but not completely flattened neck muscle activity.

Stimulus parameters proved to be a critical determinant of the evoked effects observed by LoPresti and Goodman (1968). Current intensities beyond an optimal level changed the behavioral state of the organism from one of inhibition to one of excitation. Stimulation frequency was also apparently a major influence. Several brain sites stimulated at both high [100 pulses per sec (100 pps)] and low (20 pps) frequencies induced sleep only at the low frequency. Interestingly, several stimulated sites produced sleep under both conditions. These effects correspond with Sterman and Clemente's (1962) findings with basal forebrain stimulation of the cat. The latter workers used such evidence to support their contention that the basal forebrain is normally involved in sleep regulation (and behavioral inhibition)

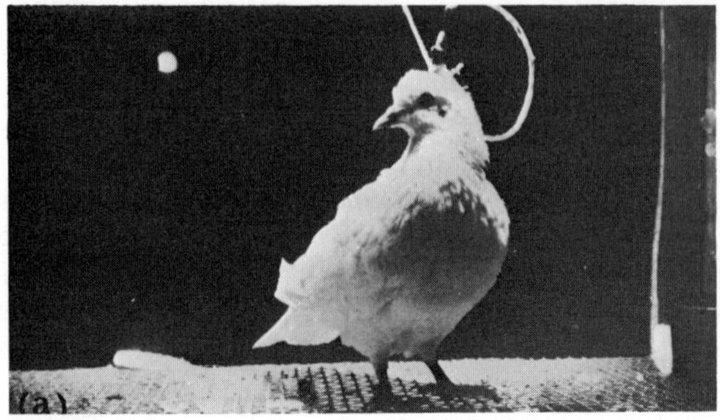

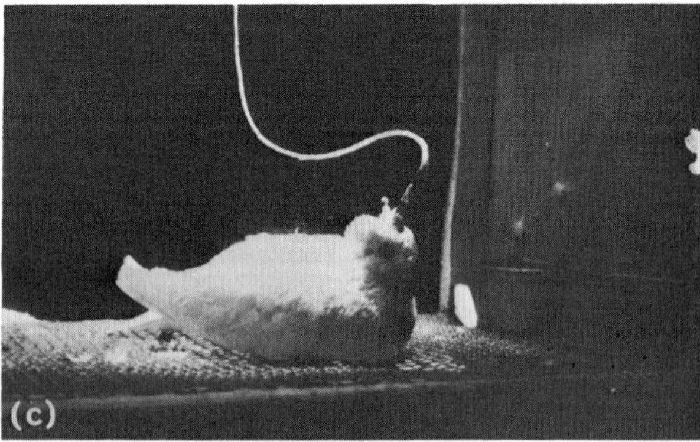

Fig. 3. The sequence of behavioral stages associated with effective sleep induction induced by electrical brain stimulation. (a). Arrest posture; (b). rest posture; (c). sleep posture.

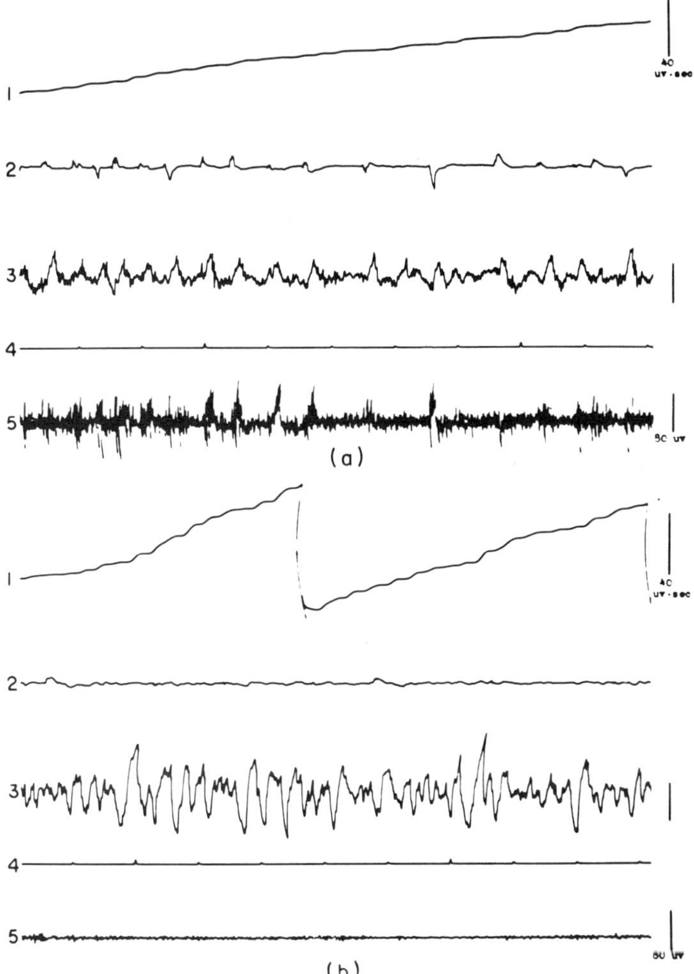

Fig. 4. Polygraphic record of sleep induction with electrical brain stimulation (100 pps, 1 msec-pulses, 20 μA, 30-sec duration) in pigeon. (a). Alert state prior to stimulation; (b). sleep after termination of brain stimulation; 1., Integration record of EEG slow wave activity (1–10 Hz); 2., EOG; 3., EEG; 4., Time, sec; 5., EMG.

and that sleep is not simply induced experimentally by slow wave recruitment in the brain.

These authors observed that a slight shift in the electrode position was often sufficient to change the evoked behavioral and electrographic effects (see Fig. 5). Stimulated brain sites from which drowsiness was induced were often adjacent to sleep-inducing sites. Two general areas of the pigeon forebrain were identified as sleep-inducing by LoPresti and Goodman (1968):

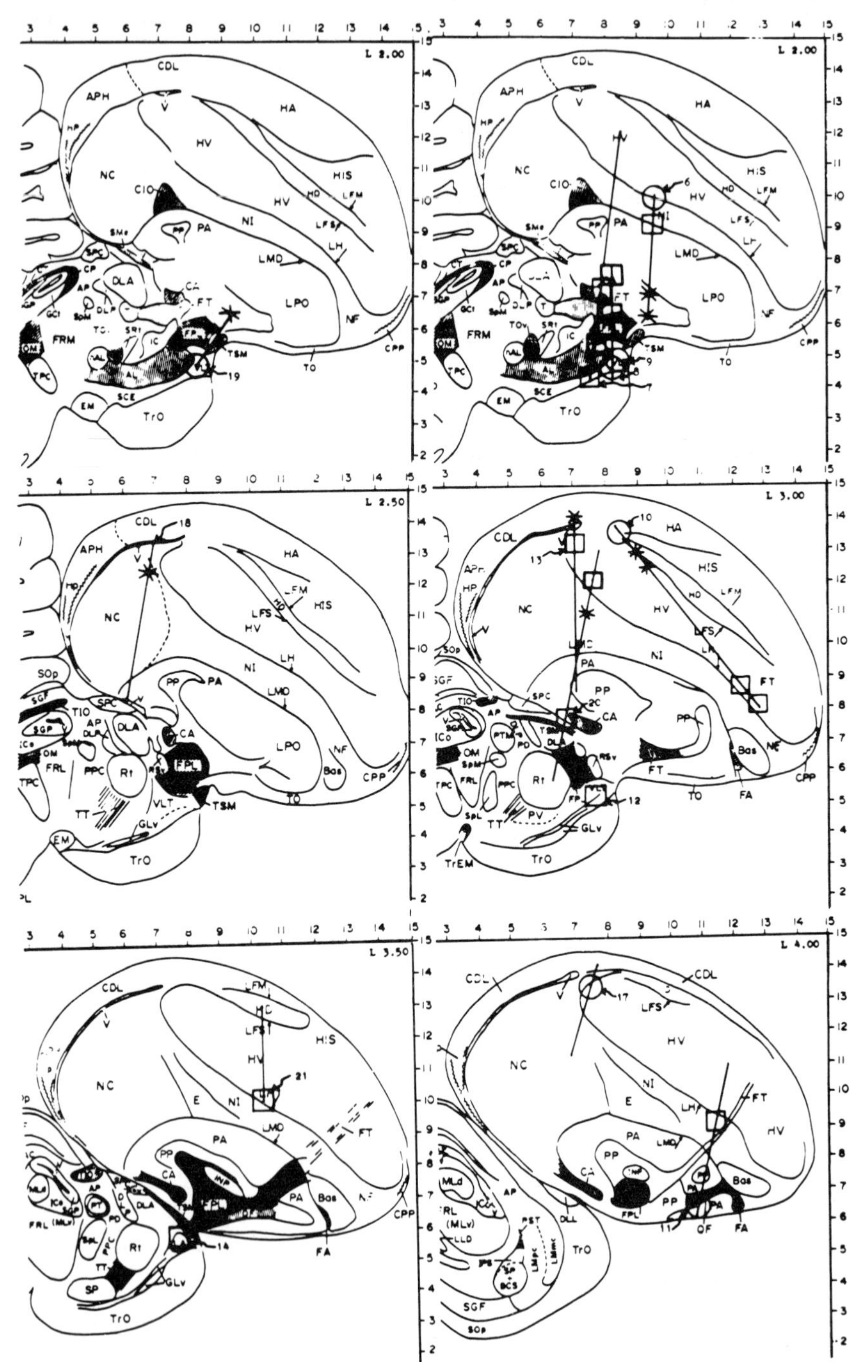

the first below the lamina medullaris dorsalis, in the basal forebrain (extending from an area just rostal to the preoptic level to the hypothalamic level) and the second a dorsal forebrain area extending over the range of the lamina hyperstriatica, one subgroup involving dorsal posterior hyperstriatum and another the neostriatum intermedium. Stimulation in this latter brain area also produced sleep in the study reported by Goodman and Brown (1966). The former brain area, it was suggested, corresponds to the area from which Putkonen (1967) induced sleep-associated behaviors in the chicken and, perhaps, is related to the area in which Sterman and Clemente (1962) produced sleep in cats.

Any attempt at parcelation of the forebrain relative to sleep or arousal control based upon brain stimulation results to date is obviously premature. Further anatomical and physiological study is required.

B. Brain Lesions

Cate (1936) reviewed a number of studies investigating the effects of cerebral ablation in birds. Of particular relevance to sleep, however, Ferrier (1886) described the Rolando-Flourens syndrome or a permanent lethargy resulting from total cerebral ablation in the pigeon. Schrader (1889), however, found that within several days such animals regained the ability to display waking activity, contrary to the previously reported permanent sleep–waking loss.

In a study in pigeons, Brunelli, Magni, Moruzzi, and Musumeci (1972) also described the sleep and waking behavior in animals with total destruction of the cerebral hemispheres. Making an assessment of sleeping and waking based on monitored behavioral and electrocardiographic (EKG) activity, their thalamic pigeons showed a recovery of sleep–waking alternation within several hours after surgery rather than the several days required by Schrader's subjects. Brunelli *et al.* explained this discrepancy as perhaps due to the extreme care they used to avoid injury to the underlying diencephalon during the surgical procedure. It appears from these results that while the thalamic pigeon regains its waking ability within several hours, it shows a much greater than normal tendency to sleep; while not explicitly mentioned, one gains the impression that the frequency and duration of sleeping and waking bouts (cycling) is affected by telencephalic destruction.

Fig. 5. Schematic sagittal views of the pigeon brain showing sites where electrical brain stimulation induced sleep (○), drowsiness (□) or activation (∗). The number in the upper right-hand corner of each section indicates the distance of the sagittal plane from the midline; the numbers at the top, bottom and sides are millimeter scalings. The numbers with arrows pointing to electrode tracks are subject identification numbers. [After Karten & Hodos (1967).]

Waking activities such as self-cleaning (particularly, preening) are regained within 24 hr, whereas feeding activities (including pecking movements, seizing and swallowing food, and drinking) appear within 2–3 weeks. Thus, such animals do eventually become self-sufficient.

Arousal was studied in thalamic pigeons (Brunelli *et al.*, 1972); sensory stimulation applied during sleep was able to produce wakefulness which was often long-lasting.

The sleep and waking effects of electrical stimulation of the nucleus reticularis pontis oralis, a brainstem structure implicated in cortical EEG desynchronization in mammals (Moruzzi, 1964), was also investigated. Short duration, high-frequency reticular stimulation was able to produce wakefulness in the sleeping bird. In longer-surviving thalamic animals either sensory or reticular stimulation blocked feeding behavior; brain stimulation intensity threshold for arousing a pigeon from sleep (1.8 V) was higher than that required to block feeding activity in the same waking animal (1.2 V). Interestingly, by changing from single train to repetitive train reticular stimulation, the arrest of feeding behavior was changed to sleep. Based upon the accompanying differentiation of electrocardiographic effects produced by this difference in stimulation parameters, these investigators concluded that two different, perhaps adjacent brainstem systems may be costimulated during pontine reticular stimulation, an *activating* and a *deactivating* one. These systems were not anatomically identified.

In a study that initially attempted to deliver putative CNS neurotransmitter substances (ACH, NE, 5HT) via a vertically repositionable cannula in the pigeon forebrain and to observe their effects on behavior, Church and Goodman (1971) were initially "plagued" with the finding that lowering of the brain cannula (probe), in some locations produced profound changes in behavioral activity, i.e., reduced responsiveness and sleep. Electrographic and behavioral measures were employed to evaluate sleep and awake states in these animals. Penetration lesions produced by lowering unilateral cannulae (1 mm diameter) within seconds produced a dramatic change: Pigeons became inactive and showed fluffed feathers, puffed chest and closed eyes. Spontaneous activity remained drastically reduced from 5–7 days after the cannulae were lowered into the basal forebrain structures (below lamina medularis dorsalis), which involved the paleostriatum augmentatum and primitivum, the parolfactory lobe, lateral septal nucleus and nucleus accumbens. These birds could be aroused by handling or sound, but they resumed their depressed state within 15 min thereafter. Behavioral and electrographic data showed evidence of sleep during much of the day.

A second group of animals with smaller brain probes (.6 mm diameter) bilaterally lowered into similar sites and several into sites more caudally situated within the thalamus (medial to nucleus rotundus, involving ventrolateral thalamic nucleus, posteroventral nucleus, nucleus intercalatus, nucleus subrotundus, and thalamic reticular nucleus), produced more severely depressed birds. They were harder to arouse, resuming their sleep behavior within 1–2 min. Even though the cannulae were removed from the brain after the initial penetration, this condition persisted until the birds were sacrificed, 2–4 weeks later. Spontaneous locomotor activity was not just decreased but virtually absent.

One pigeon in this second group was also severely depressed by probe lesions but exhibited brief periods of spontaneous activity including feeding. The results of a 24-hr period of behaviorally and electrographically monitored sleep–awake activity are shown in Table I. Total sleep time during a 24-hr period rose from 46 to 75%, while NREM and REM sleep were maintained at the same percentages but increased in terms of absolute time.

The lesions in this bird were approximately 1 mm in diameter, and involved the lateral forebrain bundle, ansa lenticularis, thalamic reticular nucleus, and ventrolateral thalamic nucleus.

Small bilateral basal forebrain lesions (1 mm diameter) produced by electrolytic destruction identified within paleostriatum (augmentatum and/or primitivum) have been reported to produce reduced locomotor activity (Ruskin & Goodman, 1971) and temporary somnolence (7–8 days) (Zeier, 1968) in pigeons.

Zeier (1971) also reported prolonged somnolent behavior states lasting for several days to a week following bilateral archistriatal lesions in pigeons. This effect was produced by lesions in lateral limbic rather than those in either medial limbic or somatic motor archistriatum.

The telencephalon, diencephalon, and lower brainstem have been implicated in the control of sleep and waking activity by the use of total and subtotal regional lesions as well as circumscribed electrical stimulation. Part of the difficulty in characterizing the sleep and waking system of the avian brain rests in our limited knowledge regarding neuroanatomical circuitry. Recent information regarding forebrain neuroanatomy (see Cohen and Karten, Chapter 2 of this volume) should be quite useful when com-

Table I. *Summary of 24 Hr Electrographic and Behavioral Record of Sleep and Waking in a Normal Control and a Bilaterally Probe-Lesioned Pigeon, Expressed in Absolute Time (Hours, Minutes) and Percentages*[a]

Activity	Lesioned	Control
Total wakefulness	6 hr (25%)	13 hr (54%)
Total sleep time (TST)	18 hr (75%)	11 hr (46%)
NREM (percentage of TST)	17 hr 10 min (97%)	10 hr 30 min (97%)
REM (percentage of TST)	50 min (3%)	30 min (3%)

[a]From Church and Goodman (1971).

bined with further neurobehavioral studies in evaluating the role of the forebrain in sleep. Furthermore, similar neuroanatomical and neurobehavioral studies are required in connection with avian brain stem mechanisms. Some limited parallels have been tentatively established or speculated upon between avian and mammalian mechanisms; further experimental studies in this area are obviously required.

C. Pharmacological Treatments

Studies in birds concerning the neurochemical bases of sleep and waking behavior have only recently begun and at a pace lagging behind similar investigations in mammals. These few studies in birds have tended for the most part to be concerned with brain amines and putative neurotransmitter substances. Those studies briefly summarized below deal with the application of some of these substances either to widespread or localized brain areas and some grossly observed behavioral changes.

Key and Marley (1962) tested a number of different sympathomimetic compounds representing several different amine categories based upon structural formulas. They found electrocortical and behavioral sleep induced in alert chicks up to 4 weeks of age by systemic injections of either adrenaline, noradrenaline, isoprenaline, cobefrin (amines with hydroxyl groups in the 3,4 position on the benzene ring with another beta carbon on the side chain), dopamine (one from a group of amines with hydroxyl groups in 3,4 position on the benzene ring but lacking any on the side chain), or phenyephrine (one from a group of amines with an hydroxyl group in either the 3 or 4 position on the benzene ring, together with another on the beta carbon of the side chain). This 4-week period in the life of the chick was followed by a transitional period during which the sleep induction effects of these compounds changed to that of behavioral and electrographic arousal (perhaps associated with a relatively slow developing blood–brain barrier in chicks).

Another group of amines which included pipadrol (one from a group of amines with no hydroxyl group on the aromatic nucleus but possibly possessing one of the beta carbon atom of the ethylamine side chain), amphetamine, phenmetrazine, and methyl phenidate (amines with no hydroxyl groups on either the aromatic nucleus or the side chain) when administered systemically produced an alerted EEG pattern and behavior; alerting produced by amphetamine was followed in the young chick by ataxia then by immobility attended by high-pitched vocalizations for hours. This led either to death or recovery within a 24-hr period.

Spooner and Winters (1967), also investigating effects of centrally active amines on the EEG and behavior, found that systemic administration of noradrenaline, adrenaline, and serotonin induced electrocortical slow-wave activity and behavioral sleep and that amphetamine, dopamine, and its precursor DOPA produced alert EEG patterns and behavior; however, changes in blood pressure did not correlate with sleep or waking effects. They therefore concluded that behavioral and electrographic patterns produced by these pharmacological agents resulted from a direct action on the nervous system rather than from cardiovascular changes.

Carbachol, a substance implicating cholinergic control of sleep and waking in mammals (Hernandez-Peon, Chavez-Ibarra, Morgane, & Timo-Iara, 1963), was also tested by Spooner and Winters. After systemic administration, they observed an induced low-frequency, high-amplitude EEG pattern along with lowered blood pressure, but with a behavioral posture which they suggested was in contrast to that induced by noradrenaline or serotonin; carbachol injection resulting in birds standing erect rather than crouching (but eyes were closed).

Marley and Nistico (1972) reported the results of an investigation in which they compared the effects of catecholamines and adenosine derivatives delivered into the brain intraventricularly (third ventricle) and directly to brain structures by infusion in adult fowls (*Gallus*) (sufficiently mature to have a well-developed blood–brain barrier). Noradrenaline, alphamethylnoradrenaline, isoprenaline or dopamine delivered directly to the hypothalamus, a brain area that normally contains neurohumoral substances such as noradrenaline, adrenaline, serotonin (Fuxe & Ljunggren, 1965), and acteylcholine (Aprison & Takahashi, 1965), induced electrocortical and behavioral sleep and lowered body temperature.

Noradrenaline, alphamethylnoradrenaline, and isoprenaline were also effective in producing sleep and temperature effects when delivered into the ventricle of the same animals but required 5 to 20 times the amount for equal effectiveness. Smaller doses were required to induce sleep than to lower the body temperature, thus providing evidence against an explanation which suggests that sleep results from a body temperature drop but rather supporting a view that sleep may result from direct CNS action. Dopamine was ineffective when administered intraventricularly; however, when a monoamine oxidase inhibitor, mebanazine, was injected prior to dopamine, dopamine produced profound sleep and lowered body temperature. Noradrenaline effects were also prolonged by mebanazine pretreatment. (Both noradrenaline and dopamine are metabolized by monoamine oxidase.)

Dopamine, when applied to either the hypothalamus or the paleostriatum augmentatum, produced a contralateral head deviation involving a jerking motion which was enhanced by mebanazine and attenuated by haloperidol, a dopamine antagonist. Sleep was not induced by either dopamine or noradrenaline when applied to the paleostriatum augmentatum, an area within the brain which normally contains a relatively high concentration of dopamine (Juorio & Vogt, 1967; Karten & Dubbeldam, 1972).

Clearly needed in the study of avian brain–behavior relationships is a combination of efforts: (*a*) the mapping of the various populations of chemical-bearing brain neurons; (*b*) the experimental manipulation of these neural groupings and the observation of physiological and behavioral consequences; (*c*) the advancement of neuroanatomic information regarding neural circuitry of the avian brain; and (*d*) the further characterization of sleep and waking activity as it is more normally found in the behaving organism. This enormous task must be pursued with great vigor in order that we may some day better understand the physiological mechanisms and behavioral significance of sleep and waking processes in birds.

References

Aprison, M. H. & Takahashi, R. Biochemistry of the avian central nervous system. II. 5-hydroxytryptamine, acetylcholine, 3, 4-dihydroxyphenylethylamine and norepinephrine in several discrete areas of the pigeon brain. *Journal of Neurochemistry*, 1965, **12**, 221–230.

Berger, H. Uber das Elektrenkephalogramm des Menschen. *Archives Psychiat.*, 1929, **87**, 527–570.

Berger, R. J. & Walker, J. M. Sleep in the burrowing owl *(Speotyto cunicularia hypugaea)*. *Behavioral Biology*, 1972, **7**, 183–194.

Brunelli, M. Magni, F. Moruzzi, G. & Musumeci, D. Brain stem influences on waking and sleep behaviors in the pigeon. *Archives Italiennes de Biologie*, 1972, **110**, 285–21.

Buchwald, N. A., Wyers, E. J., Okuma, T. & Heuser, G. The "caudate spindle." I. Electrophysiological properties. *Electroencephalography and Clinical Neurophysiology*, 1961, **13**, 509–518.

Cate, J. ten. Physiologie des Zentralnervensystems der Vogel. *Ergebnisse der Biologie*, 1936, **13**, 93–173.

Church, S. M. & Goodman, I. J. Effects of probe penetration lesions on sleep–awake behavior in pigeons. Paper presented at Society for Neuroscience meeting, Washington, D. C., 1971.

Dement, W. & Kleitman, N. Cyclic variations in EEG during sleep and their relation to eye movements, body motility, and dreaming. *Electroencephalography and Clinical Neurophysiology*, 1957, **9**, 673–690.

Ferrier, D. *The functions of the brain*. London: Smith Elder, 1886.

Fuxe, K. & Ljunggren, L. Cellular localization of monoamines in the upper brain stem of the pigeon. *Journal of Comparative Neurology*, 1965, **125**, 355–382.

Goodman, I. J. & Brown, J. L. Correlations between ICS evoked behavior and reinforcement in brain stimulation studies with pigeons. Paper presented at American Psychological Association Convention Symposium on *The Neural Basis of Avian Learning and Motivation*. New York, September, 1966.

Harwood, D. & Vowles, D. M. Forebrain stimulation and feeding behavior in the Ring Dove *(Streptopelia risoria) Journal of Comparative and Physiological Psychology*, 1966, **62**, 388–396.

Hernandez-Peon, R. Chavez-Ibarra, G. Morgane, P. J., & Timo-Iara, C. Limbic cholinergic pathways involved in sleep and emotional behavior. *Experimental Neurology*, 1963, **8**, 93–111.

Hess, W. R. *The function organization of the diencephalon*. New York: Grune and Stratton, 1957.

Hishikawa, Y., Cramer, H., & Kuhlo, W. Natural and melatonin-induced sleep in young chickens—A behavioral and electrographic study. *Experimental Brain Research*, 1969, **7**, 84–94

Jouvet, M. Neurophysiology of the states of sleep. *Physiological Review*, 1967, **47**, 117–177.

Juorio, A. V. & Vogt, M. Monoamines and their metabolites in the avian brain. *Journal of Physiology*, 1967, **189**, 489–518.

Karten, H. J., & Dubbeldam, J. L. The organization and projections of the paleostriatal complex in the pigeon (*Columba livia*). *Journal of Comparative Neurology*, 1973, **148**, 61–90.

Karten, H. J. & Hodos, W. *A stereotaxic atlas of the brain of the pigeon* (*Columba livia*). Baltimore: The Johns Hopkins Press, 1967.

Kawakami, M. & Sawyer, C. H. Induction of behavioral and electroencephalographic changes in the rabbit by hormone administration or brain stimulation. *Endocrinology*, 1959, **65**, 631–643.

Key, B. J. & Marley, E. The effect of sympathomimetic amines on behavior and electrocortical activity of the chicken. *Electroencephalography and Clinical Neurophysiology*, 1962, **14**, 90–105.

Klein, M., Michel, F. & Jouvet, M. Etude polygraphique du sommeil les oiseaux. *Comptes Rendus des Seance de la Societe de Biologie*, 1964, **158**, 99–103.

LoPresti, R. W. & Goodman, I. J. Intracranial induction of sleep in the avian forebrain. *Psychophysiology*, 1968, **5**, 199.

Marley, E. & Nistico, G. Effects of catecholamines and adenosine derivatives given into the brain of fowls. *British Journal of Pharmacology*, 1972, **46**, 619–636.

Moruzzi, G. Reticular influences on the EEG. *Electroencephalography and Clinical Neurophysiology*, 1964, **16**, 2–17.

Ookawa, T. & Gotoh, J. Electroencephalogram of the chicken recorded from the skull under various conditions. *Journal of Comparative Neurology*, 1965, **124**, 1–14.

Paulson, G. The avian EEG: An artifact associated with ocular movement. *Electroencephalography and Clinical Neurophysiology*, 1964, **16**, 611–613.

Penaloza-Rojas, J. H., Elterman, M & Olmos, N. Sleep induction by cortical stimulation. *Experimental Neurology*, 1964, **10**, 140–147.

Peters, J. J., Vonderahe, A. R. & Schmid, D. Onset of cerebral electrical activity associated with behavioral sleep and attention in the developing chick. *Journal of Experimental Zoology*, 1965, **160**, 255–262.

Putkonen, P. T. S. Electrical stimulation of the avian brain. *Annales Academiae Scientarium Fennicae* (Medica), 1967, **130**, 9–95.

Rojas-Ramirez, J. & Tauber, E. S. Paradoxical sleep in two species of avian predator (Falconiformes). *Science*, 1970, **167**, 1754–1755.

Rossi, G. F. Sleep inducing mechanisms of the brain stem. *Electroencephalography and Clinical Neurophysiology*, 1963, Suppl. **24**, 113–132.

Ruckebusch, Y. The relevance of drowsiness in the circadian cycle of farm animals. *Animal Behavior*, 1972, **20**, 637–643.

Ruskin, R. S. & Goodman, I. J. Changes in locomotor activity following basal forebrain lesions in the pigeon. *Psychonomic Science*, 1971, **22**, 181–183.

Schrader, M. E. G. Zur Physiologie des Vogelhirns. *Pflugers Archives*, 1889, **44**, 175–238.

Spooner, C. E. & Winters, W. D. The influence of centrally active amine induced blood pressure changes in electroencephalogram and behavior. *International Journal of Neuropharmacology*, 1967, **6**, 109–118.

Sterman, M. B. & Clemente, C. D. Forebrain inhibitory mechanisms: Sleep patterns induced by basal forebrain stimulation in the behaving cat. *Experimental Neurology*, 1962, **6**, 103–117.

Sterman, M. B., Knauss, T., Lehmann, D, & Clemente, C. Circadian sleep and waking patterns in the laboratory cat. *Electroencephalography & Clinical Neurophysiology*, 1965, **19**, 509–517.

Tradardi, V. Sleep in the pigeon. *Archives Italiennes de Biologie*, 1966, **104**, 516–521.

Van Twyver, H. & Allison, T. A polygraphic study of sleep in the pigeon *(Columba livia)*. *Experimental Neurology*, 1972, **35**, 138–153.

von Holst, E. & von St. Paul, U. On the functional organization of drives. *Animal Behaviour*, 1963, **11**, 1–20.

Walker, J. M. & Berger, R. J. Sleep in the domestic pigeon (*Columba livia*). Behavioral Biology, 1972, **7**, 195–203.

Zeier, H. Changes in operant behavior of pigeons following bilateral forebrain lesions. *Journal of Comparative and Physiological Psychology*, 1968, **66**, 198–203.

Zeier, H. Archistriatal lesions and response inhibition in the pigeon. *Brain Research*, 1971, **31**, 327–339.

Behavioral Adaptation on Operant Schedules after Forebrain Lesions in the Pigeon

Hans Zeier[1]
Swiss Federal Institute of Technology

Operant conditioning is widely used in behavioral investigations. In such test situations the animals tend to optimize their chances; if motivated by hunger they try to obtain a maximal amount of food reward for a minimal amount of instrumental work. The drive components that induce the animal to perform an instrumental response with maximal frequency tend to become modified by something akin to a computation of "expenses" and success, leading to adapted behavior. Such behavioral adaptation to various schedules of reinforcement was originally described by Skinner (1938), later by Ferster and Skinner (1957) and many others.

This adaptation process can be treated mathematically. In accordance with the expected utility theory, Shimp (1969) formulated a mathematical model for steady-state operant behavior. The theory states that the experimental animal chooses whichever alternative momentarily has the greatest weighted probability of reinforcement, formulated as

$$\max[P(E_i) \cdot V(E_i)], \tag{1}$$

where $P(E_i)$ is the momentary probability that response A_i, if it occurs, is reinforced, and $V(E_i)$ the value of reinforcement E_i after response A_i. The model successfully predicts the steady-state relative frequencies of the two alternatives in two-key, discrete-trial probability-learning experiments, in concurrent variable interval schedules, and in concurrent variable interval schedules modified to study magnitude and delay of reinforcement and

[1] The research upon which this paper is based was supported by the Swiss National Foundation for Scientific Research, Grant 3.608.71.

conditioned reinforcement, as well as the relative frequencies of interresponse times in one-key variable interval schedules.

In an attempt to predict mathematically acquisition as well as steady-state operant behavior, we developed a slightly different model in which the preceding one could be included. Starting from the optimization principle stated at the beginning, we may assume that the animal, by a regulation process, chooses the alternative having the greatest incentive. In other words, the animal optimizes his energy budget. This assumption can be formulated as follows:

$$\max(\text{expected reinforcement value} - \text{energy spent}) \\ = \max[P(t) \cdot V(r) - r(t)^2/2], \qquad (2)$$

where $P(t)$ is the subjective momentary reinforcement probability, and V the reinforcement value as a function of the response rate r at time t. This model successfully describes acquisition and steady-state operant responding on interval schedules, and can, for example, predict the experimental data of Catania and Reynolds (1968).

These mathematical models support the idea that behavioral adaptation on operant schedules is some kind of optimization process. For experimental studies of behavioral adaptation on operant schedules, pigeons seem to be advantageous subjects since they perform particularly well in such test situations. They peck a key for food reward with a rate of up to five times per second. They easily learn to perform rather automatically with a maximal response rate on ratio schedules. However, on schedules requiring response inhibition such as interval schedules or schedules with differential reinforcement of low rates, pigeons usually perform with response rates far above optimal levels; it seems to be rather difficult for them to withhold responses. Therefore, one might expect, at least in the pigeon, that brain lesions or drugs would disrupt operant behavior on reinforcement schedules requiring response inhibition rather than on schedules requiring maximal response rates.

I. Hyperstriatal and Neostriatal Lesions

The hyperstriatum and neostriatum, although called "striatal," are a specific to the avian telencephalon. Therefore, they cannot be directly compared with the mammalian brain. Neither the hyperstriatum nor the neostriatum are homogenous structures; each can be subdivided into many regions belonging to various systems. Nauta and Karten (1970) consider these structures as part of the "external striatum" and as equivalents of the mammalian neocortex. This anatomical situation renders it difficult to compare behavioral deficits caused by extensive hyper- or neostriatal lesions with mammalian findings.

Reynolds and Limpo (1965) trained pigeons on a multiple schedule consisting of a fixed-interval (FI) component of 4 min with a red key, alternated with a fixed-ratio (FR) component of 55 responses with a green key. Each component was presented 10 times per session. When both performances were consistent from day to day, surgical lesions were made in the entorhinal area, the accessory hyperstriatum, and most of the dorsomedial hyperstriatum. After surgery the birds were tested on the same multiple schedule. The performance of each bird on the fixed-ratio component was normal, with a characteristic high, sustained rate. However, the performance of the lesioned birds on the interval component was strikingly changed: There was a lack of the characteristic pause at the start of the interval, and sudden breaks from high rates into pauses during the interval. Performance on the ratio component showed that hyperstriatal lesions do not impair the pecking response nor render food ineffective as a reinforcer. Furthermore, the sensory discrimination between the stimuli associated with the schedule components were not affected by this lesion. The alterations on the FI component were sustained during the whole observation period. The abrupt breaks from a high rate into a pause during the interval show that these effects are not simply due to a disinhibition of responding.

An inhibitory as well as an excitatory role of the hyperstriatum was postulated by Macphail (1969). He suggested that there is a mechanism for response facilitation in the anterior hyperstriatum accessorium and a mechanism for response inhibition in the posterior hyperstriatum accessorium. Small bilateral lesions of the anterior hyperstriatum in the pigeon did not affect acquisition or reversal of a brightness discrimination. However, pigeons with such lesions were faster in extinguishing a food-reinforced runway response and more affected by punishment in a passive avoidance test than normal control animals. Experiments on active avoidance were inconclusive because the range of control and experimental scores was too wide.

In a more recent study, Macphail (1971) electrolytically lesioned the hyperstriatum of the pigeon; most damage was done to the hyperstriatum dorsale and ventrale, with some involvement of the hyperstriatum intercalatus superior. In a few cases there was also some minimal damage to the hyperstriatum accessorium and to the neostriatum, but the hippocampus was always spared. After surgery the animals learned to peck a key illuminated by a red or green light for food reward. When pecking well at both colors, a variable interval (VI) schedule, having a mean interreinforcement interval of 60 sec was gradually introduced. A daily session consisted of 25 red and 25 green components in a Gellerman (1933) sequence. Each component lasted for 60 sec. The intercomponent interval was 5 sec, during which time the key and house-lights were extinguished. Experimental as well as control animals achieved a stable VI performance, on the average, within 22 sessions. There were also no significant differences in response rates between any of the session components. During the final 12 sessions of the experiment, discrimination training was introduced; only the responses made during the green component were reinforced according to the VI schedule (positive component), whereas responses made on the red key could no longer generate reinforcements (negative component). Lesioned animals showed a slight impairment of response inhibition in the negative component, primarily during the most difficult stages of the discrimination (first two components of each day, and the first few days). However, there were no significant differences between experimental and control animals during the positive component. Both groups showed behavioral contrast (an increase in the rate of responding to a positive stimulus consequent to the introduction of a negative stimulus), which increased over the first 3 days then remained relatively stable. Macphail concluded that hyperstriatal lesions in pigeons cause a general deficit in the ability to withhold responses as a consequence of nonreward, but that this deficit is probably not a result of reduced emotional responding to nonreward. He compares this general deficit with that produced in mammals by certain limbic lesions (Gerbrandt, 1965), and suggests that some functional system common to both classes of animals has been disrupted. However, for such a comparison to be made, more data about the structural organization of the hyperstriatum as well as about behavioral deficits relative to more restricted lesions would be needed.

Lesions in the neostriatum have so far failed to produce any significant changes in operant behavior. Lesions involving a rather restricted region within the caudal neostriatum did not alter the performance on a VI (Zeier, 1968), FI, or DRL (differential reinforcement of low rates) schedule (Zeier, 1969). These negative results are not surprising, since Phillips (1964) found no changes in "wildness" of ducks having a similar type of lesion, and Zeigler (1963) did not find any apparent change in the general activity levels of pigeons after neostriatal lesions. In this case also, more anatomical data about the nature and subdivisions of the neostriatum might give a hint as to which kind of behavioral deficit we must look for.

II. Archistriatal Lesions

The archistriatum is a large and heterogenous nuclear mass in the caudal ventrolateral portion of the bird's telencephalon. It was considered homologous with the mammalian amygdaloid complex. Zeier and Karten

(1971), however, dispute this simple generalization and suggest a subdivision of the archistriatum into four major regions (see Fig. 1): the archistriatum anterior, intermedium, posterior, and mediale. The archistriatum posterior and mediale are connected with the hypothalamus through the tractus occipitomesencephalicus pars hypothalami (HOM), and thus may correspond to the mammalian amygdala. Furthermore, the archistriatum posterior gives rise to the stria terminalis (ST) which can be traced into the

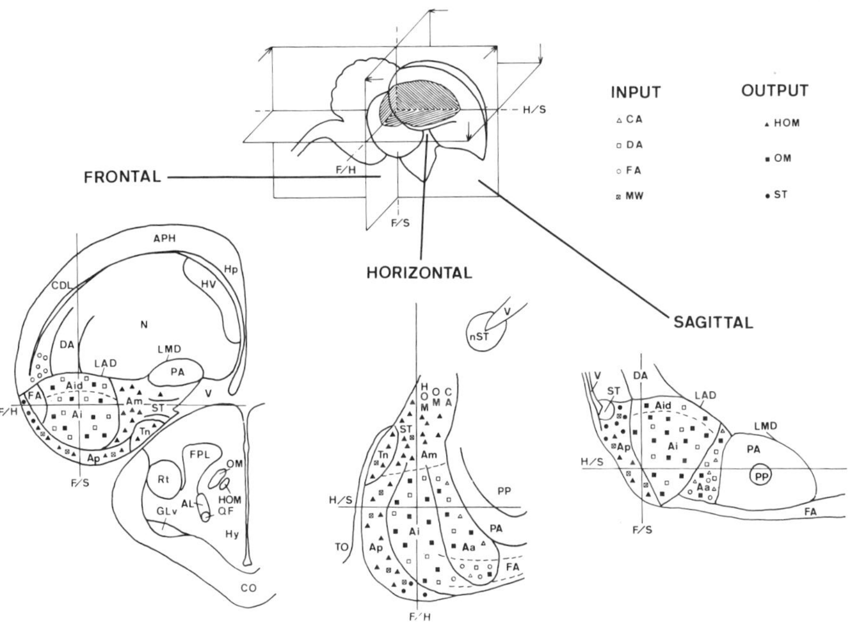

Fig. 1. Subdivisions of the archistriatum depicted in frontal (F), horizontal (H) and sagittal (S) planes with some of their afferents (input) and efferents (output). [From Zeier & Karten (1973).]

Abbreviations

Aa, archistriatum anterior; Ai, archistriatum intermedium; Aid, archistriatum intermedium, pars dorsalis; AL, ansa lenticularis; Am, archistriatum mediale; Ap, archistriatum posterior; APH, area parahippocampalis; CA, commissura anterior; CDL, area corticoidea dorsolateralis; CO, chiasma opticum; DA, tractus archistratalis dorsalis; FA, tractus fronto-archistriatalis; FPL, fasciculus prosencephali lateralis; GLv, nucleus geniculatus lateralis, pars ventralis; HOM, tractus occipitomesencephalicus, pars hypothalami; Hp, hippocampus; HV, hyperstriatum ventrale; Hy, hypothalamus; LAD, lamina archistriatalis dorsalis; LMD, lamina medullaris dorsalis; MW, Medial wall; N, neostriatum; nST, nucleus striae terminalis; OM, tractus occipitomesencephalicus; PA, palaeostriatum augmentatum; PP palaeostriatum primitivum; OF, tractus quinto-frontalis; Rt, nucleus rotundus; ST, stria terminalis; Tn, nucleus taeniae; TO, tractus opticus; V = ventriculus.

nucleus striae terminalis, a cell group positioned around the ventral anterior aspect of the forebrain ventricle. The archistriatum anterior and intermedium, however, are connected through the nonhypothalamic tractus occipitomesencephalicus (OM) with the thalamus, optic tectum, tegmentum, lateral reticular formation, lateral pontine nuclei, sensory nuclei of the brainstem, and to a small degree even with the rostral levels of the spinal cord. The OM bundle consists of thicker fibers than the hypothalamic projection, and its course and distribution strongly resemble the picture of Bagley's bundle of ungulates, a fiber bundle often considered as a variant form of part of the pyramidal tract of primates. According to these findings, the archistriatum posterior and mediale can be considered "limbic," whereas the archistriatum anterior and intermedium appear to be "somatic–sensorimotor" in nature. The most massive telencephalic afferents to the archistriatum, the tractus archistriatalis dorsalis (DA) and, to a lesser degree, the tractus fronto-archistriatalis (FA) terminate in the somatic motor archistriatum. Terminals of the anterior commissure (CA) appear to be distributed only in the archistriatum anterior, whereas the remaining portions of the archistriatum receive no commissural connections. In the archistriatum posterior, on the other hand, terminals can be found after lesions placed in the medial wall of the pallium (hippocampal and septal area).

This complex anatomical situation renders the evaluation of behavioral effects of archistriatal lesions rather difficult. In a study by Zeier (1971), pigeons were pretrained on a 1 min VI and a 10 sec DRL schedule of reinforcement, respectively, for 21 days. They then sustained bilateral surgical or electrolytic lesions in the somatic motor, lateral limbic, and medial limbic archistriatum. After surgery, they were retested for 21 days on the same schedule, and then sacrificed. For histological study of the lesions, pairs of adjacent sections were stained with a silver method (Fink & Heimer, 1967), selectively impregnating degenerated nerve fibers and endings, or with the Nissl method. The histological data from both methods were used as criteria for classifying the experimental animals into corresponding lesion groups. Cases with combined or asymmetrical lesions were discarded.

Lesions in the limbic archistriatum produced pigeons that were tamer and easier to handle, whereas somatic–motor archistriatal lesions produced more fearful and occasionally more aggressive animals. However, there were already considerable individual differences in aggressiveness before surgery —males usually being more aggressive than females. Lesions in the lateral limbic archistriatum produced the most striking behavioral changes. Postoperatively, some of them showed somnolent states, refused to take any food and failed to react to tactile stimuli. However, they resumed pecking in the operant conditioning situation after 2–3 days, and recovered completely within 1 week. In the meantime they were fed by hand. Lesions in the somatic–motor archistriatum often caused eating difficulties. Such animals apparently had a motor deficit, because they could barely keep the grains within their beaks; they also had to be force-fed, although they usually performed in the test apparatus and tried to eat the reinforcement. This motor eating disturbance may be related to the feeding mechanisms postulated by Zeigler, Green, and Karten (1969), which utilize a neuronal circuit starting in the main sensory trigeminal nucleus innervated by afferents from the oral region, and, after a synapse in the nucleus basalis, reach the anterior somatic–motor archistriatum and the caudolateral neostriatum. We may expect that a lack of sensory trigeminal information renders it difficult for the animal to keep the grains within its beak.

On the 1-min VI schedule, the pigeons performed regularly within a few preoperative daily 20-min sessions and usually achieved their 20 reinforcements per session. The response rate of each animal was stabilized after about 2 weeks, but it remained quite different among individuals. However, the response rate of all animals was far above optimal level, and only 2–5% of the responses generated a reinforcement. Figure 2 demonstrates the postoperative–preoperative performance difference on the VI schedule plotted in percentage of the preoperative level achieved during the last five training sessions. The controls ($N = 8$, unoperated animals and animals with neostriatal lesions calculated together) remained relatively close to the preoperative level throughout the entire observation period. The groups with lesions in the somatic–motor archistriatum ($N = 4$) and with increased food deprivation ($N = 4$) showed an increased response rate. The pigeons with lesions in the lateral limbic archistriatum ($N = 4$) showed postoperatively a marked initial depression in response rate, caused by their somnolent state. However,

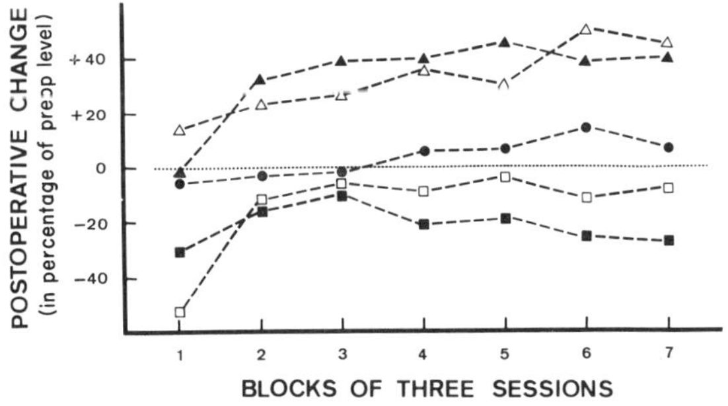

Fig. 2. Postoperative–preoperative performance difference on the VI schedule plotted in percentage of the preoperative level achieved during the last five training session. (▲), Somatic–motor archistriatum; (□), lateral limbic archistriatum; (■), medial limbic archistriatum; (△), reduced weight; (●), control. [From Zeier (1971).]

they recovered and returned to the preoperative level after a few days. The animals with lesions in the medial limbic archistriatum ($N = 4$) performed with a slightly subnormal response rate throughout the entire observation period.

On the 10-sec DRL schedule, the animals also reached relatively stable performance within about 2 weeks, but once again the differences in efficiency among individuals were quite marked. They achieved the typical DRL behavior in that they alternately emitted a few responses, turned away from the pecking key, moved around in the test chamber, approached the key again, and so on. According to the definition of Sidman (1956), the pecks with interresponse times of 2 sec or less were classified as burst responses, the remaining pecks with longer latencies as approach responses. On the average, about 10–20 % of the approach responses had an interresponse time of 10 sec or more, and were therefore reinforced. Figure 3 shows the postoperative changes in approach responses, burst responses and reinforcements of the different groups, plotted as percentage of the preoperative level. The animals with lesions in the medial limbic archistriatum ($N = 4$) showed a relatively stable increased rate of approach responses and, therefore, a decreased rate of reinforcements throughout the entire postoperative observation period. The animals with lesions in the somatic–motor archistriatum ($N = 8$) showed exactly the opposite phenomenon. Weight reduction ($N = 4$), on the other hand, caused a more fluctuating effect. The number of burst responses did not significantly change postoperatively, except for an initial increase for the group with lesions in the medial limbic archistriatum.

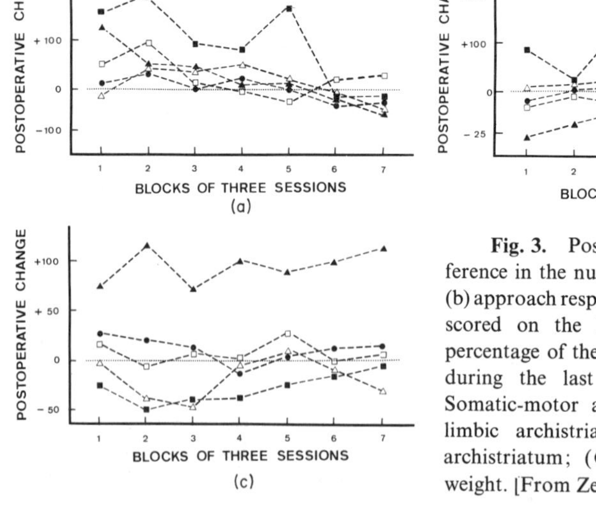

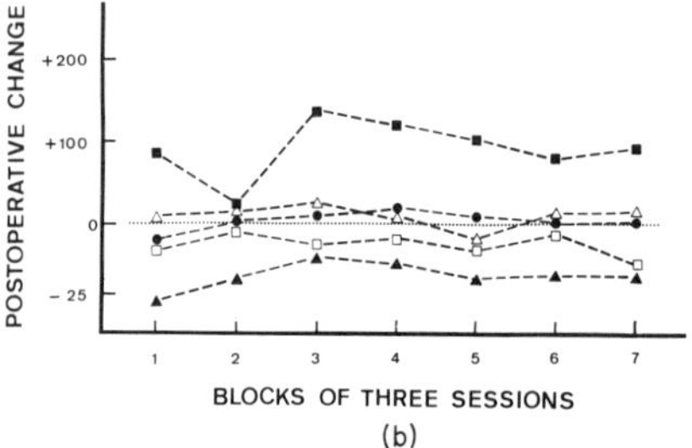

Fig. 3. Postoperative-preoperative difference in the number of (a) burst responses, (b) approach responses, and (c) reinforcements scored on the DRL schedule, plotted in percentage of the preoperative level achieved during the last 5 training sessions. (▲) Somatic-motor archistriatum; (□), Lateral limbic archistriatum; (■), medial limbic archistriatum; (●), control; (○), reduced weight. [From Zeier (1971).]

The observed alterations in operant behavior were opposed to each other in that lesions in the somatic–motor archistriatum (comparable to the mammalian neocortex) and the limbic archistriatum (comparable to the mammalian amygdaloid complex) resulted in opposite effects. Lesions in the somatic–motor archistriatum impaired the VI performance, in that these pigeons pecked with an increased response rate, although such behavior did not generate more reinforcements. However, the same lesions improved the DRL performance in that the pigeons made fewer approach responses, while the number of burst responses remained unchanged; they thereby gained more reinforcements. On the other hand, lesions in the medial archistriatum impaired the DRL performance and slightly improved the VI performance. Lesions in the posterior limbic archistriatum failed to produce significant long-lasting changes in operant behavior, although they were followed by a strikingly transient postoperative depression and increased tameness. The observed changes in operant behavior on both schedules, after archistriatal lesions, cannot be explained in terms of motivational changes, since increased food deprivation impaired efficiency on both schedules. A memory deficit or a change in the timing capability of the animal are ruled out through the findings of an earlier experiment (Zeier, 1969), where lesions involving the lateral limbic and somatic–motor archistriatum had no effect on typical FI schedule timing behavior.

The increased response rate on the VI schedule after somatic–motor archistriatal lesions may be a consequence of a motor disinhibition. This explanation correlates with anatomical findings (Zeier & Karten, 1971) inferring that this region has connections to the lateral reticular formation, optic tectum, medulla and upper spinal cord. However, the effect of this lesion on the DRL schedule suggests that the somatic–motor archistriatum also acts upon pecking behavior in a facilitatory way. The pecking response on the two schedules appears to be of a quite different nature. On the VI schedule, the pigeon usually stood in front of the pecking key, made continuous, automatic pecking responses at a relatively constant rate, and hardly showed any other behavior. On the DRL schedule, however, the pigeon turned away from the pecking key when it failed to get a reinforcement. The subsequent approach responses appeared to be much more complex than the VI responses or the burst responses, the latter being usually related to frustration (Stamm, 1964) or lack of stimulus feedback (Kramer & Rilling, 1970). It appears that the somatic–motor archistriatum inhibits pecking responses and facilitates the more complex approach responses, whereas the medial limbic archistriatum rather inhibits approach responses and facilitates monotonous pecking responses. Such an interaction between limbic gate-setting functions and motor effector systems may be possible in the bird through the close anatomical proximity between limbic and somatic–motor structures.

It seems premature to compare these findings to data obtained from mammals, since the somatic–motor archistriatum cannot yet be compared to a specific cortical structure of the mammal. However, deficits on DRL performance and passive avoidance tasks caused by lesions in the amygdaloid complex have been reported in mammals by several authors (Horvath, 1963; Pellegrino, 1968; Ursin, 1965). Furthermore, Goddard (1964) suggested that the amygdala is primarily involved in the active suppression of motivational approach behavior. Improvement of DRL behavior, on the other hand, was found in the monkey after ablation of the prefrontal cortex (Stamm, 1964). Disturbance of emotional behavior and sleeplike states after lesions in the lateral limbic archistriatum fit well into the syndrome observed in amygdaloidectomized mammals (reviewed by Gloor, 1960, and Goddard, 1964). Therefore, we may assume that the limbic archistriatum performs amygdaloid functions. However, in order to compare its different regions directly with the subnuclei of the mammalian amygdala, still more data are needed.

III. Paleostriatal Lesions

The paleostriatum of the bird corresponds to the mammalian corpus striatum. Nauta and Karten (1970) suggested a homology between the avian paleostriatum primitivum and the mammalian globus pallidus, based on the observation that each, in its respective class, has a selectively high content of iron compounds and projections to the mesencephalic tegmentum over the ansa lenticularis. They justified a homology between the avian paleostriatum augmentatum and the mammalian caudoputamen, since both structures have particularly large amounts of dopamine and acetylcholinesterase, as well as massive efferent connections to the large-celled, most internal striatal component: the paleostriatum primitivum and the globus pallidus, respectively.

After extensive paleostriatal lesions, pigeons usually display somnolent or cataleptic states and refuse to take any food (Zeier, 1968). This effect may be due to an interruption of the tractus quinto-frontalis which is involved in the above-mentioned neural feeding mechanism, as postulated by Zeigler *et al.*, (1969). Pigeons pretrained on a 1-min VI schedule therefore showed after paleostriatal lesions a marked depression in their instrumental response rate. However, they recovered quickly and, after a few days, performed on the VI schedule with a response rate slightly above the preoperative level (Zeier, 1968). Hence, we may assume that the paleostriatum also plays some inhibitory role in operant pecking behavior. More localized lesions would be needed, however, in order to evaluate these mechanisms and to compare them to mammalian findings.

IV. Conclusions

Pigeons adapt their pecking behavior to various schedules of reinforcement in a characteristic way. As a general rule, they peck with a response rate above the optimal level rather than take the risk of missing reinforcements. The pecking response per se seems to be performed fairly automatically and stereotypically. Therefore brain lesions disrupt operant behavior on schedules requiring response inhibition, rather than on schedules requiring maximal response rates. In the pigeon, several forebrain structures seem to play an inhibitory function on the pecking response and some structures seem to act in a facilitatory way. However, the schedule of reinforcement is also an effective variable for determining the effects of brain damage on operant behavior.

References

Catania, A. C. & Reynolds, G. S. A quantitative analysis of the responding maintained by interval schedules of reinforcement. *Journal of the Experimental Analysis of Behavior*, 1968, **11**, 327–383.

Ferster, C. B. & Skinner, B. F. *Schedules of reinforcement*. New York: Appleton, 1957.

Fink, R. P. & Heimer, L. Two methods for selective silver impregnation of degenerating axons and their synaptic endings in the central nervous system. *Brain Research*, 1967, **4**, 369–374.

Gellerman, L. W. Chance orders of alternating stimuli in visual discrimination experiments. *Journal of Genetic Psycholgy*, 1933, **42**, 206–208.

Gerbrandt, L. K. Neural systems of response release and control. *Psychological Bulletin*, 1965, **64**, 113–123.

Gloor, P. Amygdala. In J. Field (Ed.), *Handbook of physiology*, Vol. 2. Washington, D.C.: American Physiological Society, 1960.

Goddard, G. V. Functions of the amygdala. *Psychological Bulletin*, 1964, **62**, 89–109.

Horvàth, F. E. Effects of basolateral amygdalectomy on three types of avoidance behavior in cats. *Journal of Comparative and Physiological Psychology*, 1963, **56**, 380–389.

Kramer, T. J. & Rilling, M. Differential reinforcement of low rates: A selective ctitique. *Psychological Bulletin*, 1970, **74**, 225–254.

Macphail, E. M. Avian hyperstriatal complex and response facilitation. *Communications in Behavioral Biology*, 1969, **4**, 129–137.

Macphail, E. M. Hyperstriatal lesions in pigeons: Effects on response inhibition, behavioral contrast, and reversal learning. *Journal of Comparative and Physiological Psychology*, 1971, **75**, 500–507.

Nauta, W. J. H. & Karten H. J. A general profile of the vertebrate brain, with sidelights on the ancestry of cerebral cortex. In F. O. Schmitt (Ed.), *The neurosciences, second study program.* New York: Rockefeller Univ. Press, 1970.

Pellergrino, L. Amygdaloid lesions and behavioral inhibition in the rat. *Journal of Comparative and Physiological Psychology*, 1968, **65**, 483–491.

Phillips, R. E. "Wildness" in the mallard duck: Effects of brain lesions and stimulation on "escape behavior" and reproduction. *Journal of Comparative Neurology*, 1964, **122**, 139–156.

Reynolds, G. S. & Limpo, A. J. Selective resistance of performance on a schedule of reinforcement to disruption by forebrain lesions. *Psychonomic Science*, 1965, **3**, 35–36.

Shimp, C. P. Optimal behavior in free-operant experiments. *Psychological Review*, 1969, **76**, 97–112.

Sidman, M. Time discrimination and behavior interaction in a free operant situation. *Journal of Comparative and Physiological Psychology*, 1956, **49**, 469–473.

Skinner, B. F. *The behavior of organisms.* New York: Appleton, 1938.

Stamm, J. S. Function of cingulate and prefrontal cortex in frustrative behavior. *Acta Biologiae Experimentalis (Warszawa)*, 1964, **24**, 27–36.

Ursin, H. Effect of amygdaloid lesions on avoidance behavior and visual discrimination in cats. *Experimental Neurology*, 1965, **11**, 298–317.

Zeier, H. Changes in operant behavior of pigeons following bilateral forebrain lesions. *Journal of Comparative and Physiological Psychology*, 1968, **66**, 198–203.

Zeier, H. DRL-performance and timing behavior of pigeons with archistriatal lesions. *Physiology and Behavior*, 1969, **4**, 189–193.

Zeier, H. Archistriatal lesions and response inhibition in the pigeon. *Brain Research*, 1971, **31**, 327–339.

Zeier, H. & Karten, H. J. The archistriatum of the pigeon: Organization of afferent and efferent connections. *Brain Research*, 1971, **31**, 313–326.

Zeier, H. & Karten, H. J. Connections of the anterior commissure in the pigeon (*Columba livia*). *Journal of Comparative Neurology*, 1973, **150**, 201–216.

Zeigler, H. P. Effects of forebrain lesions upon activity in pigeons. *Journal of Comparative Neurology*, 1963, **120**, 183–194.

Zeigler, H. P., Green, H. L. & Karten, H. J. Neural control of feeding behavior in the pigeon. *Psychonomic Science*, 1969, **15**, 156–157.

The Neural Basis of Avian Discrimination and Reversal Learning

Laurence J. Stettner
Wayne State University

I. Introduction

It is clear that birds as a class possess a high degree of learning ability. It is also clear that we have as yet barely begun to investigate the relationship between these abilities and avian brain structures. Until recently, in fact, we lacked the basic neuroanatomical information about the avian brain to be able to make an all out effort in this area. Thanks to recent anatomical and electrophysiological work we now possess some of the basic information that has been sorely lacking. Visual pathways have been identified and traced into the forebrain primarily by Karten and his associates (Karten, 1965; Karten & Revzin, 1966; Revzin & Karten, 1967; Karten & Hodos, 1970). Karten (1967, 1968) has also mapped an auditory pathway at the thalamic and forebrain levels, and Erulkar (1955) has studied the auditory system electrophysiologically. These very fundamental neuroanatomical studies have at last begun to sketch a framework that can serve as a guide for functional studies of the avian thalamus and forebrain. Furthermore, and at least as important as the specific information, we have insights about vertebrate brain organization that these studies have suggested (Nauta & Karten, 1970). Our thinking has finally been freed from mammalian neocortical domination and the point has been brought home that the term "striatum" applied to the avian forebrain is purely descriptive—the bulk of these cells are neither homologous nor functionally analagous to the mammalian "corpus striatum."

These closely allied anatomical and conceptual advances have, I believe, placed us on the threshold of a period of intense investigation of the neural substrates of avian learning. The basic stock-in-trade of this effort will almost surely be investigations of the effect of brain lesions on discrimination learning. It is extremely important to be aware at the outset of the psychological (behavioral) complexities that can influence the outcome of such studies, *even in those cases where we are working with relatively well-defined anatomical areas*. Differences in training procedure, or the animal's prior learning experience, or both can often be critical in determining whether or to what degree the learning in question will be influenced by any particular brain lesion (Meyer, 1958).

A particularly good illustration of this problem in mammals has been analyzed by Elliot and Trahiotis (1972) in their review of the effects of removal of "auditory cortex" on frequency discrimination.[1] One set of investigators (Allen, 1943, 1945; Meyer & Woolsey, 1952) found that removal of temporal lobe cortical areas in the cat led to permanent deficits in frequency discrimination. Others, however (Butler, Diamond & Neff 1957; Goldberg & Neff, 1961), found that complete relearning of frequency discrimination was possible in cats that had suffered the same or even more extensive lesions. Following up on the hypothesis that this apparent contradiction could be based on differences in test procedures, Thompson (1959, 1960) directly investigated the effect of different learning paradigms. His results are particularly instructive since in all cases the sensory requirements of the tasks were the same—to discriminate between tones of 1000 and 1500 Hz. He utilized three different training procedures, all of which were basically "avoidance" learning situations in which the animal had to respond to the "positive stimulus" by running in a circular cage. A failure to respond to the positive stimulus within 2 sec led to shock; no shock was ever associated with the negative stimulus. In the first task, the stimuli (1000- and 1500-Hz tones) were presented independently in irregular order, with silence between trials. Thus, on any given trial, an animal had to "decide" whether he was listening to a 1000- or a 1500-Hz tone and to behave accordingly. In the second test procedure, the animal was presented with an array of eight tones on each trial—they were either all of *one* frequency (1000 or 1500 Hz) or alternated between the frequencies. Thus, on any trial, the subject had to "decide" whether it was listening to a sequence of the same tones or listening to an alternating sequence. In the third test procedure, the animal was continually exposed to

[1] The discussion in this section is essentially based upon material presented in Elliot and Trahiotis' review. The author wishes to express his appreciation to Dr. Elliot for making the manuscript available to him prior to publication.

1000-Hz tone pulses; the onset of a trial necessitating a response to avoid shock was signaled by an alternating sequence of 1000 and 1500 Hz.

All of these procedures required the ability to differentiate between the two tones, but they apparently differed considerably in their total information-processing requirements. With the first procedure, animals simply failed to learn (within 1500 trials) preoperatively. The other two procedures were learned, with the third procedure (repetitive 1000-Hz tones between trials) leading to faster acquisition than the second. After surgical removal of all cortex bounded by the suprasylvian and rhinal sulci, performance on the second procedure was deleteriously affected, whereas all animals relearned successfully (*with savings*) under the third procedure; it is clear that the test procedure can make all the difference. It cannot be too strongly emphasized, moreover, that this is not a question of "sloppy" or "inferior" test procedures. Each procedure was logically and methodologically sound. Furthermore, the different results are often not open to obvious interpretation. It may be easy to appreciate why the first procedure, which requires an "absolute" judgment by the animal as to which tone it is hearing would be more difficult than the other procedures, in which it simply has to detect that there is a difference in tones that are presented in a close temporal sequence. However, since both of the other tasks require only a "relative" judgment of this type, it is less clear as to why differences should exist.

The interpretation of these results and of exactly what role auditory cortex plays in the mammalian brain are questions that continue to engage researchers. The point for us here is that different test procedures can be critical in determining results of lesion studies *even in those cases where the sensory and motor demands made on the animal appear to be identical.* Different procedures seem to have different types of information-processing and response integration requirements, and these may be differentially affected by destruction of any particular set of brain cells. Perhaps the avian brain will be more willing to give up its secrets to us and this preamble will turn out to be over cautious. However, we must be aware of these possibilities, as we evalutate the contributions of the forebrain.

II. Discrimination Procedures

There are, of course, an indefinitely large number of behavioral parameters that could be relevant to interpretation of neural studies. We will discuss only some of the major different learning procedures and highlight the variables that are most relevant to the studies that will be reviewed in what follows.

A. Classical Conditioning

A fundamental distinction must be made between so-called "classical" (Pavlovian) conditioning procedures for establishing evidence of discrimination and instrumental or operant procedures. In the classical procedure a neutral stimulus is temporally paired with a reflex-eliciting (unconditioned) stimulus repeatedly until the neutral stimulus becomes a "conditioned" stimulus (CS) and elicits reflex behavior similar to that produced by the unconditioned stimulus (US). In the bird literature, the procedure has been to use electric shocks, which elicit increases in heart rate and respiration, as the US. A light or tone paired with the shock soon comes itself to elicit similar changes (Cohen & Durkovic, 1966). Discrimination or "differentiation" is established by presenting two stimuli (e.g., a white light and a red light), an "S^+" followed by shock and an "S^-" never followed by shock. Development of a consistently different response in one or more measures (e.g., respiration or heart rate) following the two stimuli is evidence of discrimination. Of major concern to us is the nature of this discrimination. The subject may continue to respond to *both* stimuli in a "conditioned" manner, but have a higher level response to S^+ than S^-. Or he might not respond at all in a conditioned manner to S^- while continuing his response to S^+. Similarly, lack of discrimination can occur through an equally high level of responding to both stimuli or through a total lack of responding to both. It is important to note the specific pattern of response levels that occurs in interpreting lesion effects, since not only sensory processing deficits but also alteration of basic response tendencies may lead to lack of differentiation.

B. Instrumental Learning

Instrumental learning procedures afford a wide range of possibilities which essentially involve either the use of rewards ("positive reinforcers") or punishments ("negative reinforcers") that are made contingent upon the animal's behavior in such a way that a desired pattern of responding is brought about.

There are a number of logical schemes that have been suggested to classify instrumental procedures (Bitterman, 1962; Hilgard & Bower, 1966) and many distinctions can be made according to the nature of the "incentive" used, nature of the response required, type of stimulus presentation, temporal and spatial relationships between stimuli presented and response required, etc. Fortunately, however, one type of situation has been almost universal in studies of instrumental learning in birds: The bird is placed in a small experimental chamber (Skinner box) and required to peck at a small window in the wall of the chamber (pecking key) in order to gain access to grain from a food bin (hopper) mounted at the base of one wall of the chamber.[2] The animal's interest in the food is maintained by keeping it on a limited feeding schedule for some time prior to and during the experiment. The feeding is usually arranged so as to maintain the animal at the same fixed percentage of his free-feeding (ad lib) weight. Discrimination learning in this situation is introduced in one of two fundamental ways. In the first, a single response key is present, but different stimuli are presented at different times. The animal is reinforced (allowed to gain access to the food hopper) for pecking at the key in the presence of one of the stimuli (S +) but not in the presence of the other (S −). Discrimination is established when there is a consistent difference in its pecking behavior according to whether S + or S − is presented. This training of differential response tendencies to two stimuli presented one at a time in temporal sequence is termed "successive" discrimination. (The term "single-key" training is often used to refer to this procedure in bird literature.)

The other method is to have two (or more) pecking keys in the experimental chamber (these keys being invariably located spatially as left and right, or left, center and right, on the same horizontal plane) and to present different stimuli *simultaneously*. Only pecks at the response key associated with S + lead to reinforcement. Discrimination is established when there is a consistent choice of the S + key over the S − key.

There are some other differences that usually obtain between successive and simultaneous discrimination, in addition to the fundamental one of the mode of stimulus presentation. In the single-key situation, the stimuli usually stay "on" for some period of time, e.g., 1 min, so that the subject can make multiple pecks at the stimulus on any one presentation ("free responding" and "free operant" are terms used to describe this situation). The *simultaneous* situation is often arranged so that a single response to either key terminates the presentation of the stimuli for a period of time (the intertrial interval). ("Discrete trials" is the term most often used to describe this procedure.) Furthermore, in the single-key situation, the animal is generally not rewarded for every peck at S + . Reinforcements occur on an intermittent basis according to some predetermined schedule, e.g., one reinforcement for 15 pecks, one reinforcement every 15 sec, etc.[3] When the discrete-trial situation

---

[2] It is, of course, possible to use aversive stimulation in this situation, also. A pigeon can be required to peck at a key to escape or avoid an electrical shock. Mammalian studies suggest that some brain structures (e.g., the amygdala) may play a special role in these situations. Eventually, training with aversive stimuli would seem necessary in order to test the degree of motivational specificity of avian brain structures.

[3] These considerations of reinforcement schedules are of the most elementary and fragmentary nature. The development and study of the effects of various schedules of reinforcement has been a major undertaking among behavioral scientists. See the chapter by Zeier in this volume for further discussion in this area.

is utilized in simultaneous discrimination learning, typically each response to the S + is reinforced.

The procedures just discussed represent the fundamental framework of training conditions under which any particular discrimination problem may be investigated. Any experimenter must make a choice on each of the above-noted dimensions: single-key versus simultaneous stimulus presentation, free-operant versus discrete trials, intermittent reinforcement versus "continuous" reinforcement. The choice is often arbitrary, since there are arguments both ways for each of the procedures. For example, simultaneous discrimination with discrete trials allows the animal to compare the stimuli directly and allows the experimenter to obtain a simple measure of discrimination in terms of percentage of trials on which S + is chosen over S −. On the other hand, successive discrimination with free-operant responding provides the animal with multiple opportunities to acquire response feedback on each stimulus presentation, and gives the experimenter a large number of responses and therefore a more stable measure to use as evidence for discrimination. It is possible to devise procedures that combine some of the desirable features of both types of training. Zeigler (1965a), for example, utilized a simultaneous procedure in which the stimuli both remain projected for 15 sec and responses to S + are reinforced on a fixed interval schedule.

Having described any discrimination learning experiment in terms of the general procedural dimensions, we are then ready to inquire into the specific nature of the discrimination problem and the conditions used. The nature of the deprivation schedule used (e.g., animals maintained at 70 or at 90% of their ad lib weight) and the specifics of the reinforcement procedure (amount of eating time allowed as well as schedule of reward) must, of course, be specified as they can be major determiners of performance. The nature of the discrimination problem presented is of course the fundamental content in which we are interested. In case we miss the forest for the trees, it should be noted that the various types of instrumental discriminations presented to birds have utilized visual cues almost exclusively. [Many discriminative procedures are readily transferrable to the use of auditory cues and there *are* some behavioral studies in this area (Heise, 1953; Jenkins & Harrison, 1960). Use of olfactory, taste, tactile, or temperature cues are more difficult but techniques for studying all of these do exist in the literature of animal learning and all are feasible.] The specific types of discrimination generally used include position (left versus right key), brightness (brightly lit versus dimly lit key), hue (red versus green), line orientation (horizontal versus vertical or other angles), size (large square versus small square) and shape (circle versus triangle). Combinations of these, of course, are also possible. These different types of discrimination vary, of course, in the degree of difficulty. Furthermore, within any type of discrimination there can be different degrees of difficulty according to the degree of similarity of S + and S −. Of course, we must not assume that we can necessarily infer discrimination difficulty from strict physical parameters or from our perceptions, but excellent techniques do *exist* (Blough, 1966) for determining with great precision how "fine" a discrimination an animal can make on any particular stimulus dimension, and the relative difficulty of discriminating between stimuli can be determined in the same way.

Finally, it should be noted that in all of what has been said so far, we have dealt only with what might be termed "simple discrimination," in which an animal is repeatedly presented with two and only two different stimulus conditions and simply attaches responding to condition A and not to B. Many more complicated contingencies are possible, and have been successfully utilized with birds (Stettner & Matyniak, 1968). Two very common types are conditional discrimination and oddity discrimination. In a conditional discrimination, an animal must vary his response to the two "primary stimuli" depending on the presence of other stimulus conditions. For example, it may be confronted with a simultaneous choice between a horizontal and vertical line on each trial, *but*, a red overhead light may be present on half the trials and a green overhead light on the other half. Reinforcement of response to the lines is made *conditional* upon the presence of the red or green light: when red is on, a response to the horizontal line is reinforced; when green is on, only responses to vertical are rewarded.

In oddity learning, three stimuli are presented on each trial. Two are the same and one is different, e.g., two reds and a green or vice versa. The reinforced key is always the one with the odd stimulus. Again, this means that the subject must sometimes choose one stimulus, sometimes the other, depending on other aspects of the stimulus situation.

C. Reversal Learning

Reversal learning may begin with a "simple discrimination" according to any one of the procedures discussed above. After the animal has acquired the initial discrimination according to the experimenter's criterion (e.g., 18 out of 20 correct choices in a discrete-trial procedure, or 90 % of all pecks made to S + over the last two training sessions in a single-key situation) the reinforcement contingencies are switched—if red was S + initially, green now becomes S + and red S −, etc. The experiment may be terminated when the animal has reached criterion on this new contingency; that is when the subject has switched its behavior from responding to one stimulus predominantly to responding to the other. An experiment terminated at this point is a single reversal experiment. Usually, a single reversal takes more trials to reach criterion or involves more errors than the original learning. (Even the single reversal may reveal differences in performance between individuals of the same or different species that might not be evident in the initial discrimination.)

Alternatively, the experiment may be continued for a second reversal back to the original reinforcement contingencies, then a third reversal, a fourth, etc. This is the multiple-discrimination reversal experiment (MDR). After going through a long sequence of reversals in an MDR experiment, birds typically learn more rapidly on the later reversals than they did on the earlier ones, sometimes learning faster on later reversals than they did in acquiring the initial discrimination problem.

In MDR training there is an additional fundamental procedural question that must be dealt with in addition to all of the ones present in the simple discrimination learning. The question is one of when to initiate each new reversal. The two basic strategies are either to run each reversal to a set criterion of learning or to run each reversal for a fixed time (or fixed number of trials). In the former case, each animal will have a somewhat different running pattern and different number of total trials; in the latter, animals may show no evidence whatsoever of learning during the early reversals. Many ingenious minor variations are, of course, possible. Disjunctive criteria, such as either 90% correct responses or 3 days of training prior to reversal can be set, or one can give an extra day of training after criterion on some (or all) of the reversals; one may always start a new reversal on the day following the time when criterion is reached or always start reversal on the same day that criterion is reached, etc. All of these factors can and do affect reversal performance. Furthermore, it should be emphasized that reversal learning appears to be at least as sensitive as simple discrimination learning to variables such as deprivation schedule, amount of reward, nature of discrimination, degree of similarity of stimuli, etc.

III. Neurobehavioral Studies

We are now prepared to review experimental studies on the neural basis of discrimination and reversal learning in birds. This review will be limited to studies that are *directly* concerned with relating brain processes or structures to discrimination or reversal learning. Furthermore, it will be limited to those studies that have been carried out with sufficient methodological precision to provide useful information according to contemporary standards for neurobehavioral research.

There are several types of studies in the literature that appear to have the potential to provide this information. First are lesion studies, which assess the effect of the destruction of particular areas of brain tissue on discrimination or reversal learning. As already noted, this type of study is a basic stock-in-trade and comprises the primary focus of this review. Closely allied to such studies are those involving transfer of learning from one eye to the other in monocularly trained animals after various possible transfer-mediating structures have been destroyed. Third are studies of chemical modifications of brain tissue as they effect discrimination and reversal.[4] Fourth are purely behavioral studies which, however, attempt to relate differences in learning performance to differential brain development in different avian species. A very important type of study that we do not as yet have but that has been described and elegantly prepared for by Cohen (1969) is to record electrically from brain cells at different stages in the learning process. This promises to be of great interest in the future. Finally, there are studies on the effect of electrical perturbation of brain cells at various stages of learning; these studies are reviewed in this volume by McCollum and Goodman.

We have, in fact, a relatively limited amount of material to hang upon the framework we have prepared. Lesion studies are the primary body of information in the area, although the number of these is quite small compared to the number of similar studies extant in the mammalian literature. (In fact, at the present time, more studies of brain lesions and learning in mammals are published every *month* than the sum total of all the avian studies in this area thus far.) We have only a few pioneering studies in the other categories. Thus, we will review the lesion studies first, and then consider additional information and suggestions from the other studies as they appear relevant to the overall picture.

A. Lesion Studies

All the lesion studies reviewed in what follows should be assumed to have a number of common characteristics unless otherwise specifically noted. The *train–operate–test* method was used; that is, animals are trained on the discrimination problem, then operated upon, then retrained. Performance on retraining is assessed by comparing preoperative and postoperative data and by comparing performance of experimental animals with controls. (Note that this method, strictly speaking, makes an assessment of *memory* of a learned task through a relearning procedure. It is thus a conservative procedure.) All the studies (unless otherwise noted) included at least some sham operated

[4] Studies of this type have not as yet been able to give us information specifically related to the neural basis of avian discrimination and reversal learning. However, Mayor (1969) has demonstrated that such processes are susceptible to blocking by puromycin and Boguch (1968; 1970) has developed techniques to assess changes in the neurochemistry of the pigeon brain that occur as a result of discrimination training.

animals and/or animals with lesions in areas different from those in the experimental animals. Histological verification of lesion sites should also be assumed, as should the presentation of testimonials to the effect that operated animals ate, drank, and pecked "normally" during the postoperative tests.

Lesion studies of discrimination acquisition will be reviewed before we look at reversal studies. An important differentiation should be recognized at the outset and kept in mind between studies involving relatively circumscribed lesions in areas that are well-defined components of the avian sensory pathways (see Cohen and Karten, Chapter 2 of this volume, for a discussion of sensory pathways) and relatively large and irregular lesions which have been made in the wulst and hyperstriate regions where anatomical definition is much less precise. The former will be reviewed first, and Fig. 1, which indicates the known components of the avian visual pathways will be useful in following them.

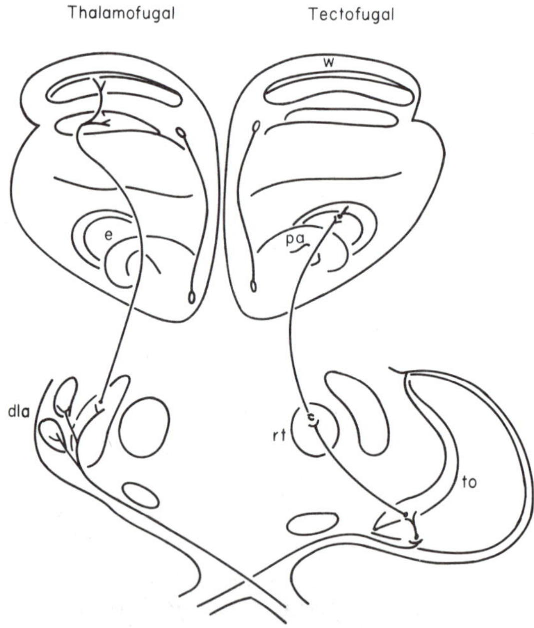

Fig. 1. Schematic representation of visual pathways in the avian brain, illustrated by showing the two principal pathways leading from the retina to the cerebral hemisphere in the owl. All retinal efferents, forming the optic tract cross in the optic chiasm to the opposite side. The right half of the diagram shows the path over the optic tectum (to) and nucleus rotundus thalami (rt) to the ectostriatum (e), a ventral component of the external striatum. On the left, the conduction route over the nucleus dorsolateralis anterior thalami (dla) is shown; this pathway is distributed to various cell layers of the "Wulst" (W), a complex part of the pallial mantle. The drawing is oriented in the frontal plane. pa, paleostriatum augmentatum. [From Nauta & Karten (1970).]

1. Discrimination

a. Visual pathway sites. Despite the obvious anatomical relationship of the optic tectum to visual processes and its easy accessability to surgery, only one study relating it directly to discrimination learning exists. Cohen (1967a) studied the effect of bilateral and unilateral tectal lesions in pigeons on a brightness-discrimination test. The task was a successive discrimination (where responses to S + were reinforced on a VI 60-sec schedule; bright and dim lights were counterbalanced as S + and S −). Animals underwent surgery prior to any training. He found that it took significantly longer for lesioned subjects to learn the task, with no difference between bilateral and unilateral damage. The effect is somewhat impressive in light of the fact that damage was relatively small, representing about 20 % of the stratum opticum layer of the tectum. In all cases animals *were* able to learn the discrimination.

After Karten (1965) determined through anatomical study that there was a major projection of nerve fibers from the tectum to nucleus rotundus of the thalamus, Hodos and Karten (1966) studied the effects of rotundal lesions on brightness and pattern discriminations. They used a simultaneous discrimination procedure; each bird learned four problems, then underwent surgery, then was retrained on the same four problems. The brightness problem was a bright versus dim square, the pattern problems were: (1) a horizontal versus a vertical bar, (2) an apex up versus an apex down triangle, and (3) an apex up versus an apex down triangle, wherein the triangles were composed of a dotted outline as opposed to a solid shape as in (2). They found deficits in the brightness discrimination and *severe deficits* in pattern discriminations. Control animals, which showed good savings on relearning included sham operations and electrolytic lesions in other areas of the thalamus as well as in the midbrain and the forebrain. They noted that animals were able to peck accurately, and ultimately, to discriminate successfully, so that the deficit produced was not a blindness nor possibly not even a sight impairment per se, but some sort of "processing" deficit.

Subsequently, Karten and Hodos (1970) traced anatomical projections in the pigeon from the nucleus rotundus into the ectostriatal area of the forebrain, and then studied the effect of ectostriatal lesions on intensity and pattern discrimination problems as in the previous study, using simultaneous stimulus presentation and a modified discrete trial technique. On each trial, the stimuli remained projected until the animal had either pecked seven times at S +, in which case reward followed, or four times at S −, in which case the stimuli went off and a period of 5 sec of darkness followed.

The animals were trained on all four problems concurrently during the preoperative period, and were operated upon when they had achieved a 90 % correct response criterion over 3 days on any three out of four problems. After ectostriatal lesions, deficits occurred in retention and relearning of the brightness discriminations and severe deficits appeared in the pattern discriminations. The deficits also appeared to be more profound for the up–down triangle discrimination (where four animals showed some learning but failed to reach criterion postoperatively) than for the horizontal versus vertical line discrimination, where all animals did eventually relearn to a 90 % correct level. Hodos and Karten also found a high correlation between the extent of the ectostriatal lesion actually produced and the extent of the deficit as represented by a negative savings score (comparison of trials to learn originally with trials to learn postoperatively where the latter is greater). As they note, the pattern of results in this experiment was very similar to that found after rotundal lesions, both as to the occurrence of the deficit and as to the ultimate "recovery of function."

Hodos (1969) investigated *color* discrimination after electrolytic lesions of the nucleus rotundus in pigeons. He also utilized operated control animals with lesions in the telencephalon or other areas of the diencephalon. The discrimination procedure used successive presentation of stimuli but with a multiple key procedure. First a colored circle appeared on a center key; after ten pecks at it, the center key light went out and two side keys were lit up. When the center key was green, a peck on the left key was reinforced with a 3–5 sec access to grain; if the center key was red, yellow or blue, the bird had to peck at the right side key in order to be rewarded. (The hues were matched for apparent brightness on the basis of data from another group of pigeons.) The animals were thus trained to discriminate green from red, from yellow, and from blue. Surgery was scheduled for each bird after it reached a criterion of 90 % choices on all problems for two successive days. Retention after surgery was excellent for birds with control lesions. Animals with bilateral damage to the nucleus rotundus showed a retention loss, returning to chance levels postoperatively, *but* they relearned rapidly, at the same rate as they had preoperatively. Thus, the effect of rotundal damage was less marked on these color discriminations than it was on pattern and intensity discrimination. Hodos notes that this may be related to the fact that the color discrimination was "easier" (learned more rapidly preoperatively); another possibility, of course, is that this result is due to differences in the systems which process color information.

An additional finding of particular interest in this study was the performance of one bird who had extensive rotundal damage and extensive bilateral damage in the dorsal area of the thalamus in the region of the thalamic input from the retina. Thus, the animal had lost the thalamic components of both the tectofugal and thalamofugal visual systems. This animal not only showed a return to chance level postoperatively, but showed a large deficit in relearning the color discriminations, although it did eventually reach criterion. Comparison with the rotundal group suggests that the relearning deficit is due to the dorsal thalamic damage whereas the initial "forgetting" is a function of the destruction of nucleus rotundus. (Note the implication that the former effect would not show up with the train–operate–test method if the nucleus rotundus were left intact.) The fact that relearning was ultimately successful despite apparent destruction of all thalamic visual input is most intriguing.

Two more recent studies have begun to round out our picture by shedding considerable light on the functional role of the thalamo-fugal visual pathway. Hodos, Karten, and Bonbright (1973) studied the effect of lesions of the nucleus opticus principalis thalamii (OPT) in the dorsal thalamus, and of the area to which it projects, nucleus intercalatus hyperstriati accessorii (IHA) within the hyperstriatum accessorium of the wulst, on pattern and intensity discrimination. Utilizing the same basic procedures as in those described above for the tecto-fugal pathway (Hodos & Karten, 1966, 1970), they found essentially no serious postoperative retention or relearning impairment for any of the problems when lesions were confined to OPT or IHA.

Since, however, efferent fibers of nucleus rotundus that project to ectostriatum pass through the OPT complex, many pigeons in this study sustained significant retrograde degeneration of nucleus rotundus as well as destruction of OPT. These showed varying degrees of postoperative deficit according to the extent of rotundal damage. Two animals had extensive bilateral destruction of OPT *and* of nucleus rotundus. These animals showed a total postoperative retention deficit on all three pattern problems and on the intensity problem, and also had a very severe postoperative relearning deficit on every problem, eventually attaining criterion after more than 20 times the amount of training necessary to learn preoperatively. Thus, lesions in the thalamo-fugal pathway, by themselves, have little effect as compared with tecto-fugal lesions in rotundus or ectostriatum. However, combined rotundal and OPT lesions lead to a much greater deficit than for rotundal lesions alone. Again, however, relearning did eventually occur. As Hodos, Karten, and Bonbright suggest, "Since no other visual inputs to the telencephalon are known, we must conclude that brain stem structures are sufficient to maintain visual discrimination, given adequate retraining procedure [p. 463]." We might add an additional note of conjecture, based on performances of the animal with combined OPT and rotundal damage in Hodos's (1969) color discrimination study, that color information processing is carried out more efficiently at brainstem levels than is pattern processing.

The results of the preceding study led Hodos (1971) to hypothesize that functions of the thalamo-fugal pathway might better be revealed by utilizing discrimination procedures which are more demanding and allow one to "test the limits" of the animal's abilities. Thus, he obtained preoperative thresholds for pigeons on a light-intensity discrimination. The training procedure utilized a three-key chamber. A white light was projected from a slide projector upon the center key; when it was unfiltered (full intensity) the pigeon was required to peck at the right side-key for reward; if it was dimmed by insertion of a neutral density filter in the projector a left-key peck was required. After training the animal on an "easy" problem utilizing a .8 log unit filter, thresholds were obtained by utilizing filter densities of .45, .30, .20, .10, and .05. After thresholds had stabilized over a 5-day period, lesions were made in the OPT, nucleus rotundus, or both.

The results were highly confirmatory of Hodos's hypothesis and of prior studies. Animals with OPT lesions showed an immediate small elevation of threshold (approximately .1 log units) postoperatively, but this elevation was apparently *permanent*—threshold values were still elevated above preoperative levels after up to 30 postoperative sessions. Pigeons with rotundal lesions showed a larger postoperative threshold elevation, but eventually returned to preoperative levels. (This mirrors their performance on pattern problems—initial deficit followed by relearning.) Animals with combined OPT and nucleus rotundus lesions showed combined effects: an immediate large threshold elevation (an inability to perform correctly with the .8-log unit filter, the "easiest" in the series used) greater than that found for either lesion alone, followed by gradual improvement to a stable threshold level that was .10 to .15 log units above preoperative levels—comparable to that found for OPT lesions without rotundal involvement.

b. Other sensory pathway sites. Avian ascending auditory pathways have also been traced up to the forebrain level (Karten, 1967, 1968). Krasnegor (1970) investigated the effects of electrolytic lesions in the telencephalic site that receives auditory projections, an area in the medial portion of the neostriatum designated as field L. He trained his animals on both an intensity and a frequency discrimination task and obtained preoperative thresholds. His method was to use a two-key chamber and require pecks to the left when the standard auditory stimulus was present and to the right when the variable stimulus (stimulus which differed from the standard) was present. Reinforcement (access to grain) occurred upon the 30th peck on the appropriate key. White noise was used to establish the intensity difference thresholds and pure tones were used in the frequency discrimination tests. Lesions were placed in field L according to the coordinates provided by Karten and Hodos's (1967) pigeon atlas. After 5–7 days for recovery, the birds with field L lesions all showed elevated thresholds on both problems postoperatively. With continued postoperative testing, preoperative thresholds for intensity difference were regained for all of the animals. On the frequency problem, two of the animals stabilized with a moderate degree of sensory loss. This initial study provides evidence of forebrain involvement in "basic" auditory processes, confirms the functional significance of the anatomical pathway described by Karten, and opens the way for further study not only of auditory discrimination but of differential effects of forebrain lesions on auditory and visual processes.

A great deal of experimental research and theoretical deliberation has been undertaken with respect to the widespread role in learning and motivational processes of what were originally conceived of as olfactory areas of the mammalian brain (Brady, 1958; Olds, 1958; Adey & Tokizane, 1967). Wenzel and Salzman (1968) explored the possibility that the olfactory system in birds might play a role in nonolfactory learning. They studied the effects of bilateral section of the olfactory nerve and of bilateral destruction of olfactory bulbs on the acquisition of color discrimination. A sham operated group and a group with bilateral hyperstriate damage, roughly equivalent in size to the olfactory bulbs, were used for comparison.

In this experiment *all* training of the animals, including adaptation to the experimental chamber, learning to eat from the grain hopper, etc., took place after surgery. The discrimination procedure utilized two keys; animals were first trained to peck at the left key when color "A" was projected on both keys and then trained to peck at the right key when color "B" was projected. After these were established independently, A and B were presented randomly at variable intervals of 20–65 sec, and the birds had to respond appropriately to the left or right as before in order to be reinforced. Wenzel and Salzman found *no effect* on the acquisition of accurate performance during the discrimination phase, but the olfactory damaged group showed a marked deficit on acquisition of eating from the hopper, of initial key peck shaping, and of initial learning to peck at the second key. Hyperstriate-lesioned birds did not show these deficits. The nature of this involvement of the olfactory structures in the acquisition of eating in a strange place and in acquiring a response requiring an instrumental act and reinforcement is broadly consistent with the kind of effects found in various areas of the mammalian "limbic" system.

c. Wulst and hyperstriatum. We now turn to studies involving the Wulst and other hyperstriatal regions. In assessing these it is important to recognize that attempts to ablate large portions of any one segment will tend to invade others and that the precise pattern of tissue destroyed may vary considerably between different studies reporting the effects of Wulst or hyperstriate ablation. Figure 2 illustrates the major hyperstriatal regions and their relationships to one another; it should be used as a guide for the studies reviewed in this section.

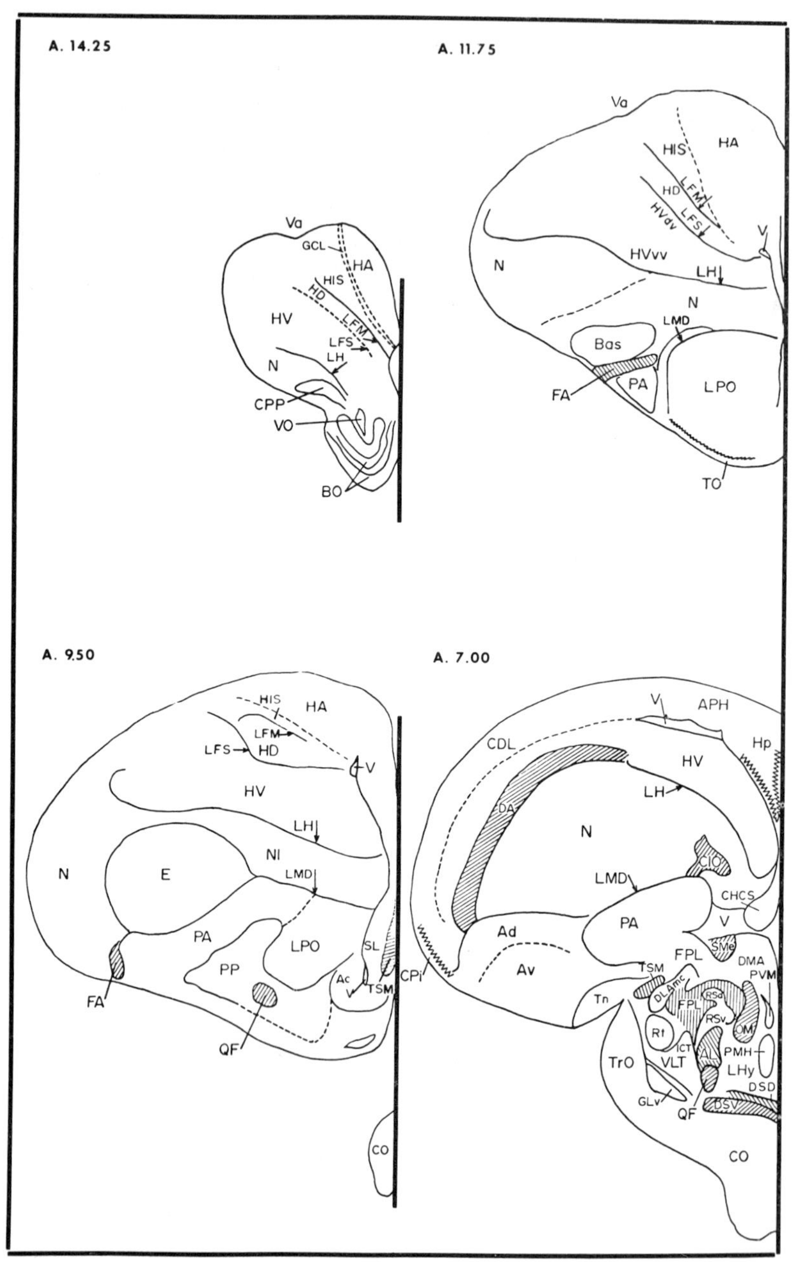

A. 14.25
Va
GCL
HA
HIS
HD
HV
LFM
LFS
LH
N
CPP
VO
BO
A. 11.75
Va
HIS
HA
HD
LFM
HVdv
LFS
V
N
HVvv
LH
N
LMD
Bas
PA
LPO
FA
TO
A. 9.50
HIS
HA
LFM
LFS
HD
V
HV
LH
N
E
NI
LMD
PA
LPO
PP
SL
Ac
V
TSM
FA
QF
CO
A. 7.00
V
APH
CDL
Hp
HV
LH
DA
N
CIO
LMD
CHCS
PA
V
Ad
SMe
FPL
DMA
Av
CPi
TSM
PVM
Tn
DLAmc
FPL
RSd
Rt
RSv
OM
ICT
TrO
VLT
AL
PMH
LHy
DSD
GLv
QF
DSV
CO

We now turn to Zeigler's (1963) landmark investigation into the effect of lesions in different areas of the forebrain on visual discrimination learning in pigeons. This study applied modern neurobehavioral methods to the avian brain for the first time, and it was the first neurobehavioral study aimed at the hyperstriatum. Zeigler used a single key discrimination situation, wherein responses during the first 40 sec of stimulus projection had no effect; thereafter a peck at the S + produced reward (10 sec of access to food) whereas a peck at S − simply turned off the stimulus for 20 sec. The problems were a brightness discrimination and a circle versus triangle form discrimination. Zeigler divided his groups so that some animals were subject to surgery prior to any discrimination training ("acquisition group"). Others were operated on after training and were then retrained ("retention group"). Zeigler's design called for one group of animals with lesions restricted to the hyperstriatum (including hyperstriatum ventrale and Wulst) and another group that would have lesions of roughly comparable size in the neostriatum. He expected, however, that there would be involvement of adjacent areas in the second group, which was thus termed the "mixed" group. The behavioral results appeared to separate the birds into three categories. The first was the hyperstriate group, which consistently evidenced deficits in acquisition and relearning of both brightness and pattern problems. The deficit was moderate

---

Fig. 2. Four sections of the pigeon brain drawn from the plates contained in the Karten and Hodos (1967) atlas. The sections range from anterior 14.25 to anterior 7.00 in terms of the stereotaxic coordinates of the atlas. The key areas for the studies discussed in this section are hyperstriatum accessorum (HA), hyperstriatum dorsale (HD), hyperstriatum intercalatus superior (HIS) and hyperstriatum ventrale (HV). HV is further subdivided into hyperstriatum ventrale dorsoventrale (HD dv) and hyperstriatum ventrale ventro-ventrale (HVvv) in the second section. HA, HD, and HIS collectively comprise the Wulst. Note that lesions of the same depth might easily cut across different combinations of these areas according to where they are placed on the anterior–posterior and medial–lateral planes within the Wulst–hyperstriatal complex. The area labeled GCL (Granule cell layer) corresponds to the location of visual projections from the nucleus opticus principalis thalamii (OPT) of the thalamus to the hyperstriatum accessorium (Karten, personal communication). This region is referred to as nucleus intercalatus hyperstriate accessorii (IHA) by Hodos *et al.* (1973)

Additional Abbreviations

AC, nucleus accumbens; Ad, archistriatum, pars dorsalis; APH, area parahippocampus; Av, archistriatum, pars ventralis; Bas, nucleus basalis; BO, bulbus olfactorius; CDL, area corticoidea dorsolateralis; CHCS, tractus cortico-habenularis et cortico-septalis; CIO, capsula interna occipitalis; CO, chiasma opticum; CPi, cortex piriformis; CPP, cortex prepiriformis; E. ectostriatum; FA, tractus fronto-archistriatalis; H, hippocampus; LFM, lamina frontalis suprema; LFS, lamina frontalis superior; H LH, lamina hyperstriatica; LMD, lamina medulla dorsalis; LPO. lobus parolfactorium; N, neostriatum; NI, neostriatum intermedium; PA paleostriatum augmentatum; PP, paleostriatum primitivum; QF, tractus quintofrontalis; SL, nucleus septalis lateralis; SMe stria medullaris; TO, tuberculum olfactorium; TSM. tractus septomesencephalicus; V, ventriculus; Va. vallecula; VO, ventriculus olfactorius.

for brightness discrimination (all animals could learn it after surgery or relearn it after surgery within 12 sessions at the most) but quite profound for the form discrimination, where three out of four birds lesioned prior to learning failed to acquire the discrimination at all within the 30 sessions allotted. Zeigler also notes that the deficit in the "retention" group is essentially a *relearning* deficit (day one performance after surgery does not differ significantly from controls), hence the effects of the lesions appear to be on the acquisition process.

The second group included subjects with mixed lesions whose performance was within the control range. The third group was composed of animals with mixed lesions who *did* show discrimination deficits. By comparing the actual lesion sites in these two groups, Zeigler was able to "partial out" which areas were implicated as responsible for the deficit and which were not. He found that subjects whose lesions involved primarily neostriate, archistriate, or corticoid areas, or the simultaneous involvement of any of these, showed no deficits. When lesions involved the paleostriate region, deficits were found. (It is likely, in the light of the subsequent anatomical and behavioral work of Karten and Hodos, cited earlier, that damage to the rotundal–ectostriatal pathway was responsible for these deficits.)

Bachrach (1963) studied the effect of lesions of the hyperstriatum accessorum (HA) of the Wulst and of hyperstriatum ventrale (HV) on color, intensity and pattern discrimination. The stimuli were red–green, bright-circle–dim-circle, horizontal-stripes–vertical-stripes, and circle–triangle. He utilized a modified simultaneous discrimination procedure wherein both stimuli remained projected until 15 pecks were made to the positive stimulus and a reinforcement was delivered. Animals were trained until they were making 90% of the pecks to S + for two consecutive days. Following this there was a 10-day rest interval and a retest (to serve as comparison data for postoperative performance). This was in turn followed by surgery and a postoperative test after another 10-day interval.

Retention of the color discrimination was unaffected by Wulst lesions and the effect on the intensity problem was either nonexistent or very slight. Retention of the pattern problems *was* affected however. Performance of animals with Wulst lesions started at a level below that of terminal preoperative performance (but still above chance) and took as long or somewhat longer to reach criterion levels as in original acquisition. In the HV group, retention of the color discrimination was again unaffected, but definite retention and moderate relearning deficits were evidenced on both the intensity and pattern problems. No histological verification is available so we cannot be sure as to precisely what areas of the Wulst and hyperstriatal masses were involved in the two groups. However, regardless of the actual extent of the lesions, a differential effect upon color discrimination and pattern discrimination was found in both groups in animals with the same lesions. Bachrach notes that the color-discrimination problem was considerably easier than the others and that the differential effect may be due to problem difficulty; we must note that this same differentiation of effect between color problems and pattern or intensity problems occurs with lesions in thalamic visual centers.

Pritz, Mead and Northcutt (1970) also studied the effect of hyperstriate lesions on visual discrimination learning in pigeons. They used a brightness problem, a color problem (red versus green) and a pattern problem, (+ versus ×). Stimuli were presented successively for 30 sec on each trial: Every fifth response was rewarded during S +, and pecks during S − had no effect. Each animal was assigned to be run on only one of the problems and a very stiff preoperative learning criterion (which essentially required a minimum of 5 days of 90% S + responses) was imposed. Surgery was carried out by aspiration: The intended lesion in the experimental group was total Wulst ablation (HA, HIS, HD), and control subjects were subjected to either sham operations *or* aspiration of neostriate or corticoid tissue roughly equivalent in volume to the Wulst ablation.

Results again indicated no effect of the hyperstriate lesions upon color discrimination, but brightness and pattern discrimination were affected. (No deficits appeared in any of the control groups.) The effect on brightness was statistically significant but relatively small and all animals relearned and ultimately performed at the same high criterion level as the controls. For the pattern discrimination, the effect was greater in terms of differences in number of days to relearn; performance even after relearning criterion was reached continued to show more variability than that of the controls, and one bird entirely failed to relearn. It is of interest to note that the behavioral pattern of the form discrimination deficit was the same as that found in Zeigler's hyperstriate retention group. Controls and lesioned animals started out "even" on relearning, but the controls relearned after a few sessions whereas the lesioned subjects needed many more. With respect to the histological findings, the authors note that ". . . it was extremely difficult to restrict this surgical destruction to the Wulst alone. In all cases, varying amounts of hyperstriatum ventrale were also injured."

Macphail (1971) found also that hyperstriate lesions affected discrimination performance. He utilized a series of electrolytic lesions, producing more uniform and circumscribed neural damage than has typically been the case with aspiration. The lesions were produced in the anteriolateral portion of the hyperstriatal complex and involved primarily the hyperstriatum dorsale of the Wulst and hyperstriatum ventrale; the hyperstriatum accessorium was minimally, if at all, damaged. Macphail studied performance on a successive red–green discrimination, with each stimulus presentation lasting 60 sec and 5 sec between stimulus presentations. Twenty-five stimulus presentations of each color were made each day. After animals had been trained to peck well at both colors (being reinforced initially for all responses, then on a 15 sec variable interval schedule, VI 15, and finally a VI 60), discrimination training was begun by continuing to reinforce responses to S + but no longer reinforcing responses to S – . There was a small but statistically significant effect on performance in terms of responding to S – ; hyperstriate animals made more pecks to it during certain stages of training than did controls. [This is an effect which is (1) both small enough in itself to be overlooked if one tested differently or simply looked at days to reach criterion and (2) of a type that could very well lead to larger differences in discrimination performance with more difficult problems.] In a second experiment with the same animals, Macphail noted no effect on acquisition of a left and right discrimination.

Cohen (1967b) studied the effect of hyperstriate lesions on differential conditioning of respiratory and heart rate responses in pigeons. A classical conditioning procedure was utilized in which he initially paired either a red or green light with shocks. When a consistent acceleration of heart rate or respiration appeared after the onset of the colored light (which preceded the shock by 6.5 sec), conditioning was considered established. Then differentiation (discrimination) training was introduced. Trials were randomly interspersed wherein a red and green light were presented on different trials and shock continued to follow the light which had been used in the initial conditioning, but no shock followed the other light. Control animals showed evidence of response differentiation on all 4 days of training. They essentially responded (with heart rate or respiratory acceleration) in a conditioned manner to both stimuli, but showed a consistently larger response to S + than to S – . Experimental subjects were marked by some evidence of deficits in differentiation for both tasks. The experimental animals' S – performance on heart rate differentiation was *lower* than that of the controls; the failure of differentiation was reflected in the fact that their S + response was not elevated as much as the controls was elevated. For respiratory conditioning the controls had a higher response to S + and a lower one to S – than did the experimentals. It should be noted that all animals continued to respond to both stimuli in a conditioned manner; furthermore, they continued to respond through an extinction period of 100 nonreinforced presentations of the former S + following differentiation. Lesions were made by aspiration and consistently involved extensive bilateral damage to the Wulst area; although the total damage was much slighter, all subjects also sustained some bilateral damage in the hyperstriatum ventrale. Subjects who had unilateral hyperstriate lesions did not show any deficit in differential conditioning.

The results of this study indicate that hyperstriatal areas are in some way involved in control of respiratory and heart-rate conditioning, but, as Cohen notes, it is difficult to tell in this case whether the failure to achieve differentiation was a result of a deficit in processing the sensory information or was due to alteration of control of response characteristics. This is a particularly important point in view of the lack of effect on color discrimination reported in instrumental situations.

In a number of other studies, where lesions were essentially of the Wulst with little or no damage to other hyperstriate regions, no effect on discrimination was obtained. Zeigler (1965a) noted no effect on color or brightness discrimination, and no effect on pattern discrimination when a sufficiently long postoperative recovery time (3 to 6 months) was allowed. Zeigler (1965b) also noted that performance on a single color *oddity* discrimination was not effected by Wulst lesions in pigeons. Stettner and Schultz (1967) also reported no effect on acquisition of a simultaneous horizontal versus vertical line discrimination in Bobwhite quail, after lesions that produced extensive damage to the Wulst, overlying cortical tissue and the "hippocampal and parahippocampal" areas. *Cortical* ablations have also failed to produce any evidence of discrimination deficit in studies by Layman (1936), Zeigler (1963), and Stettner (1966).

2. *Reversal*

a. Ablation studies. There have been a small number of investigations of reversal learning which have thus far been remarkably consistent in reporting deficits in reversal after damage to portions of the Wulst. Zeigler (1965) found deficits in spatial reversal and pattern reversal after Wulst ablations in pigeons. He used a modified simultaneous discrimination procedure wherein both stimuli remained projected for 15 sec, and the next response to S + led to reward, whereas the next one to S − simply led to extinguishing of the stimulus lights for 10 sec. He did not run each reversal to criterion, but reversed his animals *daily*, once initial acquisition was achieved.

Stettner and Schultz (1967) found marked deficits in pattern reversal learning after extensive combined Wulst (all portions) and corticoid ablations in Bobwhite quail. They employed a traditional simultaneous one-peck-per-trial discrimination situation and ran each reversal to criterion. They also found that deficits were the same in animals that had undergone 25 reversals to criterion preoperatively as in those subjected to surgery prior to any training.

Schade-Powers (1969) studied the effects of Wulst lesions on multiple-reversal learning in pigeons. Her training situation utilized a novel "magazine response" procedure that has been shown to produce more rapid discrimination acquisition and clearer evidence of improvement in reversals than standard key-peck techniques. In this procedure a single peck-key was mounted above two feeding apertures each with a grain hopper mounted behind it. A white light on the center key began a trial. A peck at the white light lit up the two food apertures, one with a red light and one with a green light. The bird then exercised its choice by sticking its head inside of the food hopper and tripping of a photo-cell switch. A correct choice operated the hopper mechanism, swinging the food tray into position to provide access to the grain. An incorrect choice simply extinguished the lights until the next trail.

Utilizing this procedure throughout, acquisition and 30 reversals of a red–green discrimination were run preoperatively. Reversals were not run to criterion; there was a fixed period of 2 days per reversal, ten trials per day. The birds showed considerable improvement over the reversal series.

Ablations were made by aspiration with the lesions focused in the anterior portion of the Wulst in one group of animals and in the posterior portion of the others. Histological analysis showed a variable pattern of ablation, with all lesions impinging somewhat upon the main hyperstriatal mass as well as the Wulst. Hyperstriatum accessorium (HA), hyperstriatum intercalatus superior (HIS), hyperstriatum dorsale (HD), and hyperstriatum ventralis superior (HV) were all damaged to varying extents. In analyzing her anatomical findings, Schade-Powers was able to group her animals into two categories: an anterior group which primarily had a lesion of HD and HIS and a posterior group where the greatest damage was to HA and HV. *All* of the operated animals behaved as if naive at the beginning of the reversal series. However, as the reversal series progressed the posterior group began to have more success than those with the anterior lesions. The groups did not differ in the initial portion of each reversal (first five trials of Day 1), but from this point on the *anterior group* made more errors than the posterior animals on each of the later reversals. (This was true even in those cases where the posterior lesions were larger than the anterior ones.) This suggests that the lesions did not produce a difference in retention, but rather a different rate of learning within each reversal.

The variation in location and size of lesions in this study was such that in no sense can it be considered conclusive (one complicating factor for example, is that in all cases large portions of all Wulst segments were spared on at least one side), but the suggestion that HIS and HD ablation is primarily responsible for the reversal deficit is most intriguing and suggests further study. This study clearly points the way toward a systematic investigation of the effects on reversal of bilateral electrolytic lesions restricted to each of the Wulst areas.

Macphail (1971), in Experiment 2 of his study already noted, reported a deficit in spatial reversal learning after electrolytic lesions primarily damaging hyperstriatum dorsale of the Wulst as well as hyperstriatum ventrale. He utilized a standard simultaneous one-peck-per-trial problem, and ran each reversal problem to a criterion of 15 consecutive correct trials (though it only took one session to reach criterion on each reversal for this easy problem). Macphail's results are especially interesting in that a highly significant effect on reversal was obtained with much smaller lesions than previously used, despite the very easy reversal problem. There seems no doubt that reversal learning of a variety of forms is highly susceptible to Wulst lesions. It should also be noted that the ablations, though leading to marked deficits, do not "abolish the ability to reverse."

Again, results from corticoid ablations are negative; Stettner (1966) reported that pattern reversal performance of four quail subjected to extensive corticoid destruction was entirely within the control range.

b. Analysis of deficit. Interest in the reversal-learning deficit has focused on the nature of the deficit, especially when we consider that discrimination performance per se is unaffected in the same animals. Reversal learning in animals has been a subject of extended behavioral investigation and considerable theoretical speculation (e.g., Bitterman, 1965, 1969, Gonzales, Behrend, & Bitterman, 1967; Gossette, 1968; Mackintosh, 1969; Matyniak, Wheeler, & Stettner, 1971). A simplified schematic of one way of conceptualizing the different stages of reversal learning is presented in Fig. 3. The most obvious place to look for the source of reversal deficit is in the stage of perseveration to the previously rewarded stimulus. Indeed, this was the basis of Macphail's (1971) study of responding to S− after hyperstriate lesions—the higher rate of response to S− in lesioned animals supports the hypothesis that these animals do have some difficulty in withholding responses after reinforcement, but the effect was small. Also, Macphail (1969) found in another study that small lesions of the hyperstriatum accessorium actually *increased* rather than decreased tendencies to withhold responding after nonreward. This line of investigation is particularly intriguing (and enticing), of course, because of the extensive mammalian work on response perseveration and reversal deficits produced by limbic lesions primarily in the hippocampal and septal areas. We have also looked at our reversal data in this light, with somewhat surprising results. We have analyzed reversal performances for all our subjects by breaking down total errors made in each reversal (in a discrete trial, simultaneous situation) in terms of errors prior to reaching a 50% level of response over any given ten trial block and error from this point until criterion was reached. Figure 4 tells the story for four control subjects and two experimental subjects with Wulst lesions which showed marked reversal deficits (quail). Although there is some difference in the 0–50% score, indicating greater perseveration, the bulk of the difference is in the 50% = criterion scores.

A further analysis suggests the basis for this. Prior to reaching criterion, animals often respond exclusively to one side of the apparatus regardless of the visual stimulus. We arbitrarily defined any block of ten trials with

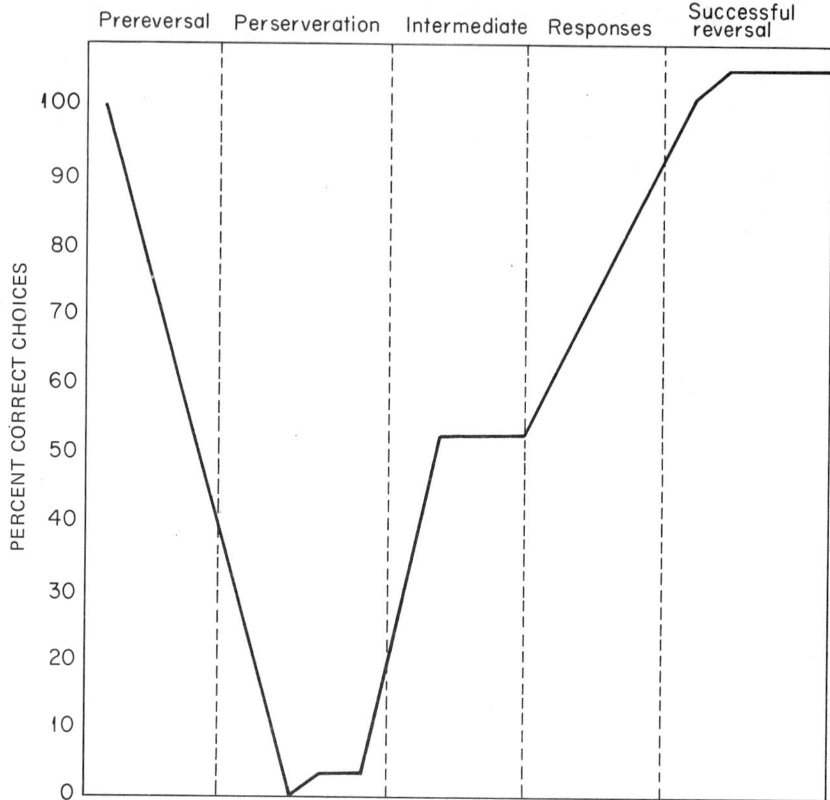

Fig. 3. Hypothetical stages of reversal learning. This representation assumes that any given reversal is run to criterion before a new one is programmed. Thus, a high level of correct responses characterizes the prereversal period. At the beginning of each reversal animals will still be responding to the *previously* correct (now incorrect) stimulus. This is the stage of perseveration of the previously rewarded response. When the subject reaches chance level or above he has passed out of the perseveration stage; he may remain at chance level or somewhere between chance level and criterion level for a varying length of time with a varying pattern of responses—hence the term intermediate stage. If successful in the problem he reaches criterion level on the new response for the final stage. A new reversal introduced at this point would start the cycle all over again. The number of trials occupied by any stage is, of course, variable and in some cases may be too short to detect clearly in any given animal's record.

eight responses or more to one side (remember that S+ and S− are randomly switched from side to side) as a "biased block." A biased block analysis for the same animals shown in Fig. 4 is shown in Fig. 5. It is clear that biased blocks account for the bulk of the reversal deficit and that the biased blocks

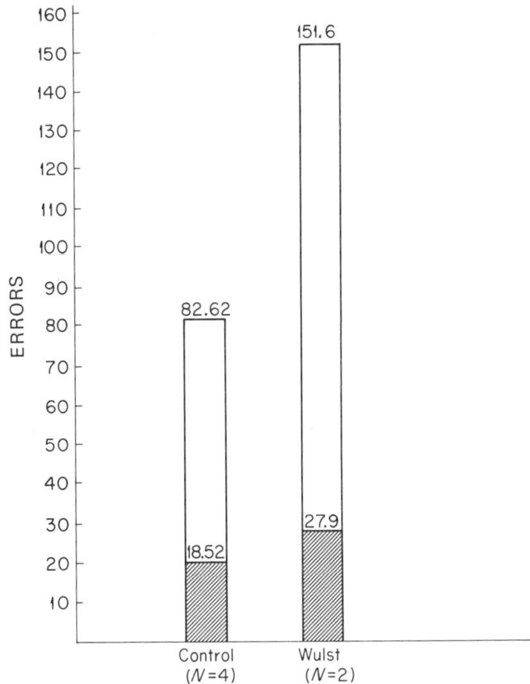

Fig. 4. Performance of control and Wulst-lesioned Bobwhite quail on successive reversals of a horizontal versus vertical stripes discrimination. Errors are broken down into those made during the perseveration stage of each reversal (0 to 50 % correct, shaded bars) and those made during the intermediate stage (50 % to criterion, open bars).

tendency develops in the reversal situation. It is our belief that the best framework for analyzing this effect is provided by the two-stage theory of discrimination learning developed over the past decade by Sutherland and Mackintosh (1971). Detailed discussion of this analysis is unfortunately beyond the scope of this chapter, but it essentially suggests that learning "what to attend to" is a critical stage in discrimination learning, particularly in visual reversal learning. Following from the above, we can suggest the working hypothesis that the Wulst-ablated animals in our study had great difficulty in paying attention to the relevant stimulus in the initial stages of each reversal. This deduction from the pattern of reversal errors is consonant with the Hodos *et al.* (1973) analysis of the function of the hyperstriatal component of the avian thalamo-fugal system. Noting that Cohen and Pitts (1967) reported that stimulation of the hyperstriatum in awake pigeons resulted in specific orienting head movements and that Cohen (1967b) noted

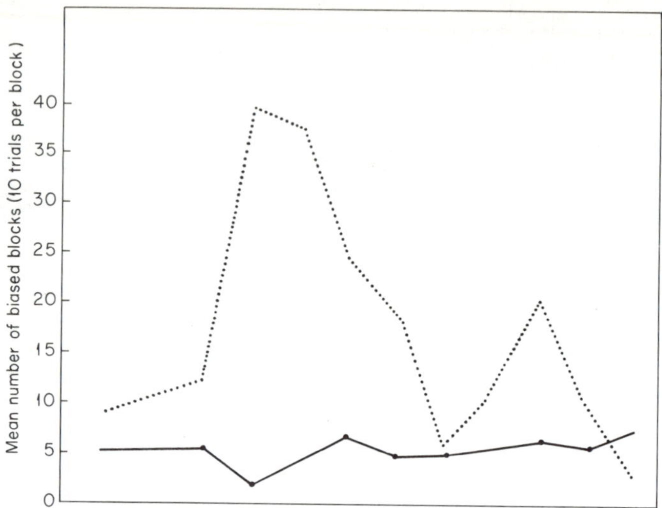

Fig. 5. Number of "position-biased" blocks during successive discrimination reversal of a horizontal versus vertical stripes problem. (——), Control, $N = 4$, (···), Wulst-lesioned. N-2. (See text for further explanation).

a diminution in visual orienting responses[5] after hyperstriatal lesions, they hypothesize that ". . . the findings of Stettner and Schultz of deficits in reversal learning may be the result of an impairment of their subjects ability to shift their attention from one aspect of the stimulus configuration to another [p. 464]." This similarity of interpretation stemming from different kinds of evidence certainly suggests that the selective attention interpretation has potentially a high degree of resolving power in interpreting Wulst deficits and should be carried further and tested in new situations.

B. Interocular Transfer Studies

An approach to the study of the organization of brain function that has proven quite fruitful in mammalian work is to try to identify structures and pathways that are critical for transfer of information from one hemisphere

[5] It should also be noted that Zeigler and Cohen (1965) found a deficit in neck-stretch response and associated movements which lasted several months in pigeons with hyperstriatal lesions. (These neck-stretch responses were classified as "responses to novelty" on the basis of observation by the experimenter). One might predict that allowing a long post-operative recovery period and waiting for these responses to return to normal levels could obliterate (or at least reduce) the reversal learning deficits.

of the brain to the other (Sperry, 1961; Meyers, 1962). The avian brain is particularly intriguing in this respect, since anatomical and electrophysiological evidence suggests that visual input is initially totally segregated in different halves of the brain; there appears to be complete decussation of the optic chiasm in the bird, so that all projections from a given retina go to the tectum or the thalamus on the opposite side. Consistent in a general way with this anatomical picture are reports suggesting that the completely normal, neurologically intact, pigeon has some difficulty with certain types of interocular transfer (Levin, 1945; Mello, 1965). The obvious first step in this area is to try to identify the pathways critical for visual information transfer by interrupting different commisural systems. The ability to transfer is then assessed by training the animals to make a discrimination with only one eye exposed, and testing for the presence of learned discrimination with only the untrained eye exposed.

The major commissural pathway studied in mammals has been the corpus callosum, a structure not present in birds. The bird has a number of commissural systems, however, that could serve as transmitters of visual information. Cuenod and Zeier (1967) studied the effect of cutting two different commissural systems on the interocular transfer of a simultaneous, two key, discrimination in pigeons. They sectioned either the tectal and posterior commissures (adjacent to one another in the region of the dorsal posterior portion of the tectum) or the anterior commissure (located at the base of the telencephalon). All animals were trained on two discriminations in succession with different eyes so that two transfer tests could be carried out for each subject. For example, an animal might be trained with the left eye exposed to discriminate blue from green, then, after a transfer test on blue–green with the right eye, it would be trained with that eye on a red–yellow discrimination and tested for transfer back to the left eye. They found that all their birds were capable of positive interocular transfer after complete section of the anterior commissure, although the transfer was less than "perfect," in that most birds made some errors with the untrained eye. With the tectal and posterior commissures, their histology revealed a fairly complex picture of different degrees of destruction, from total sectioning of the commissures to leaving a good part of them intact. All of the birds who had part of the commissures intact showed evidence of positive transfer, though it was less than perfect in five of the eight animals in this group and the transfer scores appeared related to the degree of commissural destruction produced. It is also important to note that transfer scores on the second discrimination were much higher for these animals than on the first. In fact, seven of the eight birds in this group showed "perfect" transfer on the second discrimination. Thus, at this point one must consider the deficits in the group to be temporary in that full recovery of transfer function appears to take place rapidly with training.

Of the three animals with complete section of posterior and tectal commissures, two showed no evidence of transfer on either discrimination and one showed perfect transfer on both. This puzzling pattern is similar to results found in an unpublished study on pigeons by Myers and Stettner, who used a pattern discrimination problem (vertical versus horizontal stripe) and a go, no-go training procedure. They also found severe transfer deficits in some animals with total tectal and posterior commissural section, but also had animals that showed excellent transfer.

In the light of these results we can offer only the tentative conclusion that neither the tectal-posterior commissure nor the anterior commissure are critical for the occurrence of transfer but both are to some extent involved in the transfer of visual information. In judging how extensively they are involved, we must keep in mind the fact that normal animals occasionally have difficulty with interocular transfer and that we cannot necessarily assume that a population of normals will all show perfect transfer on their initial transfer test.

In a more extensive study, Meier (1971) studied transfer of color and either up–down ($\vee\wedge$, $\cup\cap$) or left-right ($><$, $\supset\subset$) mirror-image pattern discrimination using performance of normal and sham-operated birds as a base line. His experimental animals had either the tectal and posterior commissures, the supraoptic decussation (DSO) or all three commissures sectioned. He found essentially little or no deficit for any of the problems after the commissures in the first case were sectioned but severe impairment for both the color or pattern problems (in most cases, there was virtually no transfer) after the DSO was cut. A particularly interesting result was obtained with respect to transfer of the right and left mirror-image patterns. Mello (1965) had previously reported that normal animals show on interocular transfer their maximum rate of pecking on a generalization test to the mirror-image of an oblique line that has been reinforced during training. (That is, animals who had been trained to peck with their left eye at a 45° line, ⊘, showed maximum pecking response to a 135° line, ⦸, when tested with the right eye.) In line with these results, Meier found that his control animals took considerably longer to reach the discrimination criterion on the left–right mirror-image problems with the second eye than they had in original monocular learning. That is, there was a marked negative transfer effect, suggesting that what transferred was a tendency to peck at the mirror image of S+, which was in this case the S−. On the other hand, animals with the DSO sectioned showed superior performance (compared to controls) on these problems with the second eye, with their data interpreted as showing *no* transfer, rather than negative transfer.

Thus, the results seem quite clear that the DSO is a primary vehicle of interocular transfer of visual information; if we were to identify a "split-brain" bird equivalent to the callosum sectioned mammal it would be one with the DSO severed. Meier notes that the DSO seems to have connections with both avian visual systems. Tectal fibers have been traced anatomically in the ventral portion of the DSO, and projections through the DSO are also indicated from the dorsal area of the thalamus. Meier essentially suggests that there is a path from hyperstriatum to the dorsal thalamic area, to the thalamus on the other side through the DSO and then upon the opposite hyperstriatum. Thus, the DSO serves in a sense as a forebrain commissure but at the thalamic level.

Meier, Maier, and Cuenod (1971), in a very intriguing experiment, have obtained evidence that further supports these ideas about the functional significance of the DSO. They taught pigeons on a simultaneous pattern discrimination with only one eye exposed (monocular training) at a time. Thus, each eye learned the problem "independently." They then made unilateral lesions of the hyperstriatum accessorium (HA) and in some cases also sectioned the DSO. Retention of the previous discrimination was not affected in any case. Monocular learning of a *new* discrimination was not affected for *either* eye in animals with unilateral HA ablation that had the DSO intact. However, in animals that had the DSO sectioned in addition to the unilateral HA lesion, there was a learning deficit with the eye contralateral to the hemisphere where the lesion was located. This result not only supports the notion that the DSO is a critical pathway in linking the hyperstriatum to visual input from the ipsilateral eye, it allows for the possibility of an elegant split brain preparation in which one can assess the role of various areas of the hyperstriatal mass in visual learning and retention. After section of the DSO, each subject can be used as its own control by training and testing both eyes independently in conjunction with selective unilateral forebrain lesions.

A different, somewhat more complex, question about interhemispheric transfer of discrimination learning was asked by Mello (1968) in an experiment involving unilateral tectal lesions. She attempted to determine whether the memory trace of a monocularly learned discrimination was formed unilaterally or bilaterally. To illustrate by a specific example: When an animal successfully transfers a monocularly learned discrimination from the right eye to the left eye, is it because the information has been passed over to the opposite tectum and "recorded" there during training (bilateral storage) or is it because the tectum on one side sends the information to the opposite one during *testing* of the other eye (unilateral storage)? Mammalian evidence from the chiasm-sectioned cat suggests that the storage is unilateral, because destroying visual association cortex in the *trained* hemisphere *after* training but before testing for transfer leads to absence of successful interocular transfer. If storage were bilateral, the "message" would have already been transmitted to the nontrained hemisphere prior to the lesions and there would be interocular transfer. Thus, Mello trained pigeons monocularly to make two color and two pattern discriminations using a single key procedure. Then (after a retention test with the trained eye for comparison purposes), she removed virtually all the tectum on the trained side (i.e., the side contralateral to the trained eye), and tested for transfer. The birds were, at this point, of course, unable to function with the originally trained eye (as expected); Mello reports that they were "blind or near-blind" in the eye contralateral to the damaged tectum. However, the untrained eye still had "its" tectum (the contralateral one) intact and so the bird could be tested with this eye to see if the discrimination had transferred. The birds did, in fact, transfer both color and pattern discrimination problems successfully. The result is clear cut in demonstrating that successful transfer can occur without the presence of the tectum on the trained side at the time of transfer. Mello considered two possibilities in the face of this result; either (1) the memory trace for discrimination is formed bilaterally at the time of training, or (2) it is formed unilaterally at some level above the tectum and sent over to the intact tectum during the transfer test. She tended to favor the former interpretation. However, in the light of the findings that it is the DSO and not the tectal–posterior commissural system that is most critically involved in transfer of visual discrimination, the possibility of a unilateral trace being formed at thalamic or forebrain levels must be considered very strongly. Experiments utilizing Mello's design but involving destruction of the nucleus rotundus and ectostriatum on the trained side would appear to be the next logical (and most fascinating) step.

Now that some of the critical structures have been identified, transfer studies promise to tell us much about the neural basis of avian discrimination learning. Perhaps the most striking aspect of the findings thus far is the lack of involvement of tectal structures in the transfer process, despite their anatomical prominence. These findings are, in fact, consistent with lesion studies showing the importance of nucleus rotundus, ectostriatum, and hyperstriatum in visual learning. The transfer findings are further evidence in opposition to the idea that the midbrain is a major visual *learning* center in avians and that telencephalization of these functions is an essentially mammalian "invention."

C. Comparative Studies

One apparently very straightforward way of investigating the neural basis of learning is to look at the neural apparatus of animals whose learning abilities differ. This approach, however, which has been pursued in various forms with varying degrees of vigor since the early years of this century, is admittedly beset with difficulties of methodology and interpretation. The essence of the problem is that species differences in performance on learning tasks (and differences in brain structure, for that matter) may reflect differences in the particular sensory–motor and motivational adaptations of the species rather than differences in learning per se. Discussion of these difficulties will not be necessary here if we simply consider that initial results showing correlations between differential learning abilities and differential brain structures are to be taken as first approximations to be explored further with a variety of techniques.

There has been one set of intriguing correlations, suggested primarily by the work of Gossette and his co-workers who have followed Bitterman's (1965) lead and tested a large number of species on reversal learning. In Gossette's initial comparative avian study, (Gossette, Gossette, & Riddell, 1966) multiple reversals of a position discrimination were studied in four different avian species: the domestic chicken, the bobwhite quail, the yellow head parrot, and the red-billed blue magpie. The first two species belong to the same family (Phasianidae), while the parrots and magpies belong to different orders (Psittaciformes and Passeriformes, respectively). Performance in order of superiority (least number of errors over a series of 29 reversals to criterion), beginning with the best, was shown by magpies, then parrots, followed by quail and chickens, with a *sharp break* between the parrots and the quail, as can be seen in Fig. 6. Essentially, magpies and parrots had very similar curves and so did quail and chickens, with a lot of daylight in between.

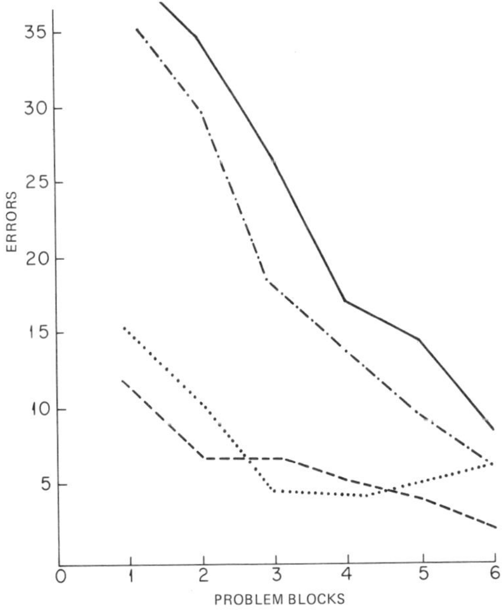

Fig. 6. Mean number of errors per problem for each bird group across six problem blocks. Each block value is the mean error per problem across five problems. Key: (—) chicken; (–·–·), quail; (· · ·), parrot; (–––), magpie. The reversals were of a left-right position discrimination. [From Gossette, Gossette, & Riddel (1966).]

Strong corroboration of marked differences in successive discrimination reversal learning between avian species was obtained in a subsequent study by Gossette (1967) that utilized 29 reversals of a brightness (black versus white) problem rather than a spatial one. In this study the subjects were magpies, mynas, parrots, and chickens, listed in order from poorest to best performance, as can be seen in Table I. Thus, the same order of performance among chicken, parrot, and magpies was found as in the prior study which had used spatial reversals, with the myna falling just behind its closest relative (of the three) in performance. However, the magnitude of the differences between species, most strikingly between the parrots and magpies, was considerably greater than in the spatial reversal study. Gossette suggests that the more difficult brightness task served to magnify the differences that had shown up on the spatial test and that, in general, the more closely related species are, the more difficult the task should be in order to reveal differences.

Although the preceding behavioral results represent only a beginning of discrimination reversal comparison among avian species, they are particularly interesting in the light of comparative brain features. Portman and Stingelin (1960) found marked variability in forebrain development between species of different avian orders. They devised a "telencephalization" index to reflect these differences, a complex measure that essentially relates the weight of the cerebral hemispheres to the weight of the brain stem. In line with Gossette's results, the quail and the chicken are among the lowest of the avian species in telencephalization (as are Galliformes, generally), whereas species of the Psittaciforme and Passeriforme orders show the highest index.

If we now look in more detail at the structure of the hemispheres of different avian species we find great variability in the "Wulst" development as perhaps the most conspicuous feature, and this difference probably accounts for a good portion of the differences in forebrain development. The Wulst not only differs in size in different avian species, but also in position. Adamo (1965) for example, described the Wulst of three species representing different forms of development: the chicken with a relatively underdeveloped Wulst; the raven in which the Wulst is very highly developed and differentiated in the rostral portion of the hemisphere; and the African

Table I. *Mean Error and Range of Errors for Four Avian Species*

	Magpie	Myna	Parrot	Chicken
Mean Error	229.2	316.7	443.0	662.3
Range	172–271	246–361	440–446	510–822

lovebird (a Psittaciforme) in which the Wulst is highly developed but in the mediocaudal portion of the hemispheres.[6] In line with these differences, she also found differences in the pattern of descending fibers revealed by degenerating fibers after Wulst lesions.

> Efferent projections from the lesioned areas utilize the septomesencephalic tract and lateral forebrain bundle although the route primarily utilized in each species of bird varies. Chickens use both tracts extensively, in ravens the lateral forebrain bundle predominates, and in African lovebird primarily the septomesencephalic tract is utilized [Adamo, 1965, p. 88].

The (admittedly incomplete) information that exists certainly seems part of a consistent thread. Avian species vary widely in their reversal learning performance; they also vary considerably in their forebrain size and structure, particularly with respect to Wulst development. The species showing superior reversal are those that have a more extensively developed Wulst. (We have already noted that Wulst lesions in quail and pigeons lead to a reversal learning deficit.) It is reasonable to conclude that Wulst cells participate heavily in information processing that is important to reversal learning, and that a more extensively developed Wulst enables the organism to carry out these processes more effectively. In no way, of course, should this be interpreted to mean that the Wulst is a "reversal learning" center. The Wulst undoubtedly is involved in performing a variety of functions, and even those processes that play a role in reversal learning most certainly will also be reflected in any number of other behavioral situations which should be revealed by further research.[7]

[6] Phillips (1968) noted an intriguing fact about the role of the Psittaciforme forebrain. He found that large lesions in the hyperstriate, neostriate or archistriate regions of the brain of the peach-faced lovebird (*Agopornis roseiocolis*) had extremely debilitating effects on their functioning. They showed muscular weakness, especially of the legs, had to be force-fed for 7 to 10 days, and so many of them died after bilateral operations that Phillips ended up doing his lesions in two stages, one hemisphere at a time, with a 7–14 day recovery period in between. This result is truly astonishing as compared with the case in pigeons or gallinaceous birds where muscular weakness is not noted and rapid recovery from extensive bilateral forebrain lesions is the rule. This difference should reinforce the point, noted by Phillips, that marked differences exist in avian forebrain development and that we should be perhaps no more surprised at this result than we would be to find greater debilitation caused by cortical ablation in a monkey or dog as compared to a rat.

[7] In line with this point, Krushinsky (1965) noted a correlation between Wulst development and performance of different avian species on his "insight" detour problem.

References

Adamo, N. J. *Structure and function of the avian wulst.* Unpublished doctoral dissertation, University of Florida, 1965.

Adey, W. R. & Tokizane, T. (Eds.), *Progress in brain research: Volume 27: Structure and function of the limbic system*, New York: Elsevier, 1967.

Allen, W. F. Results of prefrontal lobectomy on acquired and on acquiring correct conditioned differential responses with auditory, general cutaneous, and optic stimuli. *American Journal of Physiology*, 1943, **139**, 525–531.

Allen, W. F. Effect of destroying three localized cerebral cortical areas for sound on correct conditioned differential responses of the dog's foreleg. *American Journal of Physiology*, 1945, **144**, 415–528.

Bachrach, H. *Effects of end brain lesions upon visual discrimination learning in pigeons.* Unpublished master's thesis, City College of City University of New York, 1963.

Bitterman, M. E. Techniques for the study of learning in animals: Analysis and classification. *Psychological Bulletin*, 1962, **59**, 81–93.

Bitterman, M. E. Phyletic differences in learning. *American Psychologist*, 1965, **20**, 396–410.

Bitterman, M. E. Habit reversal and probability learning: Rats, birds and fish. In R. Gilbert & N. S. Sutherland (Eds.), *Animal discrimination learning*. New York: Academic Press, 1969, 163–175.

Blough, D. S. The study of animal sensory processes by operant methods. In W. K. Honig (Ed.), *Operant behavior, areas of research and application.* New York: Appleton, 1966.

Boguch, S. *The biochemistry of memory: With an inquiry into the function of the brain mucoids*, New York: Oxford Univ. Press, 1968.

Boguch, S. Glycoproteins of the brain of the training pigeon. In A. Lajtha (Ed.). *Protein metabolism of the nervous system.* New York: Plenum, 1970.

Brady, J. V. The paleocortex and behavioral motivation. In H. F. Harlow and C. N. Woolsey (Eds.), *Biological and biochemical bases of behavior*, Madison, Wisconsin: Univ. of Wisconsin Press, 1958.

Butler, R. A., Diamond, I. T., & Neff, W. D. Role of auditory cortex in discrimination of changes in frequency. *Journal of Neurophysiology*, 1957, **20**, 108–120.

Cohen, D. H. Visual intensity discrimination in pigeons following unilateral and bilateral tectal lesions. *Journal of Comparative and Physiological Psychology*, 1967, **63**, 172–174. (a)

Cohen, D. H. The hyperstriatal region of the avian forebrain: A lesion study of possible functions, including its role in cardiac and respiratory conditioning. *Journal of Comparative Neurology*, 1967, **131**, 559–570. (b)

Cohen, D. H. Development of vertebrate experimental model for cellular neurophysiologic studies of learning. *Conditional Reflex*, 1969, **4**, 61–80.

Cohen, D. & Durkovic, R. G. Cardiac and respiratory conditioning differentiation and extinction in the pigeon. *Journal of the Experimental Analysis of Behavior*, 1966, **9**, 681–688.

Cohen, D. H. & Pitts, L. H. The hyperstriatal region of the avian forebrain: Somatic and autonomic responses to electrical stimulation. *Journal of Comparative Neurology*, 1967, **131**, 323–336.

Cuenod, M. & Zeier, H. Transfer visuel interhemispherique et commisurotomie chez le pigeon. *Archives Suisses de Neurologie, Neurochirurgie et de Psychiatrie*, 1967, **100**, 365–380.

Elliot, D. H. & Trahiotis, C. Cortical lesions and auditory discrimination. *Psychological Bulletin*, 1972, **77**, 198–222.

Erulkar, S. D. Tactile and auditory areas in the brain of the pigeon. An experimental study by means of evoked potentials. *Journal of Comparative Neurology*, 1955, **103**, 421–427.

Goldberg, J. M. & Neff, W. D. Frequency discrimination after bilateral ablation of cortical auditory areas. *Journal of Neurophysiology*, 1961, **24**, 119–128.

Gonzalez, R. C., Behrend, E. R., & Bitterman, M. E. Reversal learning and forgetting in bird and fish. *Science*, 1967, **158**, 519–521.

Gossette R. L. Successive discrimination reversal (SDR) performance of four avian species on a brightness discrimination task. *Psychonomic Science*, 1967, **8**, 17–18.

Gossette, R. L. An examination of the retention decrement explanation of comparative successive discrimination reversal learning by birds and mammals. *Perceptual and Motor Skills*, 1968, **27**, 1147–1152.

Gossette, R. L., Gossette, M. F., & Riddell, W. Comparisons of successive discrimination reversal performances among closely and remotely related avian species. *Animal Behaviour*, 1966, **14**, 560–564.

Heise, G. A. Auditory thresholds in the pigeon. *American Journal of Psychology*, 1953, **66**, 1–19.

Hilgard, E. R. & Bower, G. H. *Theories of learning*. New York: Appleton, 1966.

Hodos, W. Color discrimination deficits after lesions of nucleus rotundus in pigeons. *Brain, Behavior and Evolution*, 1969, **2**, 185–200.

Hodos, W. "The Evolution of the Visual System." Invited fellow address, division of comparative and physiological psychology, American Psychological Association Convention, 1971, Washington, D.C.

Hodos, W. & Karten, H. J. Brightness and pattern discrimination deficits in the pigeon after lesions of nucleus rotundus. *Experimental Brain Research*, 1966, **2**, 151–167.

Hodos, W. & Karten, H. J. Visual intensity and pattern discrimination deficits after lesions of ectostriatum in pigeons. *Journal of Comparative Neurology*, 1970, **140**, 53–68.

Hodos, W. Karten, H. J., & Bonbright, J. C. Jr. Visual intensity and pattern discrimination after lesions of the thalamofugal pathway in pigeons. *Journal of Comparative Neurology*, 1973, **148**, 447.

Jenkins, H. M. & Harrison, R. H. Effect of discrimination training on auditory generalization. *Journal of Experimental Psychology*, 1960, **59**, 246–253.

Karten, H. J. Projections of the optic tectum of the pigeon (*Columba livia*). *Anatomical Record*, 1965, **148**, 297–298 (abstract).

Karten, H. J. The organization of the ascending auditory pathway in the pigeon (*Columba livia*). I. Diencephalic projections of the inferior colliculus (cucleus mesencephali lateralis, pars dorsalis). *Brain Research*, 1967, **6**, 409–427.

Karten, H. J. The ascending auditory pathway in the pigeon (*Columba livia*). II. Telencephalic projections of the nucleus ovoidalis thalami. *Brain Research*, 1968, **11**, 134–153.

Karten, H. J & Hodos, W. *A stereotaxic atlas of the brain of the pigeon*. (*Columba livia*). Baltimore, Maryland: Johns Hopkins Press, 1967.

Karten, H. J. & Hodos, W. Telencephalic projections of the nucleus rotundus in the pigeon (*Columba livia*). *Journal of Comparative Neurology*, 1970, **140**, 351–52.

Karten, H. J. & Nauta, W. J. H. Organization of the retinothalamic projections in the pigeon and owl. *Anatomical Record*, 1968, **160**, 373 (abstract).

Karten, H. J. & Revzin, A. M. The afferent connections of the nucleus rotundus in the pigeon. *Brain Research*, 1966, **2**, 368–377.

Krasnegor, N. A. The effects of telencephalic lesions on auditory discrimination in pigeons. Unpublished doctoral dissertation, University of Maryland, 1970.

Krushinsky, L. V. Solution of elementary logical problems by animals on the basis of extrapolation. *Progress in Brain Research*, 1965, **17**, 200–228.

Levin, J. Studies in the interrelations of central nervous structures in binocular vision: I. The lack of bilateral transfer of visual discrimination habits acquired monocularly by the pigeon. *Journal of Genetic Psychology*, 1945, **67**, 105–130.

Layman, J. D. The avian visual system: I. Cerebral functions of the domestic fowl in pattern vision. *Comparative Psychology Monographs*, 1936, **12**, No. 58.

Mackintosh, N. J. Comparative psychology of serial reversal and probability learning: Rats, birds and fish. In R. Gilbert & N. S. Sutherland (Eds.). *Animal discrimination learning.* New York: Academic Press, 1969, 137–167.

Macphail, E. M. Avian hyperstriatal complex and response facilitation. *Communication in Behavioral Biology*, 1969, **4**, 129–137.

Macphail, E. M. Hyperstriatal lesions in pigeons: Effects on response inhibition, behavior contrast, and reversal learning. *Journal of Comparative and Physiological Psychology*, 1971, **75**, 500–507.

Matyniak, K., Wheeler, G., & Stettner, L. J. Habit reversal in the crow. *Communications in Behavioral Biology*, 1971, **6**, 183–190.

Mayor, S. J. Memory in the Japanese quail: Differential effects of puromycin and acetoxy-cycloheximide. *Science*, 1969, **166**, 1165–1167.

Meier, R. E. Interhemispharischer Transfer visueller Zweifachwahlen bei kommisurotomierten Tauben. *Psychologisch Forschung*, 1971, **34**, 220–245.

Meier, R. E., Maier, V., & Cuenod, M. Visual learning following unilateral telencephalic lesions in the split-brain pigeon. *Brain Research*, 1971, **34**, 8 (abstract).

Mello, N. K. Interhemispheric reversal of mirror-image oblique lines after monocular training in pigeons. *Science*, 1965, **148**, 252–254.

Mello, N. K. The effects of unilateral lesions of the optic tectum on interhemispheric transfer of monocularly trained color and pattern discrimination in the pigeon. *Physiology and Behavior*, 1968, **3**, 725–734.

Meyer, D. R. & Woolsey, C. N. Effects of localized cortical destruction on auditory discrimination conditioning in the cat. *Journal of Neurophysiology*, 1952, **15**, 149–162.

Meyer, D. R. Some psychological determinants of sparing and loss following damage to the brain. In H. F. Harlow & C. N. Woolsey (Eds.), *Biological and Biochemical Bases of Behavior*, Madison, Wisconsin: Univ. of Wisconsin Press, 1958.

Meyers, R. E. Transmission of visual information within and between hemispheres: a behavioral study. In V. B. Mountcastle (Ed.), *Interhemispheric relations and cerebral dominance.* Baltimore, Maryland: Johns Hopkins Press, 1962.

Nauta, W. J. H. & Karten, J. J. A general profile of the vertebrate brain with sidelights on the ancestry of cerebral cortex. In F. O. Schmitt, (Ed.), *The Neurosciences: Second study program.* New York: Rockefeller Univ. Press, 1970.

Olds, J. Adaptive functions of paleocortical and related structures. In H. F. Harlow, C. N. Woolsey (Eds.), *Biological and biochemical bases of behavior*, Madison. Wisconsin: Univ. of Wisconsin Press, 1958.

Phillips, R. E. Approach withdrawal behavior of peach-faced lovebirds, Agapornis roselicolis, and its modification by brain lesions. *Behaviour*, 1968, **31**, 163–184.

Portmann, A. & Stingelin, W. The central nervous system. In A. J. Marshall (Ed.), *Biology and comparative physiology of birds, Vol. 2*, New York: Academic Press, 1960.

Pritz, M. B., Mead, W. R., & Northcutt, R. G. The effects of wulst ablations on color brightness and pattern discrimination in pigeons. *Journal of Comparative Neurology*, 1970, **140**, 81–100.

Revzin, A. M. & Karten, H. J. Rostral projections of the optic tectum and nucleus rotundus in the pigeon. *Brain Research*, 1967, **3**, 264–276.

Schade-Powers, A. The role of the avian hyperstriatum in habit reversal. Doctoral Dissertation, Bryn Mawr College, 1969.

Sperry, R. W. Cerebral organization and behavior. *Science*, 1961, **133**, 1749–1757.

Stettner, L. J. Effect of end brain lesions on discrimination and reversal learning in the bobwhite quail (*Colinus virginianus*). Paper presented at American Psychological Association Meeting, New York, 1966.

Stettner, L. J. & Matyniak, K. The brain of birds. *Scientific American*, 1968, **218**, 64–76.

Stettner, L. J. & Schultz, W. Brain lesions in birds effects on discrimination acquisition and reversal. *Science*, 1967, **155**, 1689–1692.

Sutherland, N. S. & Mackintosh, N. J. *Mechanisms of animal discrimination learning*. New York: Academic Press, 1971.

Thompson, R. F. The effect of training procedure upon auditory frequency discrimination in the cat. *Journal of Comparative and Physiological Psychology*, 1959, **52**, 186–190.

Thompson, R. F. Function of auditory cortex of cat in frequency discrimination. *Journal of Neurophysiology*, 1960, **23**, 321–334.

Wenzel, B. M. & Salzman, A. Olfactory bulb ablation or nerve section and behavior of pigeons in non-olfactory learning. *Experimental Neurology*, 1968, **22**, 472–479.

Zeigler, H. P. Effects of end brain lesions upon visual discrimination learning in pigeons. *Journal of Comparative Neurology*, 1963, **120**, 161–194.

Zeigler, H. P. *Effects of hyperstriate ablations upon discrimination learning*. Progress report to City College Research Foundation, 1965. (a)

Zeigler, H. P. *Endbrain lesions and oddity discrimination in pigeons*. Paper presented at Eastern Psychological Association Meeting, New York: 1965. (b)

Zeigler, H. P. & Cohen, D. *Hyperstriate lesions and components of general activity*. Progress report to City College Research Foundation, 1965.

Brain Perturbation and Memory Disruption: A Comparison between Classes

Richard H. McCollum
Allegheny College

Irving J. Goodman[1]
West Virginia University

It is very fitting in this volume reporting the Second Lashley Memorial Conference that the man whom the conference honored had among his scientific interests the behavior of birds and, perhaps better known, his marked concern for the central mechanisms for learning and memory. Until his death he remained in awe of the ability of organisms to retain information and was somewhat frustrated by his inability to determine the mechanisms for the memory process.

The present chapter represents an extension of the concern with the problem of trying to understand the neural bases of memory, particularly with the hypothesized processes occurring immediately after learning that affect memory. It is known that memory is disruptable by certain traumatic stimuli acting on the CNS shortly after some learning experience, making memory of the experience inaccessible at a later time. Primary emphasis here will be given to the understanding of memory that has been accumulated through study of the effects of electrical or chemical brain perturbation on the memory of some learned task.

[1] Financial support for this work was in part provided by a West Virginia University General Medical Research Grant to Irving J. Goodman.

I. Findings in Mammals

A. Memory Disruption by Various Agents

Extensive investigation has been made of the disruptive effects of electrical stimulation upon memory. The two methods of electrical stimulation reported are electroconvulsive shock (ECS), applied to large but indeterminate volumes of brain tissue (Thompson, 1958; Coons & Miller, 1960; Weissman, 1963; McGaugh & Madsen, 1964; Chevalier, 1965; Kopp, Bohdanecky, & Jarvik, 1966; Schneider & Sherman, 1968; Nielson, 1968a,b) and intracranial stimulation (ICS), applied to circumscribed and identifiable areas within the brain (Glickman, 1958, 1961; Mahut, 1962, 1964; Kasper, 1964; Goddard, 1964a; Pellegrino, 1965; Stein & Chorover, 1968; Kesner & Doty, 1968; Wyers, Peeke, Williston, & Herz, 1968; Brunner, Rossi, Stutz, & Roth, 1970; McDonough & Kesner, 1971).

Trauma-produced memory disruption, which results in the most severe loss of retention of events immediately preceding the trauma, or retrograde amnesia, has been studied by investigators using a variety of other experimental agents. Anesthesia (Abt, Essman, & Jarvik, 1961; Pearlman, Sharpless, & Jarvik, 1961; McGaugh & Alpern, 1966), anoxia (Hays, 1953; Thompson & Pryer, 1956), and carbon dioxide inhalation (Quinton, 1966) have produced retention deficits. Some investigators have found memory disruption after postrial immersion in hot or cold water (Jacobs & Sorenson, 1969) or severe hypothermia (Beitel & Porter, 1968). Bilateral brain puncture (Bohdanecky, Bohdanecky, & Jarvik, 1967) and cortical spreading depression (Pearlman, 1966; Bures & Buresova, 1957) are examples of trauma directed to the brain which produced a memory loss of a learned task. Some drug treatments, other than anesthesia, also have produced retention deficits: physostigmine administered intraperitoneally (Hamburg, 1967); cortical injection of puromycin (Flexner, Flexner, & Stellar, 1963); or anticholinesterase (Deutsch, 1969) have impaired the memory of a learned experience. There have also been a great number of studies of the effects of brain lesions on memory. However, since this latter technique does not involve short-term posttrial brain perturbation, such studies will not be considered in this paper.

Our emphasis will be on the effects of posttrial electrical stimulation on memory. Relevant findings in mammals will be reviewed first, followed by a more detailed coverage of findings in birds.

B. Different Explanations

Various investigators have posited several explanations for the retrograde amnesia resulting from electroconvulsive shock. Coons and Miller

(1960) hypothesized that when electroconvulsive shock was presented repeatedly, amnesia resulted from a fear of the goal with an approach–avoidance conflict leading to a deficit in performance. Adams and Lewis (1962a,b) presented ECS after avoidance trials, then presented more avoidance trials 3 days later. They suggested that a retention deficit occurred because ECS is an unconditioned stimulus for the convulsive response which competes with the desired response. An ECS-produced inhibition of the Pavlovian variety, which becomes conditioned to the ECS situation, has been the explanation suggested by Lewis and Maher (1965). They cite studies using ECS to produce a conditioned emotional response which can produce proactive inhibition. Lewis, Miller, and Misanin (1968) found that rats familiarized with the experimental situation before one trial passive avoidance learning showed no evidence of memory disruption due to ECS, but subjects presented with ECS after passive avoidance in typical conditions (without familiarization) showed a significant learning deficit. They hypothesized that prior familiarization permits a new memory (for a foot shock) to be integrated into the existing system, thereby protecting it, to some extent, from the inhibiting effects of ECS.

The bulk of the literature, however, supports a theory not yet mentioned. Current research has led to widespread acceptance of the hypothesis of memory *consolidation* which involves a two-stage process. Muller and Pilzecker (1900) posited the view that memory exists in a dynamic and unstable form for a short time following an experience, and assumes a relatively permanent form through stabilization in the second stage. The consolidation theory also suggests that memory, while in its unstable form, is susceptible to disruption by a strong stimulus which can prevent its becoming stabilized. The second stage is seen as one in which memory is not nearly so easily disrupted. The work of a number of investigators (e.g., Glickman, 1961; McGaugh, 1966; McDonough & Kesner, 1971) provides support for the consolidation theory of memory.

Hudspeth, McGaugh, and Thompson (1964) tested whether ECS caused a retention deficit due to disturbance of the consolidation process or to aversion of a fear-producing stimulus. In their foot-shock groups, avoidance learning was more complete with a training procedure using *longer* learning trial-to-ECS intervals, a finding which is consistent with the consolidation view; however, in nonfoot-shocked groups, the incidence of avoidance was higher with *shorter* learning trial-to-ECS intervals, a finding which supports the theory that fear of ECS causes the retention deficit. On the basis of this investigation, the authors hypothesized that ECS produces amnesia for events preceding the treatment, but that fear develops with repeated ECS treatments. Experimental findings of Chorover and Schiller (1965, 1966) support this hypothesis, in that impairment of retention was inversely related to ECS delay intervals.

Nielson (1968) has offered some evidence suggesting that rather than disrupting the consolidation of the memory, ECS alters memory *retrieval.* His subjects received ECS before and after training, and were tested for retention 4 days later. He concluded that failure of retention occurs whenever the state of brain excitability is modified (as by ECS) from the base line established during the training procedure. According to Neilson, memory retrieval depends upon matching brain excitability states during learning and recall periods.

C. Recovery of Memory Loss

Studies investigating the permanence of ECS-induced decrements in performance have revealed conflicting results. Zubin (1948) and others have shown that memory impairment in humans is not permanent; retention returned to their subjects within a few weeks after the termination of ECS treatments. Zinkin and Miller (1967) tested animals at 24, 48, and 72 hr after an ECS treatment which followed a one-trial inhibitory avoidance response and found recovery from amnesia; however, they did not use a balanced design to control for the effects of repeated testing. Cooper and Koppenall (1964) and Kohlenberg and Trabasso (1968), the former using rats and the latter using mice, found a deficit in the amount of impairment induced by ECS when their subjects were tested 48 hr after passive avoidance learning. Results conflicting with the preceding findings have been reported by Chevalier (1965) in rats given passive avoidance training in a circular runaway and by Greenough, Schwitzgebel and Fulcher (1968) and Luttges and McGaugh (1967) who used mice in a step-down passive avoidance situation, i.e., no recovery from the ECS-induced impairment could be observed.

D. Parameters of effective Stimulation

Within broad limits the disruption of performance induced by ECS has been shown not to be related to the current level of the ECS. Weissman (1963), using rats, found that so long as a tonic seizure resulted the current level of the ECS was not critical within a wide range of currents. Ray and Barrett (1969) trained mice for one trial per day for 5 days in a passive avoidance situation; the learning–ECS interval and the ECS duration were held constant and the ECS intensity varied from 5 to 80 mA. They found that high-intensity ECS produced more disruption of the learned response than low intensities and that this effect interacted with mode of ECS delivery (via eyes, ears, or cross-modal) and the intensity of foot shock. Pagano,

Bush, Martin, and Hunt (1969) investigated the effects of different ECS intensities on the duration of retrograde amnesia in rats following single trial passive avoidance learning in a step-down situation; ECS intensity was either 95 or 55 mA and was of .2-sec duration, administered .5 sec after a .8-mA foot shock of 2 sec duration. High-intensity ECS produced amnesia lasting for 48 hr, while low-intensity ECS produced a deficit in retention lasting 1 hr but no deficit when the animals were observed after 23 hr. Jarvik and Kopp (1967) demonstrated that subconvulsive threshold shock was effective in producing amnesia. These results demonstrate the effectiveness of a wide range of currents in the production of amnesia as well as the occurrence of amnesia without convulsions. McGaugh and Alpern (1966) add support to this finding in their observation of a decrement in the retention of a passive avoidance response even when convulsions were inhibited by the administration of ether prior to the ECS treatment.

E. Intracranial Stimulation

To eliminate convulsions and ambiguities concerning punishing effects of brain perturbation, and also to investigate more thoroughly the specific brain structures involved in producing memory loss, a number of studies have employed intracranial stimulation (ICS), circumscribed electrical brain stimulation. Several brain areas have been studied. Glickman (1958) using a high intensity, long duration current (three 20-sec bursts) stimulated the midbrain reticular formulation, and found retention reduced 24 hr later. Thompson (1958) investigated an alternation task with cats and found that bilateral stimulation of the caudate nucleus after each trial interfered with retention as compared to stimulation of the midbrain tegmentum or no stimulation. Mahut (1962) stimulated the reticular activating system in rats and found detrimental effects upon memory only when subjects were stimulated immediately after the learned response, but not when there was a delay of 165 sec between response and stimulation. She also found that those subjects stimulated in the nonspecific thalamic nuclei showed less retention than subjects who either received stimulation in the midbrain tegmentum, or no stimulation at all. In a later study (Mahut, 1964) it was found that low-intensity current administered immediately after a response produced a deficit in learning; when presented before the response it showed no effect, and when delivered during the choice period of the discrimination task, learning was facilitated. The deficit produced in this study occurred under the same conditions as the earlier study (Mahut, 1962) and seemed to depend partly on the temporal relationship of the task to stimulation (with a delay of 20 sec no deficit was produced).

The role of the amygdala and hippocampus in memory consolidation has been investigated in several studies. Goddard (1964b) described the amygdala as a part of the limbic system which seems essential for the establishment of avoidance behavior and the modification of approach behavior. Animals with lesions of the amygdala are "hypermotivated" in approach behaviors (to food, water, etc.) and are inhibited in acquisition of fear and avoidance behaviors toward previously neutral stimuli. Goddard (1964a) has shown that rats stimulated in amygdaloid nuclei with a continuous low-intensity current were impaired on fear-motivated learning but not on food-motivated learning. Amygdaloid stimulation was also found to impair the acquisition of a passive avoidance response by Pellegrino (1965) and by McDonough and Kesner (1971). The amygdala is considered to be important for the production of amnesia by Kesner and Doty (1968) who found that ICS applied to the amygdala or dorsal hippocampus following a simple learning task produced amnesia, but ICS similarly applied to the septum, fornix, or ventral hippocampus failed to do so.

Lidsky, Levine, Kreinick, and Schwartzbaum (1970) have reported a study that gives little support to the Goddard (1964a) results. Subconvulsive intensities of bilateral amygdaloid stimulation adjusted to avoid various overt response manifestations failed to produce any retrograde amnesia effects. This difference in findings might have been explained by Goddard's use of *unilateral* stimulation compared to the use by Lidsky *et al.* of *bilateral* stimulation. However, Bresnahan, and Routtenberg (1972) reported that current as low as 5 μA applied *unilaterally* to the medial nucleus of the amygdala resulted in memory disruption of a learned task.

Memory loss in a passive avoidance task after hippocampal stimulation has also been reported by Brunner and Rossi (1969), and Brunner *et al.* (1970). Wyers *et al.* (1968) found a retention deficit following bilateral electrical stimulation of the ventral, but not dorsal hippocampus on a maze task, and that single-pulse bilateral stimulation of the caudate nuclei or ventral hippocampal dentate regions produced retroactive interference of a passive avoidance response 24 hr later. Here the retention deficit was essentially the same with foot-shock-to-brain stimulation intervals as long as 30 sec. At present, the structures that seem to be implicated as functionally significant in producing retrograde amnesia for passive avoidance in mammals are the amygdala, caudate nucleus, hippocampus, and nonspecific thalamic area of the reticular activating system. Further experimentation is required to clarify some of the inconsistent findings.

Studies supporting the consolidation theory using ECS and ICS have investigated the effect of the duration of the foot-shock-to-ECS or foot-shock-to-ICS intervals in passive avoidance situations, with conflicting results. In studies already mentioned, Mahut (1962) found that a delay of

165 sec produced no retention deficit in rats, and none with a 20-sec delay before ICS in cats. Wyers *et al.* (1968) found retrograde amnesia in rats with a delay interval of 30 sec. McGaugh and Alpern (1966), using mice, found that ECS delivered 25 sec after a trial produced amnesia. Ray and Bivens (1968) report that at any of four foot-shock intensities, ECS administered 10 sec but not 160 sec after the learning trial disrupted learning. Electroconvulsive shock did not disrupt memory in rats when presented 30 sec or longer after foot shock in studies by Pfingst and King (1968) using a one-trial response choice technique, and by Schneider and Sherman (1968) using a step down apparatus. Studies by Chorover and Schiller (1965, 1966) found that ECS presented .5 to 10 sec after learning disrupted memory but found no retrograde amnesia with shock-to-ECS intervals of 30 or 60 sec. On the other hand, ICS has been reported to produce retrograde amnesia for a passive avoidance task in cats with delay intervals of up to 5 min (McDonough & Kesner, 1971).

II. Findings in Birds

With this background in mammals, we now can turn to the issue of memory disruption in avians. As in the mammalian studies, the greatest amount of data concerns the production of amnesia by chemical or electrical stimulation of the brain after a single learning trial. One-day-old chicks have been the subjects for the majority of the work to be mentioned, with the remainder of the data coming from work with adult pigeons.

A. Chemical Stimulation

The basic procedure for the study of memory disruption in chicks has been to present a lure (shiny bead) at which the animals would normally peck, but which they learned to avoid after one peck because the lure was coated with an aversive-tasting liquid (usually methyl anthranilate). Perturbations presented after this one-trial procedure have been used to disrupt the avoidance learning.

Cherkin and Lee-Teng (1965), found that a 5-min halothane anesthesia treatment begun immediately after the learning trial produced retrograde amnesia, but memory survived when administration of the anesthetic began 1.5 hr after the learning trial. According to the consolidation theory, memory of the learned event was stabilized within the 1.5-hr period and was thus no longer vulnerable to disruption by halothane anesthesia. The question of when *within* this period consolidation is completed remains to be answered.

Flurothyl, a chemo-convulsant, has been used to investigate RA in chicks. A study by Cherkin (1969) manipulated the training-to-treatment interval, and the intensity of duration of treatment. He found that with low doses of flurothyl, training-to-treatment intervals that could produce retrograde amnesia were shorter than those associated with higher doses and/or longer durations of the drug treatment. Disruption of memory was produced with strong treatments up to several hours after learning. These data were interpreted to suggest that weaker treatments do not entirely block the consolidation process but have the effect of slowing its rate, leading to higher retention scores than those found with stronger amnesic treatments. Cherkin suggests that the use of weak amnesic treatments may lead investigators to hypothesize shorter consolidation periods than are actually the case.

B. Electroconvulsive Shock

Several parameters of transcranial currents which effectively block consolidation of one-trial learning in chicks have been investigated by Lee-Teng and her colleagues. Using an ECS current of 28 mA, .45 sec duration and various learning–ECS intervals, Lee-Teng and Sherman (1966) found that retention deficits were most pronounced when the interval was less than 30 sec. Further testing indicated that the disruption of memory was not an all-or-none phenomenon; rather, there seems to be continuous strengthening of the memory trace during the consolidation period.

After determining that ECS-produced tonic and clonic seizures in 90% of subjects were sufficient to cause RA in chicks, the effects of subconvulsive current (SCC) were examined (Lee-Teng, 1970). It was found that current intensities from 7 to 48 mA, involving behavioral effects ranging from no seizures to full tonic and clonic seizures, did not have significantly different effects on the disruption of memory. Animals in all of the groups displayed amnesia. Using a wider range of SCC and ECS intensities, Lee-Teng found some correlation between current intensities and degree of retrograde amnesia produced in day-old chicks. Investigations have indicated that there is a consolidation period for memory of the one-trial passive avoidance response which lasts approximately 30 sec. Higher intensities and/or longer durations of current do not seem to lengthen the consolidation period, although at learning-to-treatment intervals of less than 30 sec the retention deficit was found to be a function of current intensity (Lee-Teng, 1969, 1970).

The effects of SCC on recall at different SCC-treatments–recall-intervals have also been examined by Lee-Teng (1968). The results of that investigation indicated that at SCC levels sufficient to cause memory disruption on a test

24 hr later, retention after 1, 2.5, 5, 10, and 30 min was not impaired. This was interpreted as indicating that memory may consist of a short-term and a long-term phase, such that disruption of one may not necessarily produce disruption of the other. This interpretation should be considered with reservations, though, because there is reason to believe that the SCC intensities used with neonatal animals may produce some physiological or behavioral deficits which would cause a cessation of pecking behavior at the posttreatment times tested for reasons unrelated to disruption of memory processes.

Electrocorticograms monitored following transcranial stimulation in chicks have indicated that intensities which produce RA (10 and 15 mA), also produce spike activity (Lee-Teng & Giaquinto, 1969). It is not clear, however, whether the presence of electrocortical spike activity after stimulation is a necessary or sufficient condition for the production of amnesia.

With the knowledge that both transcranial currents and flurothyl treatments can cause retrograde amnesia in chicks, Magnus and Lee-Teng (1971) attempted to compare the effect of electrical current with that of low doses of flurothyl. Since flurothyl at high doses can disrupt memory when administered several hours after learning (Cherkin, 1969) while convulsion-inducing electric currents of high intensity do not significantly impair memory when delivered after a relatively short period (Lee-Teng, 1970), it was hypothesized that transcranial electric currents may act as a low dose of flurothyl, allowing for memory stabilization over apparently short consolidation periods. The study (Magnus & Lee-Teng, 1971) used the "adding of engrams" approach, involving the assumption that subjects having two learning-treatment sessions would show more retention on a test than subjects which had received one control-treatment and one learning-treatment session. The results did not support the hypothesis in this case, however, and it could not be assumed that the current necessarily acted as a low dose of flurothyl, incompletely blocking consolidation. The authors adopted the interpretation that the two agents are not directly comparable, and possibly act on different stages in the consolidation process.

Some recent work in our laboratory (McCollum, Jason, & Goodman, submitted for publication) attempted to determine the effects of ECS on memory in an adult avian species, the pigeon (*Columba livia*). The reports of memory disruption in an avian neonate (the chick) had left unanswered the question of the comparability of ECS effects in birds and mammals, independent of a possibly confounding developmental variable.

In this study, subjects learned a passive avoidance task (response inhibition). Food-deprived subjects were trained to approach (by crossing a grid floor) and eat grain available from a feeder during an auditory signal. Training was carried out for 10 trials daily until a high performance criterion was reached. The following day, five similar trials were run but on the sixth trial, instead of food, foot shock was delivered as the subject approached the feeder. Seven seconds after this aversive experience, a 30 mA ECS (sufficient to produce tonic and clonic seizures in all subjects) was delivered transcranially to all experimental subjects. Two control groups were also run: one group received only foot shock on Trial 6 without ECS; a second group received no foot shock, but did receive ECS following Trial 6. In addition, in order to assess the contribution, if any, of different ECS treatment-recall test intervals and multiple testing to the possible recovery of memory, subjects were given two recall tests on one of two schedules: (A) 24 hr after ECS (Test 1) and 96 hr after ECS (Test 2); or (B) 96 hr after ECS (Test 1) and 120 hr after ECS (Test 2).

As might have been expected based upon results in chicks and mammals, amnesic effects (the tendency to approach the feeder despite the aversive experience) of ECS were quite marked in the experimental group. This was true for all hours of recall testing with no signs of recovery from memory loss seen from Test 1 to Test 2. Among those questions left unanswered is that regarding the duration of the disruptable or labile memory state following a learning experience. Perhaps, as in mammals and chicks, the answer depends in part upon the parameters of the disrupting stimulus and, beyond this, upon the very nature of that which is learned and to be recalled.

1. Localized Brain Stimulation

As previously mentioned, ECS studies are not without problems of interpretation due to such difficulties as separating out the effects of behavioral seizures, determining possible punishing effects of ECS and problems of neuroanatomical analysis due to widespread CNS involvement. Although studies in mammals have attempted to isolate brain structures critically involved in memory storage and/or retrieval, little effort in this direction has yet been made in birds. An effort in this direction was attempted by an investigation of circumscribed brain stimulation with ICS (McCollum, Jason, & Goodman, submitted for publication).

Using a passive avoidance task similar to that used in their ECS experiments, adult pigeons with chronically implanted electrodes received ICS (single pulse, 750 μA, .5 msec, square wave) 7 sec after the aversive foot-shock trial. An ICS was delivered bilaterally to either neostriatum, archistriatum, or paleostriatum. (No overt behavioral changes were observed.) The combined results for the ECS and this ICS experiment are shown in Fig. 1. Surprisingly, retrograde amnesia was observed in all three experimental groups. The retrograde amnesia produced by ICS was consistent over repeated testing; animals tested either at 24 and 96 hr or 96 and 120 hr after ICS showed no recovery of memory.

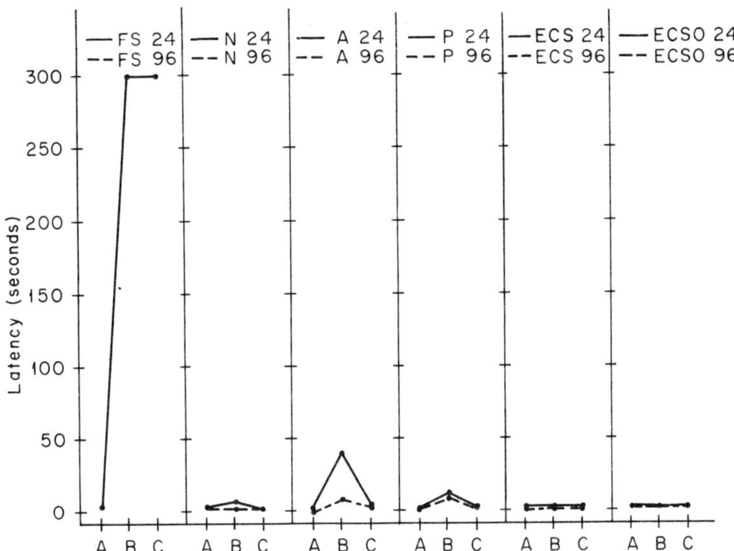

Fig. 1. Mean total running time for the first five trials on acquisition Day 6 (before treatment), A; Recall test 1 (24 or 96 hr after treatment), B; Recall test 2 (96 or 120 hr after treatment). C. *Group Treatments*: FS, foot shock only (control); ECS, foot shock followed by electroconvulsive shock after 7 sec; ECSO, electroconvulsive shock (no foot shock) (control); N, foot shock followed by ICS in neostriatum after 7 sec; A foot shock followed by ICS in archistriatum after 7 sec; P, foot shock followed by ICS in paleostriatum after 7 sec; 24, Recall test 1 interval following treatment; 96, Recall test 1 interval following treatment.

Memory disruption which resulted from ICS stimulation of sites in three different brain regions (see Fig. 2) in the pigeon is explainable in at least two different ways. Although the stimulation parameters used produced differential effects on memory depending upon the structures stimulated in the mammalian brain (Wyers *et al.*, 1968), it is possible that similar parameters used in the pigeon spread to nearby neural structures outside the immediate area of electrode placement and thus *indirectly* produce retrograde amnesia; or, it may be that memory is more vulnerable to disruption *directly* from a greater number of areas in the avian brain than in mammals.

Another experiment carried out in the study of McCollum *et al.* attempted to clarify these results by using a much lower ICS intensity (100 μA). There was also an attempt made to manipulate the magnitude of the amnesic effect by utilizing two different time intervals between the learning trial and the presentation of ICS: immediate, and 20-sec delay. All recall tests were carried out 24 hr after treatment.

The results indicated statistically significant differences ($p < .05$) in retention between subjects receiving ICS and the foot-shock control subjects.

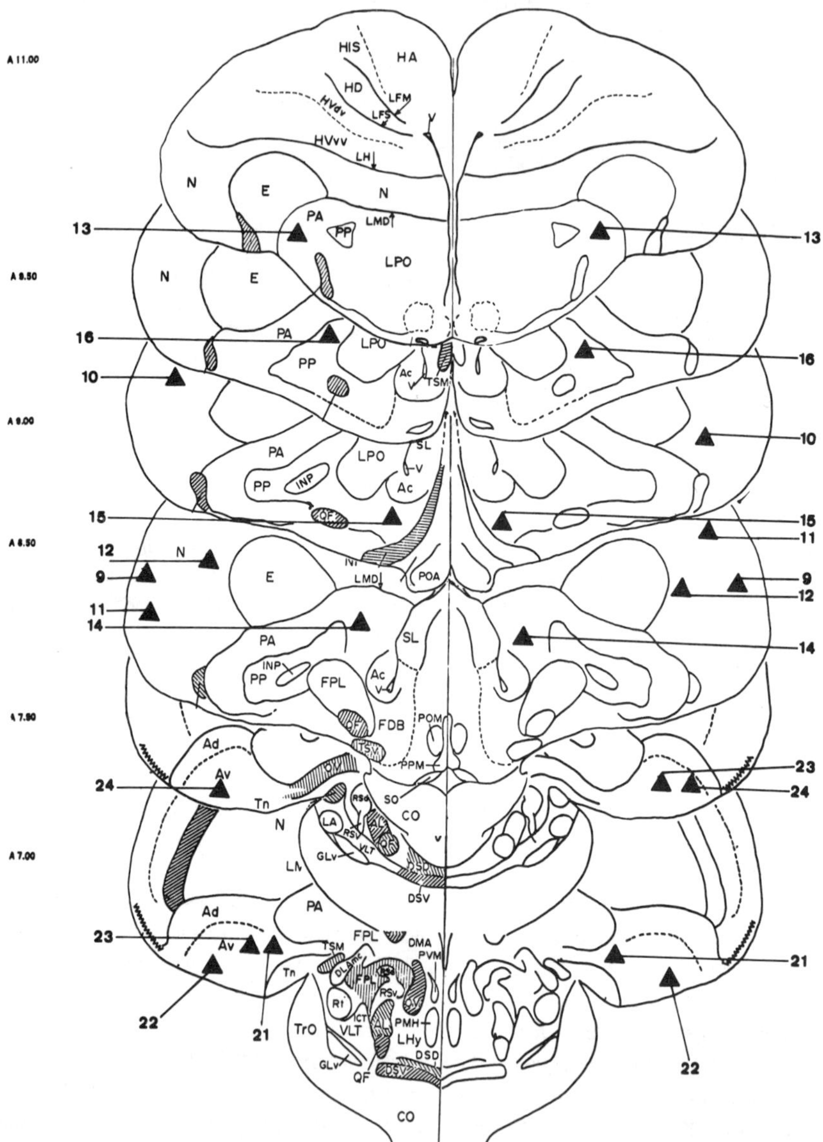

Fig. 2. Neostriatal, paleostriatal and archistriatal sites stimulated with 750-μA pulse, bilaterally in ICS experiment 1 designated on serial atlas sections [After Karten & Hodos (1970).]

The immediate and delayed ICS presentation did not produce significantly different results, suggesting that memory in adult pigeons is vulnerable to disruption for learning-to-treatment intervals perhaps longer than 20 sec. A consolidation period for memory of 20 sec has received support in the avian ECS literature (Lee-Teng, 1970). The ICS literature has reported consolidation periods ranging up to 30 sec and longer, so that the memory disruption found in the present investigation is compatible with previous reports in mammals.

It is interesting, however, that amnesia in pigeons has been produced with current intensities (bilateral, single pulse square waves of 750 μA or 100 μA) subsequently found not to produce brain afterdischarges (McCollum, unpublished observations), an electrophysiological phenomenon reported to accompany retrograde amnesia inducing currents in mammals. Further investigation into the significance of afterdischarges in producing memory disruption in mammals as well as avians is needed.

A second finding in McCollum's last experiment was that neostriatal and not archistriatal brain stimulation produced memory disruption; the control group (no ICS treatment) differed significantly from the neostriatal group but not from the archistriatal group (paleostriatal stimulation not used in this experiment) (see Fig. 3). This surprising finding leads to questions concerning the roles of the neostriatum and archistriatum in memory formation, and to questions concerning their sensitivity to disruption with low current intensities. The answers to those questions can only be speculated at this time because of the limited anatomical and neurobehavioral data available.

Recent evidence concerning neostriatum has identified it as part of the "external striatum," which has been considered by some to be homologous to the neocortex in mammals (Nauta & Karten, 1970). The neostriatum is not a homogeneous structure; parts of that region have been associated anatomically and physiologically with several different ascending sensory projection systems, descending efferent pathways and a diffuse interneuronal organization (Cohen & Karten, 1974). The McCollum study (McCollum *et al.*, submitted for publication) produced memory disruption by stimulating the lateral portion of neostriatum intermediale, some distance lateral to ectostriatum (the latter a structure implicated in vision by Karten and Hodos, 1970). The effectiveness of such stimulation in producing memory disruption is puzzling.

In the case of the archistriatum there is anatomical evidence (Zeier & Karten, 1971) supporting the presence of four subdivisions of the avian archistriatum, two of which may be comparable to the mammalian amygdala (archistriatum posterior and archistriatum mediale), associated with the hypothalamus, and two (archistriatum intermedium and archistriatum

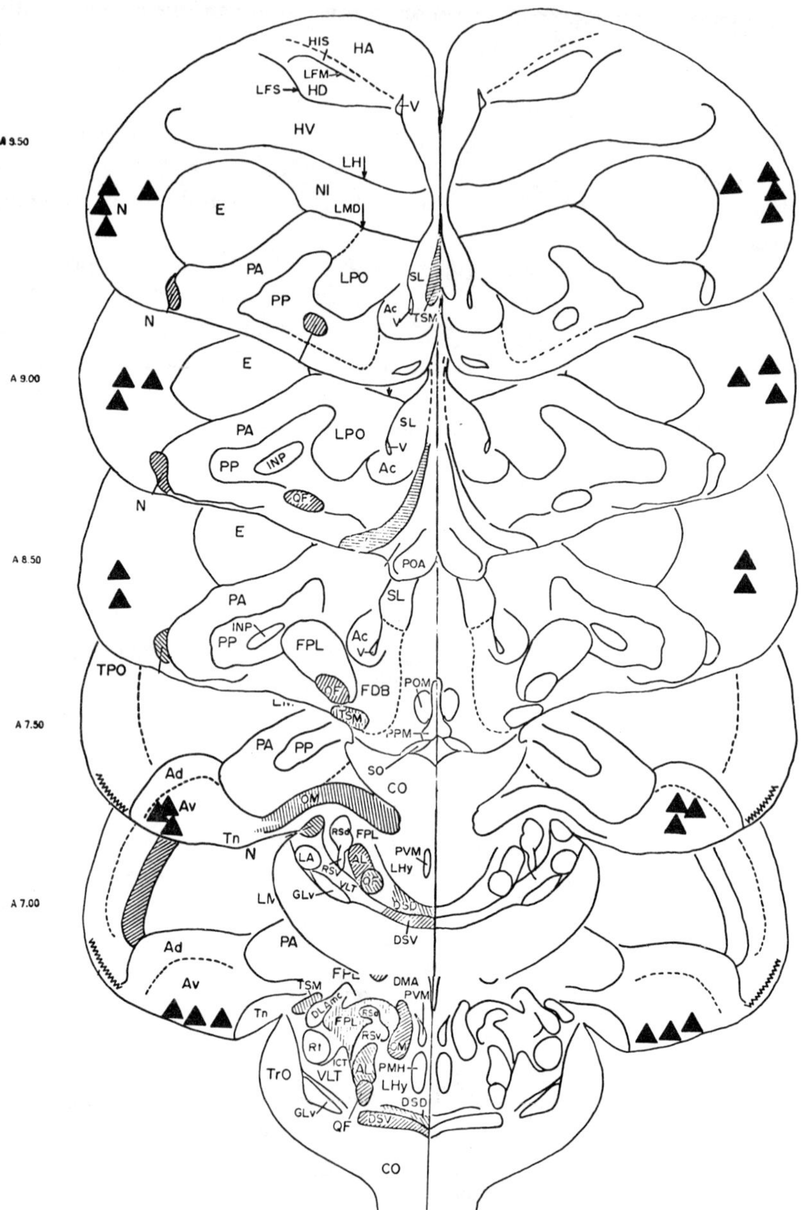

Fig. 3. Neostriatal or archistriatal sites stimulated bilaterally (100 μA, single pulse; represented on brain atlas cross-sections [After Karten & Hodos (1970).]

anterior) which appear to be associated with somatic sensorimotor function and are related to the descending tractus occipitomesencephalicus (suggested as, perhaps, a variant form of part of the pyramidal tracts of primates). Histological findings in the McCollum experiments indicated that all of the electrode tips were in either the archistriatum intermedium or the anterior archistriatum (in the somatic sensorimotor portion of the archistriatum). One might speculate that the disruptive effect on memory resulting from the use of 750 μA rather than 100-μA ICS in this archistriatal area may have resulted from current spread to the hypothalamic-associated portion of the archistriatum. Further investigation will be needed to compare the disruptive effects of ICS on memory when low-level currents are directed to the hypothalamic-associated (limbic) subdivisions of the avian archistriatum.

III. Summary

A relatively small number of studies have been made in birds regarding memory-disrupting effects of brain perturbation, compared to that in mammals. The data gathered so far indicate that anesthetics, chemo-convulsants, ECS, and subconvulsive currents disrupt memory in chicks with the consolidation period fixed around 30 sec for electrical disruption of memory traces. In adult avians, it has been demonstrated that ECS or higher intensity ICS (750 μA) as in mammals will produce long lasting amnesia for passive-avoidance learning. The use of lower intensity ICS (100 μA) has demonstrated differential effects, depending upon the area of the brain stimulated, and has led to speculation concerning the comparability of structures in avian and mammalian brains. Further ICS studies are necessary in order to clearly determine which avian and mammalian brain structures are instrumental for memory consolidation and/or retrieval. A question has also been raised concerning the necessity of initiating electrical afterdischarges from the brain in order to produce memory deficits.

The limits of the consolidation period for memory have not been found in avians, but the learning-to-stimulation intervals of 7 and 20 sec in adult avians which produced memory loss is consistent with findings of a number of ECS and ICS investigations in mammals. Although the work cited has generally provided support for the consolidation theory of memory, it is entirely possible that other theories of memory formation and retrieval could be supported with different manipulations of brain perturbation in avians and mammals.

References

Abt, J. P., Essman, W. B., & Jarvik, M. E. Ether induced retrograde amnesia for one-trial conditioning in mice. *Science*, 1961, **133**, 1477–1478.

Adams, H. E. & Lewis, D. J. Electroconvulsive shock, retrograde amnesia, and competing responses. *Journal of Comparative and Physiological Psychology*, 1962, **55**, 299–301. (a)

Adams, H. E. & Lewis, D. J. Retrograde amnesia and competing responses. *Journal of Comparative and Physiological Psychology*, 1962, **55**, 302–305. (b)

Beitel, R. E. & Porter, P. B. Deficits in retention and impairments in learning induced by severe hypothermia in mice. *Journal of Comparative and Physiological Psychology*, 1968, **66**, 53–59.

Bohdanecky, M., Bohdanecky, A., & Jarvik, M. E. Amnesic effects of small bilateral brain punctures in the mouse. *Science*, 1967, **157**, 334–336.

Bresnahan, E. & Routtenberg, A. Memory disruption by unilateral low level, sub-seizure stimulation of medial amygdaloid nucleus. *Physiology and Behavior*, 1972, **9**, 513–526.

Brunner, R. L. & Rossi, R. R. Hippocampal disruption and passive avoidance behavior. *Psychonomic Science*, 1969, **15**, 228–229.

Brunner, R. L., Rossi, R. R., Stutz, R. M., & Roth, T. C. Memory loss following post-trial electrical stimulation of the hippocampus. *Psychonomic Science*, 1970, **18**, 159–160.

Bures, J. & Buresova, O. Cortical spreading depression as a memory disturbing factor. *Journal of Comparative and Physiological Psychology*, 1957, **50**, 125–129.

Cherkin, A. Flurothyl toxicity: A remarkable species difference between chick and mouse. *Psychopharmacologia*, 1969, **15**, 404–407.

Cherkin, A. & Lee Teng, E. Interruption by halothane of memory consolidation in chicks. *Federation Proceedings*, 1965, **24**, 328. (abstract).

Chevalier, J. A. Permanence of amnesia after a single post-trial electro-convulsive seizure. *Journal of Comparative and Physiological Psychology*, 1965, **59**, 125–127.

Chorover, S. L. & Schiller, P. H. Short term retrograde amnesia in rats. *Journal of Comparative and Physiological Psychology*, 1965, **59**, 73–78.

Chorover, S. L. & Schiller, P. H. Reexamination of prolonged retrograde amnesia in rats. *Journal of Comparative and Physiological Psychology*, 1966, **61**, 34–41.

Coons, E. E. & Miller, N. E. Conflict versus consolidation of memory traces to explain "retrograde amnesia" produced by ECS. *Journal of Comparative and Physiological Psychology*, 1960, **53**, 524–531.

Cooper, R. M. & Koppenal, R. J. Suppression and recovery of a one-trial avoidance response after a single ECS. *Psychonomic Science*, 1964, **1**, 303–304.

Deutsch, J. A. The physiological basis of memory. *Annual Review of Psychology*, 1969, **20**, 85–104.

Flexner, L. B., Flexner, J. B., & Stellar, E. Memory in mice as affected intracerebral puromycin. *Science*, 1963, **141**, 57–59.

Glickman, S. E. Deficits in avoidance learning produced by stimulation of the ascending reticular formation. *Canadian Journal of Psychology*, 1958, **12**, 97–102.

Glickman, S. E. Perservative processes and consolidation of the memory trace. *Psychological Bulletin*, 1961, **58**, 218–238.

Goddard, C. J. Amygdaloid stimulation and learning in the rat. *Journal of Comparative and Physiological Psychology*, 1964, **58**, 23–30. (a)

Goddard, C. J. Functions of the amygdala. *Psychological Bulletin*, 1964, **62**, 89–109. (b)

Greenough, W. T., Schwitzgebel, R. L., & Fulcher, J. K. Permancence of ECS as a function of test conditions. *Journal of Comparative and Physiological Psychology*, 1968, **66**, 554–556.

Hamburg, M. D. Retrograde amnesia produced by intraperitoneal injections of physostigmine, *Science*, 1967, **156**, 973–974.

Hays, K. J. Anoxic and convulsive amnesia in rats. *Journal of Comparative and Physiological Psychology*, 1953, **46**, 216–217.

Hudspeth, W. J., McGaugh, J. L., & Thompson, C. W. Aversive and amnesic effects of electroconvulsive shock. *Journal of Comparative and Physiological Psychology*, 1964, **57**, 61–64.

Jacobs, B. L. & Sorenson, C. A. Memory disruption in mice by brief post-trial immersion in hot or cold water. *Journal of Comparative and Physiological Psychology*, 1969, **68**, 239–244.

Jarvik, M. E. & Kopp, R. Transcorneal electroconvulsive shock and retrograde amnesia in mice. *Journal of Comparative and Physiological Psychology*, 1967, **64**, 431–433.

Karten, H. J. & Hodos, W. Telencephalic projections of the nucleus rotundus in the pigeon (*Columba livia*). *Journal of Comparative Neurology*, 1970, **140**, 35–52.

Kasper, R. Attenuation of passive avoidance by continuous septal stimulation. *Psychonomic Science*, 1964, **1**, 219–220.

Kesner, R. P. & Doty, R. W. Amnesia produced in cats by local seizure activity initiated from the amygdala. *Experimental Neurology*, 1968, **21**, 58–68.

Kohlenberg, R. & Trabasso, T. Recovery of a conditioned emotional response after one or two electroconvulsive shocks. *Journal of Comparative and Physiological Psychology*, 1968, **64**, 270–273.

Kopp, R., Bohdanecky, A., & Jarvik, M. E. Long temperal gradient of retrograde amnesia for a well-discriminated stimulus. *Science*, 1966, **153**, 1547–1549.

Lee-Teng, E. Presence of short-term memory after immediate post-trial subconvulsive currents in chicks. Proceedings, 76th Annual Convention, APA, 1968, 329–330.

Lee-Teng, E. Retrograde amnesia gradients by subconvulsive and high convulsive transcranial currents in chicks. *Proceedings of the National Academy of Science*, 1970, **65**, 857–865.

Lee-Teng, E. & Giaquinto, S. Electrocorticograms following threshold transcranial electroshock for retrograde amnesia in chicks. *Experimental Neurology*, 1969, **23**, 485–490.

Lee-Teng, E. & Sherman, S. M. Memory consolidation of one-trial learning in chicks. *Proceedings of the National Academy of Sciences*, 1966, **56**, 926–931.

Lewis, D. J. & Maher, B. A. Neural consolidation and electroconvulsive shock. *Psychological Review*, 1965, **72**, 235–239.

Lewis, D. J., Miller, R. R., & Misanin, J. R. Recovery of memory following amnesia. *Nature*, 1968, **220**, 704–705.

Lidsky, T. I., Levine, M. S., Kreinick, C. J., & Schwartzbaum, J. S. Retrograde effects of amygdaloid stimulation on conditioned suppression in rats. *Journal of Comparative and Physiological Psychology*, 1970, **73**, 135–149.

Luttges, M. Q. & McGaugh, J. L. Permanence of retrograde amnesia produced by electroconvulsive shock. *Science*, 1967, **156**, 408–410.

Magnus, J. G. & Lee-Teng, E. The absence of residual memory consolidation following transcranial current. *Physiology and Behavior*, 1971, **7**, 113–115.

Mahut, H. Effects of subcortical electrical stimulation on learning in the rat. *Journal of Comparative and Physiological Psychology*, 1962, **55**, 272–277.

Mahut, H. Effects of subcortical electrical stimulation on discrimination learning of cats. *Journal of Comparative and Physiological Psychology*, 1964, **58**, 390–395.

McCollum, R. H., Jason, K. M., & Goodman, I. J. Effects of electroconvulsive shock or intracranial stimulation on passive avoidance learning in pigeons. (Submitted for publication).

McDonough, J. H. & Kesner, R. P. Amnesia produced by brief electrical stimulation of the amygdala or dorsal hippocampus in cats. *Journal of Comparative and Physiological Psychology*, 1971, **77**, 171–178.

McGaugh, J. L. & Alpern, H. P. Effects of electroshock on memory: Amnesia without convulsions. *Science*, 1966, **152**, 665–666.

McGaugh, J. L. & Madsen, M. C. Amnesic and punishing effects of electro-convulsive shock. *Science*, 1964, **144**, 182–183.

Muller, G. E. & Pilzecker, A. Experimentalle Beitrage zur Lehre vom Gedachtnis. *Z. Psychol.*, Suppl. No. 1 (1900).

Nauta, W. J. H. & Karten, H. J. A general profile of the vertebrate brain, with sidelights on the ancestry of cerebral cortex. In F. O. Schmitt (Ed.), *The neurosciences.* New York: Rockefeller University Press, 1970.

Nielson, J. C. Evidence that electroconvulsive shock alters memory retrieval rather than memory consolidation. *Experimental Neurology*, 1968, **20**, 3–20.

Pagano, R. R., Bush, G., Martin, G., & Hunt, E. B. Duration of retrograde amnesia as a function of electroconvulsive shock intensity. *Physiology and Behavior*, 1969, **4**, 19–21.

Pearlman, C. A. Similar retrograde amneic effects of ether and spreadi-cortical depression. *Journal of Comparative and Physiological Psychology*, 1966, **61**, 306–308.

Pearlman, C. A., Sharpless, S. K., & Jarvik, M. E. Retrograde amnesia produced by anesthetic and convulsant agents. *Journal of Comparative and Physiological Psychology*, 1961, **54**, 109–112.

Pellegrino, L. The effects of amygdaloid stimulation on passive avoidance. *Psychonomic Science*, 1965, **4**, 189–190.

Pfingst, B. A., & King, R. A. Effects of post-training electroconvulsive shock on retention-test performance involving choice. *Journal of Comparative and Physiological Psychology*, 1969, **68**, 645–649.

Pfingst, B. E. & King, R. A. Time course of consolidation as measured by response-choice behavior. *Proceedings of the 76th Annual Convention of the American Psychology Association*, 1968, **3**, 327–328.

Quinton, E. E. Retrograde amnesia induced by carbon dioxide inhalation, *Psychonomic Science*, 1966, **5**, 417–418.

Ray, O. S. & Barrett, R. J. Disruptive effects of electroconvulsive shock as a function of current level and mode of delivery. *Journal of Comparative and Physiological Psychology*, 1969, **67**, 110–116.

Ray, O. S. & Bivens, L. W. Reinforcement magnitude as a determinent of performance decrement after electroconvulsive shock. *Science*, 1968, **160**, 330–332.

Schneider, A. M. & Sherman, W. Amnesia: A function of the temporal relation of footshock to electroconvulsive shock. *Science*, 1968, **159**, 219–221.

Stein, D. C. & Chorover, S. L. Effects of post-trial electrical stimulation of the hippocampus and caudate nucleus on male learning in the rat. *Physiology and Behavior*, 1968, **3**, 787–791.

Thompson, R. The effect of intracranial stimulation on memory in cats. *Journal of Comparative and Physiological Psychology*, 1958, **51**, 421–426.

Thompson, R. & Pryer, R. S. The effect of anoxia on the retention of a discriminated habit. *Journal of Comparative and Physiological Psychology*, 1956, **49**, 297–300.

Weissman, A. Effect of electroconvulsive shock intensity and seizure pattern on retrograde amnesia in rats. *Journal of Comparative and Physiological Psychology*, 1963, **56**, 806–808.

Wyers, E. J., Peeke, H. V. S., Williston, J. S., & Herz, M. J. Retroactive impairment of passive avoidance learning by stimulation of the caudate nucleus. *Experimental Neurology*, 1968, **22**, 350–366.

Zeier, H. & Karten, H. J. The archistriatum of the pigeon: Organization of afferent and efferent connections. *Brain Research*, 1971, **31**, 313–326.

Zinkin, S. & Miller, A. J. Recovery of memory after amnesia induced by electrocovulsive shock. *Science*, 1967, **155**, 102–103.

Zubin, J. Memory functioning in patients treated with electric shock therapy. *Journal of Personality*, 1948, **17**, 33–41.

The Neural Substrate of Emotional Behavior in Birds

D. M. Vowles and L. D. Beasley
University of Edinburgh

I. Introduction

The marked differences in organization between the avian and the mammalian brain have proved of considerable interest, resulting recently in many studies to investigate the effects of brain stimulation on the behavior of birds. These studies cover a variety of species, and have involved both electrical stimulation and implantation of hormones in the brain. Perhaps the most complete description of the effects of ESB is given by Putkonen (1967); this and other papers are listed in Table I.

Many studies have been concerned with the overt behavior elicited by electrical brain stimulation (ESB) and the animal's orientation toward external stimuli (such as stuffed birds and predators) but none have followed up the pioneering work of von Holst and von St. Paul (1963) on the dynamics of drive. The aim of this study has been to map the brain of the Barbary dove (*Streptopelia risoria*), using implanted electrodes, and to ascertain which regions are involved in the overall behavior pattern of attack, defense, and flight. The quantitative effects of stimulation on the interaction of the subject with external stimuli were studied in detail, and attempts made to correlate the functions with anatomical structures.

No attempt has been made to try to reduce the proposed neural substrate to electrophysiological or biochemical activities of neurons. The aim is to describe the separate behavioral functions of those different parts of the brain which together make up the main circuitry underlying agonistic behavior. If the brain is the structure whose function is behavior, then this study is essentially one of functional anatomy. Von Holst and von St. Paul (1963) stressed the importance of developing adequate concepts and descriptive terms for such studies and this will be reiterated.

Table I. *A Summary of Publications on the Effects of Stimulation of the Avian Brain on Behavior*

Type of bird	Nature of brain stimulation	Area of interest	Authors
Pigeon	Electrical	Agonistic and reproductive behavior	Akerman (1966)
	Electrical	Autonomic responses	Cohen & Pitts (1967)
	Electrical	Agonistic and vocal behavior	Delius (1970)
	Electrical	Reward & punishment	Macphail (1967, 1968)
	Electrical	Operant Performance	Zeier & Akert (1968)
	Electrical	Feeding	Zeigler, Green & Karten (1969)
Dove	Electrical	Agonistic behavior	Harwood & Vowles (1967)
	Hormonal	Sexual behavior	Hutchison (1967, 1969 1970)
	Hormonal	Reproductive behavior	Komisarak (1967)
Domestic Chicken	Electrical	Reward & punishment	Andrew (1969)
	Hormonal	Reproductive behavior	Barfield (1965, 1969)
	Electrical	Agonistic behavior	Cannon & Salzen (1971)
	Hormonal	Sexual behavior	Gardner & Fisher (1968)
	Electrical	Various	von Holst & St. Paul (1963)
	Electrical	Vocalization	Peek & Phillips (1971)
	Electrical	Agonistic & other behavior	Phillips & Youngren (1971)
	Electrical	Miscellaneous behavior and autonomic responses	Putkonen (1967)
Mallard	Electrical	Agonistic behavior	Maley (1969)
	Electrical	Agonistic behavior	Phillips (1964)
Quail	Electrical	Vocalization	Potash (1970)
Gulls	Electrical	Agonistic behavior & Vocalization	Delius (1970)

II. Methods

A. Surgery

The surgical techniques involved in the chronic implantation of electrodes have been described in detail by Vowles and Harwood (1966) and Harwood and Vowles (1967). Concentric, bipolar electrodes were used; the outer barrel was of 27 gauge hypodermic tubing, and the core was .010-in.

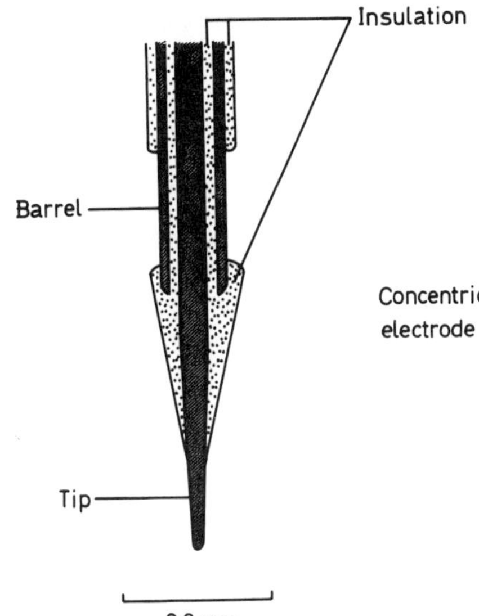

Fig. 1. A diagrammatic representation of the bipolar electrodes used for brain stimulation.

diameter stainless steel wire insulated with Diamel. The dimensions are shown in Fig. 1. Up to eight electrodes were implanted in any one bird, but in most studies only two or four electrodes were used.

In the exploratory studies, electrodes were implanted stereotaxically on 1mm three-dimensional grid references. Thus tips of electrodes were placed in loci separated by 1 mm from other electrode positions. Toward the midline of the brain and in areas such as the hypothalamus and thalamus, a .5 mm spacing was adopted. After the forebrain and brainstem had been explored systematically, other electrodes were implanted in areas of special interest. The electrode positions were determined histologically. Brains were fixed by perfusion with formal–saline. Alternate sections were stained with Glees silver stain for tracts and Nissl stains for nuclei.

The majority of subjects were adult males, although a few hens were included. They had all previously had successful breeding experience and were in full breeding condition except for some subjects that had been castrated. They were housed individually in visual isolation from each other, on an artificial day of 14 hr, at a mean temperature of 18°C.

B. Testing

Electrical stimulation was supplied from a Grass S8 stimulator via a stimulus isolation unit. A variety of stimulus parameters have been tested: The most efficient were found to be rectangular biphasic pulses, of 1 msec duration, at 100 pulses per second. The tip was negative relative to the barrel of the electrode. From the socket on the bird's head a light lead ran up to a rotating connector suspended from a counterbalanced lever. The stimulus current was monitored continuously on an oscilloscope. The stimulus was delivered via a 100 K resistor in series with the preparation, which itself had a resistance of 2–5 K, thus minimizing current changes due to fluctuations in the resistance of the preparation. The currents used were between 5 and 100 μA peak to peak.

The bird was tested in an experimental cage made of chicken wire mesh and transparent perspex. The base of two opposite sides of the cage was cut away and covered with a loose plastic curtain about 2 in. deep. This was arranged so that a model of a predator could be inserted on either side of the cage through the plastic curtains without the subject being able to see the intruder beforehand. The cage normally contained a perch, food, water, grit and a nest bowl; eggs and nest material were sometimes supplied. Every subject was very familiar with handling and with the cage before testing started.

Testing began not less than 2 weeks after the implantation of electrodes and continued for several months, in some cases for more than a year. Initially each subject was given four pilot sessions of about 30–45 min each, during which a variety of stimulus parameters were used and the bird was exposed to one of the model predators and to other doves in an adjoining cage. Those subjects that seemed to show a possible change in their agonistic responses as a result of ESB were later subjected to further detailed experimentation. No subject received more than one experimental session per day.

The two model "predators" used in this study were a white rubber model of a human skeleton and a black rubber model of a spider, each about 2 in. long. They are children's toys. Each stimulus was presented on the end of a long thin rod inserted under the plastic skirt. The spider is a "stronger" stimulus than the skeleton. Both stimuli elicit violent agonistic responses not only in doves but also in a variety of other birds and rats and monkeys. The behavior ranges from attack to flight and seems identical to that caused by a real predator such as a small rat. These responses in many ways resemble the behavior shown to other members of the same species during courtship and nest defense.

The test cage was housed in a soundproof cubicle. The bird could be viewed through a half silvered mirror and on closed circuit television; behavior was continuously recorded on a videotape recorder. During stimulation, comments were made on an audio tape recorder and on a data checklist. Between each test the videotape was played back and the behavioral items were typed into a Linc-8 computer from a special keyboard, which added real time signals and stored all the data for subsequent analysis. The application of ESB varied between 10 sec and 1 min according to the experiment involved. An interval of at least 2 min was allowed between each test and in cases where there appeared to be an aftereffect longer periods were used.

The methods of testing have been described by Harwood and Vowles (1967) and Vowles and Beazley (1973). Essentially we investigated the following:

1. whether overt, agonistic behavior was elicited by ESB in a repeatable and reliable manner, and if so whether this was stereotyped or variable, both in its behavioral component and its threshold;

2. if subthreshold ESB facilitated agonistic responses to the model predators, and if so whether this was affected by the position of the predator in the left or the right visual fields;

3. how the thresholds of behavior changed following stimulation;

4. whether the responsiveness to external stimuli varied following stimulation;

5. whether the latency of responses to ESB were facilitated by the presence of a model predator.

III. Results and Discussion

Most of the detailed results to be discussed here have already been published elsewhere, but the separate topics were then considered in isolation. In this section the results will be presented in summary, an attempt made to relate the behavioral effects of ESB to the neural structures involved, and a general synthesis attempted.

A. Characteristics of Agonistic Effects

A wide range of behavioral effects of ESB were found. These included pecking, preening, sleeping, defecation, panting, sunbathing, and many others. No reproductive behavior was observed, and vocalizations were only obtained in a very few cases. Most electrodes elicited more than one kind of behavior, sometimes, but not always with consistently different thresholds. A common combination (as reported by Delius, 1970) is pecking at food material, preening, and sleeping. When agonistic behavior occurred it was often followed by preening, possibly as a consequence of changes in feather postures, etc. Forced movements and simple motor effects were seen from many loci. This paper will be concerned on with the effects of ESB on agonistic behavior.

The movements and postures making up agonistic behavior have been described in detail by Harwood and Vowles (1967) and resemble those described for pigeons and other species by the authors listed previously. All behavioral items occurring before, during, and after ESB were recorded and analyzed; but for convenience some are grouped under the headings of aggressive, defensive, or fearful behavior. The items falling in these three categories are listed below.

Classification of Agonistic Responses

Aggressive	Defensive	Fearful
Pecking	Wing-raising	Escape
Wing-slapping	Wing-fending	Fleeing
Wing-boxing	Leaning	
	Crouching	

Other items such as alarm (an upright posture with elongated neck and sleeked feathers) or avoidance (an ambivalent attitude side-on to an intruder) are specified when they occur.

The effects of ESB on agonistic behavior fell into three general categories:

1. Group 1

Stimulation by itself produced no overt agonistic behavior, but often induced restlessness and changes of feather posture associated with emotional states. If the model predator was present ESB was found to facilitate aggressive, fearful, or defensive behavior to the external stimulus. The subject showed changes in the vigor and duration of agonistic responses to the predator, and sometimes agonistic responses were produced which had not been seen in the unstimulated state. The same electrode under the same stimulus conditions sometimes affected different types of agonistic behavior, and in some test periods could increase both fear and aggression.

A methodologically important point here is the criterion for accepting that ESB has produced a change in agonistic behavior. This is entirely statistical, and was done by comparing the reactions to the model predator in unstimulated and stimulated conditions. Since the mood of the bird changed during one experimental session as well as between sessions, it was necessary that control and test periods be intermingled during the same overall test session. Since stimulation often produces aftereffects lasting several minutes, adequate time must be left between stimulation tests in order not to distort the controls. In fact prestimulation responses, responses during

Table II. *The Effects of ESB on the Occurrence of Agonistic Responses to a Model Predator* [a]

	Number of tests	
	44	70
Behavior	Control	ESB
Looking down	20	64
Avoidance	9	33
Escape	0	25
Charge	24	33
Pecks	76	152
Wing raising plus wing slaps	3	27

stimulation, and responses after stimulation must be treated separately for analysis. A typical series of control and stimulation responses are shown in Table II. This makes clear the variability of the effects of ESB and the necessity for statistical analysis.

2. *Group 2*

Electrical brain stimulation elicited very stereotyped agonistic responses from these animals, such as leaning, wing raising, and wing-fending. The response is always stimulus bound, and has a very short latency (less than 1 sec from the onset of stimulation); the same electrode under the same conditions always gives the same response; the presence of an external stimulus has no effect on the response (except that ESB effects may be temporarily suppressed if the predator elicits a competing response); and there are virtually no aftereffects of stimulation. When subthreshold ESB is used there is no effect upon the agonistic responses to a model predator. These effects represent a subgroup of the "automatic" agonistic responses (automatisms) described for example by Putkonen (1967), Phillips (1968), and Maley (1969).

3. *Group 3*

Electrical brain stimulation elicited overt agonistic responses in the absence of appropriate external stimuli. When these responses occurred they appeared to be typical of normal agonistic responses to a real predator—and often looked as if the subject was responding to an hallucination of a dangerous intruder (avian paranoia?), although this should be taken as a description not an explanation. The responses differ from those in Group 2 in their greater variability as well as the apparent normality of their motor patterns. This variability can be shown in a variety of ways—fluctuations in threshold and latency of responses, changes in the type of response (e.g., from attack to fleeing) both within and between test sessions, failure to elicit a response during some test sessions, and the breaking off of a behavioral sequence before its completion. Not all subjects show the same range of variability. Thus in Fig. 2 the thresholds for eliciting fleeing in one prepara-

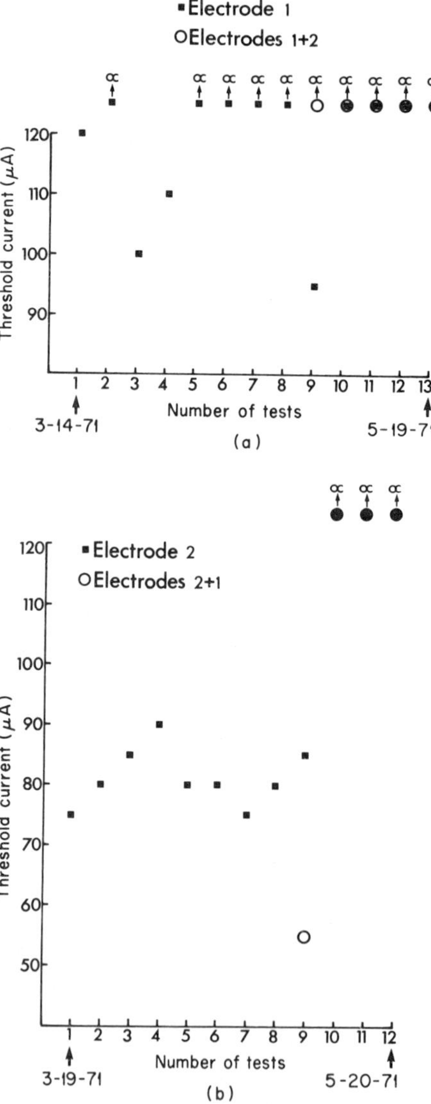

Fig. 2. The threshold for eliciting flight and fleeing at two electrode sites in Bird F.O. 15. At electrode 1 (■) fleeing was elicited on only four of the nine days of tests. At electrode 2 fleeing was elicited during every test. The days on which tests started and ended are indicated below the abscissa. The double symbol indicates tests where both electrodes were used.

tion are shown, and it is clear not only that the same behavior is elicited in almost every test, but that the threshold is very stable for electrode 1. In the same bird electrode 2, which also elicited fleeing, showed not only variations in threshold but in a number of tests no fleeing at all could be elicited.

When the same behavior is repeatedly elicited, it appears fairly automatic. However, these effects cannot be included in Group 2 because they appear less rigidly stereotyped (i.e., unlike forced movements), they have a longer latency, the behavior may outlast the stimulus by a few seconds, and, as will be seen later, subthreshold ESB facilitates agonistic responses to the model predators. The relatively invariant type of responses described here seems to represent the extreme of a range from very variable to more predictable effects.

Some other variable responses are exemplified in Table III, which shows the number of tests (using optimal ESB, as determined in pilot sessions) in which various items of agonistic behavior were elicited. Electrode 2 in both S171 and S348 were located in the basal palaeostriatum in the region of the tractus thalamo-frontalis medialis. Both electrodes elicited defensive and fearful behavior but 348/2 appeared relatively less fearful. In 171/2 defensive behavior was elicited in half and fearful behavior in the other half of the tests—thus agonistic behavior occurred on every test. However, in 348/2 no agonistic responses occurred in 3 tests, and in 17 of the remainder only leaning (a weak defensive response) occurred. These effects are not dependent upon the individual, since two other electrodes in the same subjects gave different results. In subject 171/7 the electrode was symmetrically placed on the opposite side of the brain to 2, but it elicited wing-raising in only four of 13 tests and this occurred without any context of agonistic activity. Birds commonly respond to a sudden stimulus by partially raising a wing for a brief period, and it seems likely that ESB functioned here as a startling stimulus. In subject 348 electrode 8 was in the neostriatum. Although agonistic responses occurred in only 28% of the tests they appeared to be typical of

Table III. *The Number of Tests in Which Various Items of Agonistic Behavior Were Shown in Two Subjects*

Bird/Electrode	No. of tests	Alarm	Avoidance	Wing raise	Flees
171/2	12	10	8	6	6
171/7	13	0	0	4	0

	No. of tests	Alarm	Avoidance	Lean	Wing raise	Wing slap	Flees
348/2	38	17	10	17	6	5	7
348/8	25	4	0	0	0	3	4

normal agonistic behavior, and this electrode was therefore included in Group 3.

The examples in Table III have been presented in some detail, since they raise two important questions for this kind of work. First, how repeatable must the effects of stimulation be if they are to be regarded as significant? Well-defined items of agonistic behavior such as wing-raising, or fleeing, rarely occur spontaneously in these conditions, so that even a few, infrequent occurrences during stimulation would be regarded as statistically significant; other actions, such as alarm, occur in response to unavoidable extraneous sounds, etc. Nevertheless, although statistical significance may be reached the biological significance of such infrequent events must still be decided. Second, it is important that the overall behavior pattern be considered. For example, wing-raising occurred in isolation from its normal agonistic behavioral context in 171/7. Third, two electrodes in apparently similar positions may not produce the same effects. Although this is a common finding in brain stimulation, negative results are seldom reported in detail and often an experimenter will discard those preparations which give behavioral effects in which he is not interested. The heuristics for this are quite clear, but in interpreting the results caution is required not only in attributing one particular behavioral function to a given location, but also in assuming that the unwanted behavior does not affect the behavior which is being studied.

The main characteristics of agonistic behavior affected by ESB have already been outlined under Groups 1, 2, and 3. Some further characteristics which refine this classification will be described later. But first the location of the electrodes which fall into these three groups will be described.

B. Anatomical Localization

The localization of all electrodes which affected agonistic behavior are shown in Fig. 3. The solid circles correspond to Group 1 responses, the circles containing stars to Group 2 responses, and the open circles to Group 3 responses. No attempt is made to indicate the type of agonistic patterns which are involved. In Fig. 4 the locations of all electrodes which did not influence agonistic behavior are shown. In addition, 16 electrodes with negative effects were located outside the brain tissue.

The locations of electrodes which fall in Groups 1 and 2 make a clear pattern. Group 1 loci are restricted to the forebrain, the majority lying in the neostriatum and the hyperstriatum ventrale. A few lie in the paleostriatum and septal areas. In the first two areas they are concentrated above the dorsal aspects of the ectostriatum; and a few are located just lateral to the ventricle (extending in an anteriodorsal direction in a band from Group 3, which lie more ventrally).

The Group 2 loci form a compact group in the diencephalon. They lie in a band running ventrolaterally from below the anterior dorsal thalamic nuclei in a region medial and ventral to the rotundus and lateral to the ovoidalis. It is in this region that the tractus tecto-thalamicus dorsalis runs (Huber & Crosby, 1929) which may well implicate this pathway.

The loci of Group 3 form a much more widespread pattern which runs through the telencephalon, diencephalon, and mesencephalon. Within the midbrain a few sites are found within the reticular system—a region through which course many ascending and descending fibers. In the tectal area numerous sites are found ventral and medial to the ventricle. These lie in the region of the intercollicular nucleus (ICo) and the nucleus mesencephalicus lateralis dorsalis (MLd). In the dove these nuclei are closely and intricately apposed, and with the size of electrodes used it is not possible to separate the functions of the two nuclei. However, electrodes have been found with their tips clearly restricted to one of the two nuclei, and it seems probable that both are implicated. In the thalamus a number of sites were located in the anterior dorsal thalamic nuclei (DLA)—particularly the lateralis pars medialis. Many active sites were also found in anterior hypothalamic regions—in the preoptic and supraoptic areas anterior and dorsal to the septo-mesencephalic tract (TSM). The diffuse organization of the cell bodies in these regions and the large size of the electrodes make it difficult to specify particular nuclei. Very few active sites were found in the posterior hypothalamic areas (unlike the results obtained by other workers), but very few electrodes were successfully implanted and tested in this region.

The forebrain areas involved include the archistriatum, the paleostriatum augmentatum (a band of active sites runs antieriodorsally here just lateral to the ventricle, as described also by Putkonen, 1967), the septal region and a few areas of the hyperstriatum and neostriatum. As would be expected a large number of active sites are found in the main tracts connecting the forebrain with the diencephalon and mesencephalon—particularly the tractus septo-mesencephalicus (TSM), the occipito-mesencephalic tract (OM), the anterior commissure (A.C.) and parts of the lateral forebrain bundle (FPL). It is significant that a large number of negative points were obtained from the ectostriatum and the large central core of the FPL, which connects it with the rotundus (Hodos & Karten, 1966; Powell & Cowan, 1961). The active fibers seem to lie more medially in the tractus frontothalamicus medialis (TFM), which connect the paleostriatal and neostriatal areas with thalamic regions (Huber & Crosby, 1929).

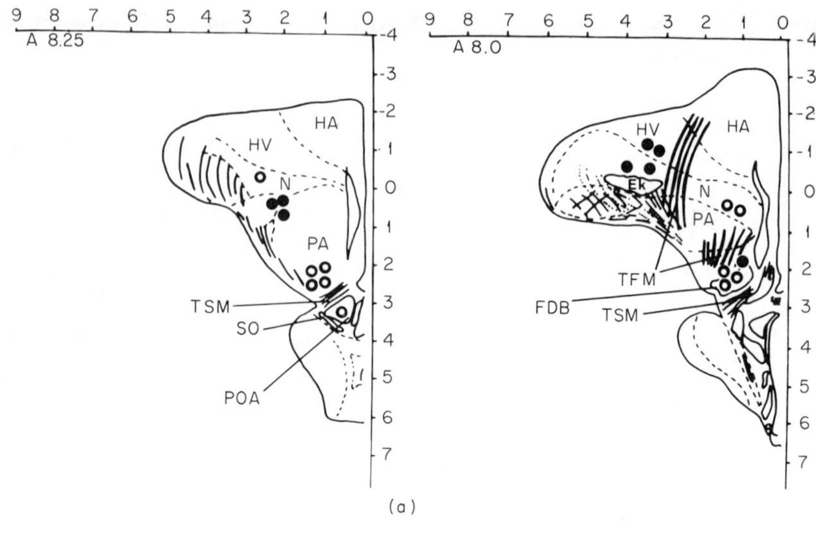

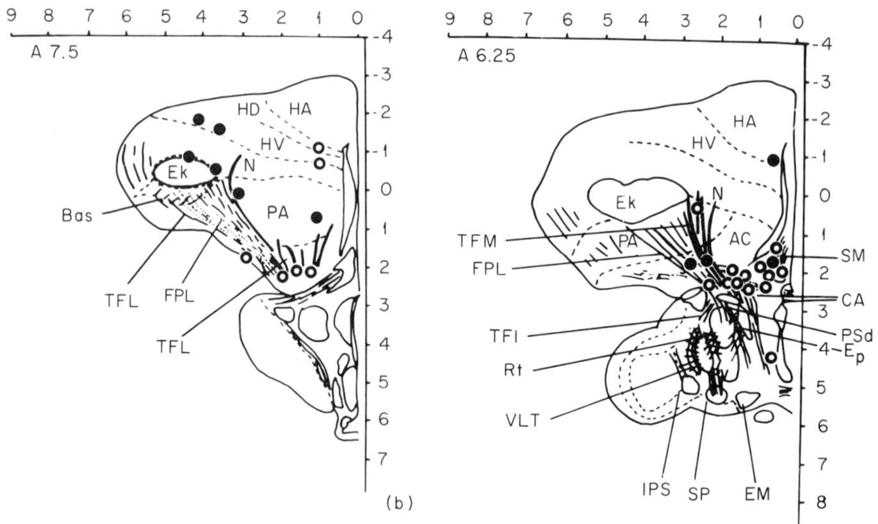

Fig. 3. The location of electrodes which produced agonistic affects in the Barbary dove. Solid circles correspond to Group 1, the circles containing stars to Group 2, and the hollow circles to Group 3 (see text). All types of agonistic affects have been grouped together. The numbers indicate the anterior position in millimeters from a zero position passing through the pineal body.

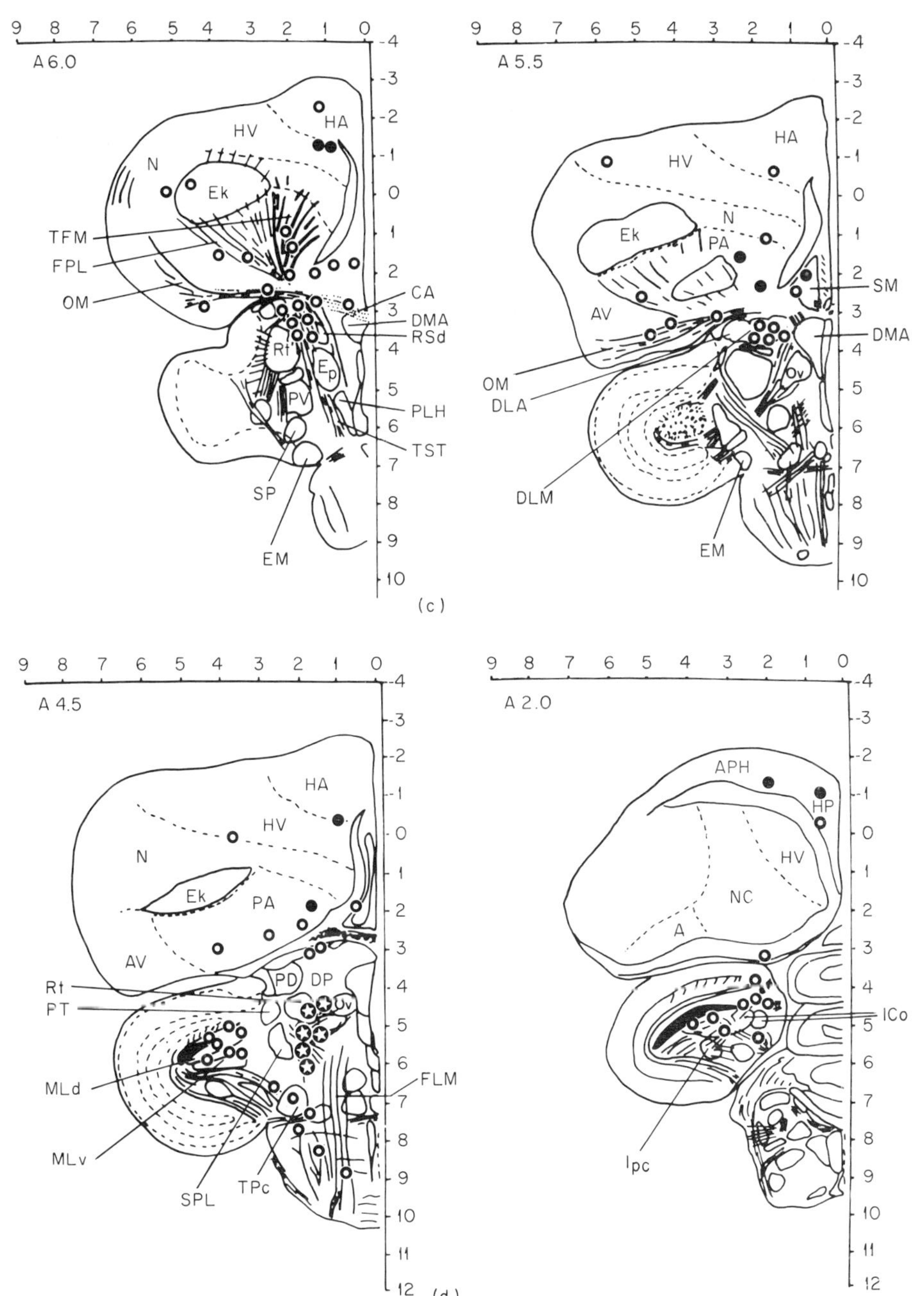
A6.0
HV
HA
N
Ek
TFM
FPL
OM
CA
DMA
RSd
Rt
Ep
PV
PLH
TST
SP
EM
A5.5
HA
HV
N
Ek
PA
SM
AV
DMA
OM
DLA
Ov
DLM
EM
(c)
A 4.5
HA
HV
N
Ek
PA
AV
Rt
PT
PD
DP
Uv
MLd
MLv
SPL
TPc
FLM
(d)
A 2.0
APH
HP
HV
NC
A
ICo
Ipc

1. Comparisons with Other Species

Most of the brain areas shown to be involved in the Barbary dove have also been implicated in other species by other workers. Figures 5–8 show the loci from which emotional behavior was obtained in domestic chickens and mallards. The results are also in agreement with those of Akerman (1966) and Cannon and Salzen (1971) in the domestic chicken and the pigeon. There is good agreement among all these studies. Many emotional responses can be obtained from the archistriatum and the occipito-mesencephalic tract (OM). Within the archistriatum the ventral regions and those areas near the OM are most involved. Perhaps the greatest concentration of active loci lies in and around that area where the tractus septomesencephalicus, the lateral forebrain bundle and the occipito-mesencephalic tracts swing downward between the base of the forebrain and the anterior diecephalic areas—implicating the septal, the basal paleostriatal, the supraoptic and preoptic regions, and their tracts. In the present study more points were found in the neostriatal areas and hyperstriatal areas, and this may be due to the rather more detailed testing methods used. The present study seems also to involve the paleostriatum primitivum and possibly the tractus thalmo-frontalis medialis more than previous work (this area is specifically mentioned by Putkonen, 1967).

Other workers have reported the involvement of dorsal anterior thalamic nuclei, and the intercollicular and dorsolateral mesencephalic nuclei. This was strongly confirmed, numerous active sites being identified in these regions in the Barbary dove.

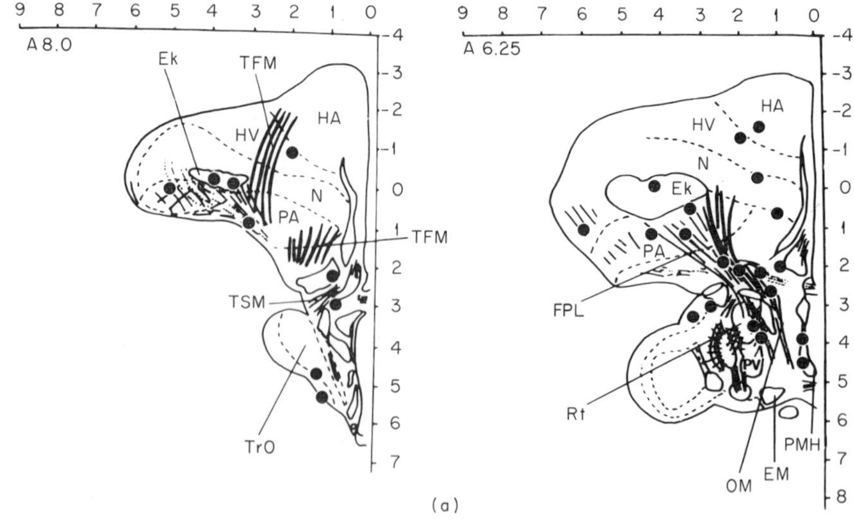

(a)

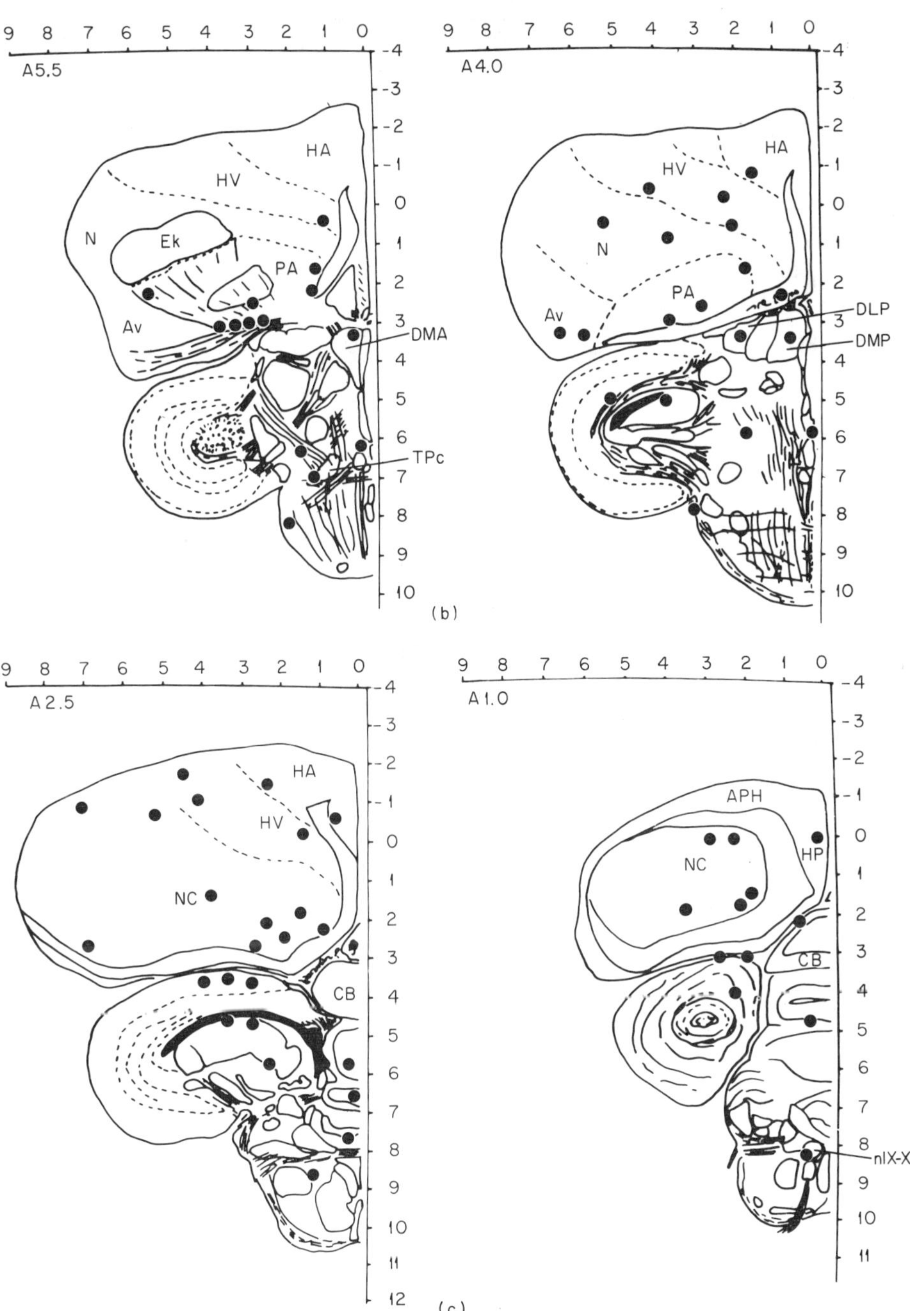

Fig. 4. The location of electrodes which produced no agonistic affects in the Barbary dove.

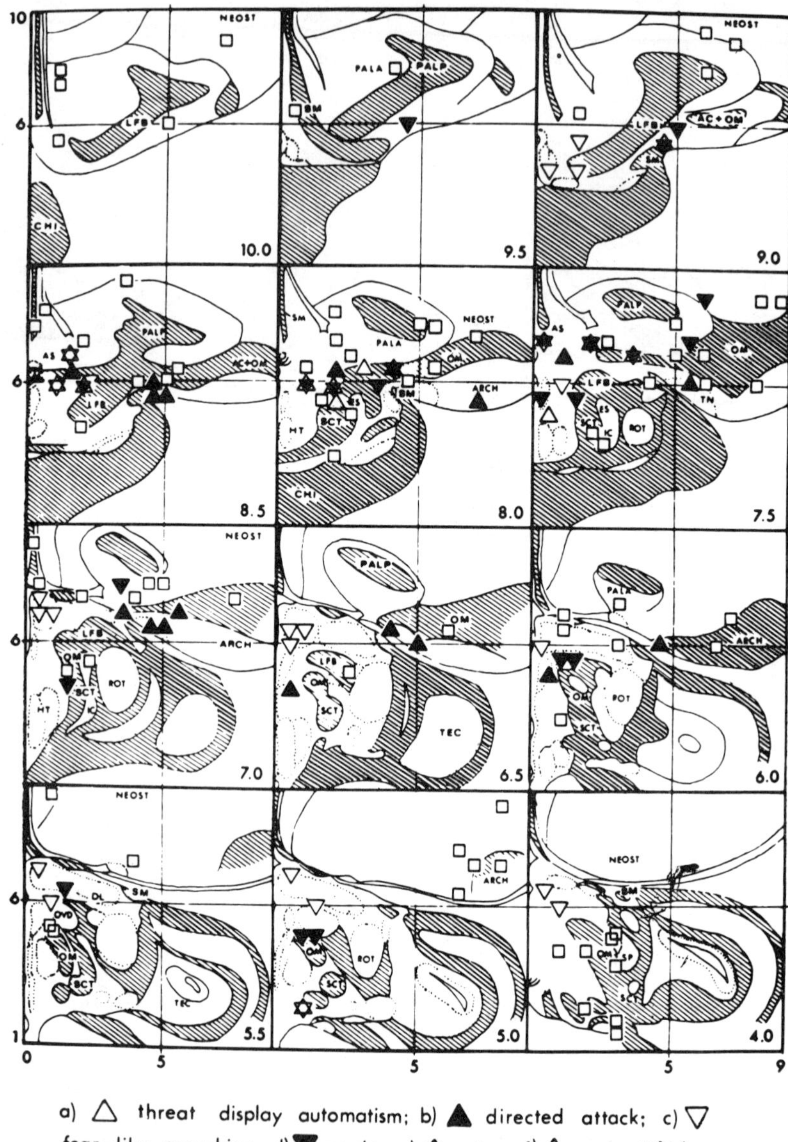

Fig. 5. Locations from which agonistic behavior was obtained in domestic chickens. [From Putkonen (1967).]

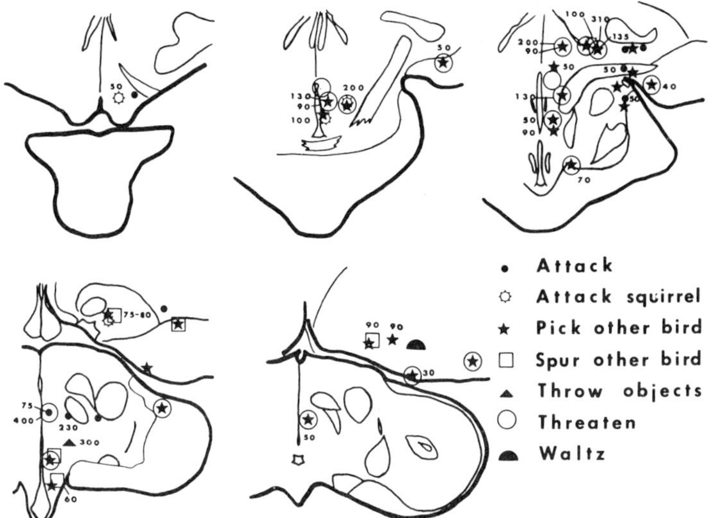

Fig. 6. Electrodes from which agonistic responses were obtained in the domestic chicken. [From Phillips & Youngren (1971).]

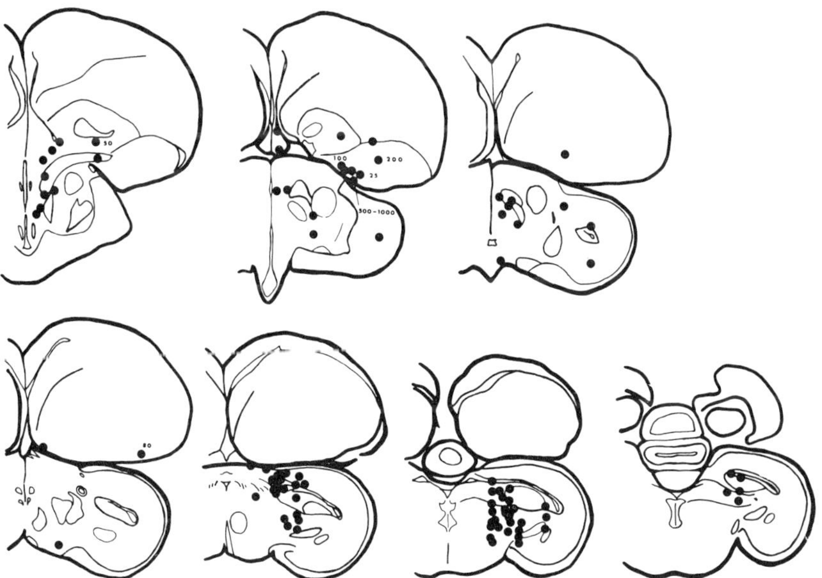

Fig. 7. Locations from which panic responses were obtained in the domestic chicken. [From Phillips & Youngren (1971).]

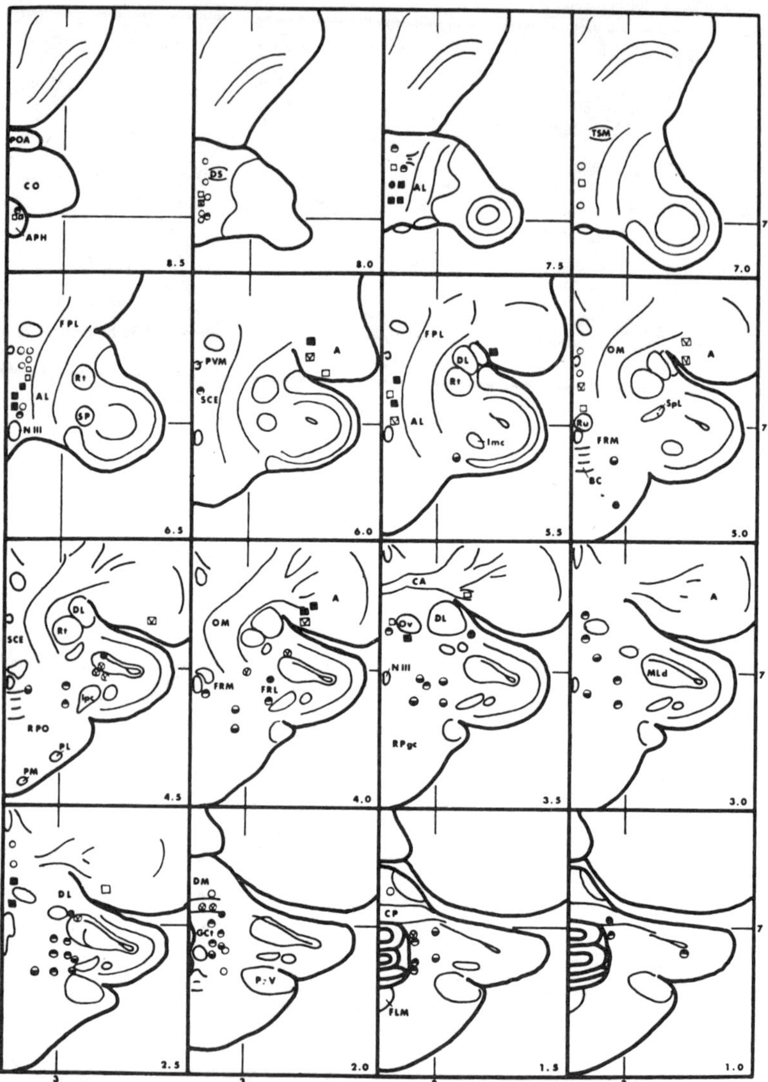

Locations of sites producing aggressive or escape reactions. Symbols: (■), directed attack; (□), threat display automatism; (☒), aggressive calling; (○), sneak-into-cover; (●), crouch-flat; (◓), directed escape; (◒), undirected escape; (⊗), alarm calling.

Fig. 8. Location of electrodes producing agonistic responses in the Mallard duck. [From Maley (1969).]

In other studies numerous active loci could be found very close to the midline both in the posterior hypothalamus and the midbrain. This was not confirmed in the present study, perhaps because too few electrodes were implanted and tested in this region.

It will be clear that all the studies quoted are in good agreement concerning the main areas involved in agonistic behavior, and these can now be regarded as well established. A more detailed discussion of the "functional" neuroanatomy will be reserved until later.

C. Types of Agonistic Responses

So far, various kinds of agonistic responses have been grouped together, irrespective of their nature. In this section the effects of ESB on aggressive, defensive, and fearful behavior will be analyzed to see if those areas of the brain involved in different kinds of behavior are disparate or overlapping. This may be a theoretically important point for understanding the motivation underlying bird displays. For example, in gulls (Tinbergen, 1959) some threat displays are apparently derived from a combination of attack and flight behavior. Von Holst and von St. Paul (1963) have described how stimulation through two electrodes which separately elicit attack and flight might together cause threat behavior.

Behavior classified as aggressive, defensive, or fearful has been defined earlier in this chapter. It should be stressed that although the behavior is shown to predators as well as to other doves it was not possible to study the effects of ESB on responses to other doves. All the subjects were in full breeding condition and kept in isolation. As a consequence they reacted to another dove with intense and prolonged courtship or attack, which overcame any effects of ESB except at extremely strong stimulus currents. This methodological flaw in the present study, is now being remedied by the use of castrated subjects.

Two criteria were used in determining the loci which gave rise to aggressive, defensive, and fearful behavior. First, only those sites were included which could be clearly located in the histological material; all sections in which, for example, damage made location of the electrode tip equivocal were rejected. When precise positive localization is desired, poor histological material and stereotaxic coordinates are not adequate guides. Second, only those preparations in which ESB gave clearly identifiable behavioral effects were included. The locations of the electrodes are shown in Fig. 9, which includes Groups, 1, 2, and 3 from the previous section. In the case of Group 3 both the overt behavior elicited by ESB alone and the responses to external stimuli which were facilitated by subthreshold ESB are included;

these are not always the same. Stars indicate when one electrode affected more than one kind of behavior, and the specific behavior is indicated.

The distribution of points involving different kinds of agonistic responses shows no simple pattern. A count of the number of loci with different agonistic effects grouped according to the area of the brain involved is given in Table IV. The relative distribution of aggressive, fearful, and defensive points appears to be independent of the specific brain areas with a few exceptions. The anterior commissure and the mesencephalic reticular areas seem particularly involved in fearful responses, while the intercollicular nuclei and the nucleus mesencephalicus lateralis dorsalis have rather few fearful points. If the hyperstriatal, hippocampal, and septal areas are considered together with the tractus septomesencephalicus (which is acceptable on anatomical grounds) they have more fearful loci than other areas; this result is strengthened by the fact that most of the defensive responses that result from stimulation in these areas are biased toward fear (e.g., leaning, crouching, and wingfending, accompanied by sleeked feathers and an elongated neck). In distinction, the dorsal anterior thalamic nuclei and the tractus thalamofrontalis medialis (which links these nuclei with the basal paleostriatum) have a relatively high number of aggressive loci, and the defensive responses there are biased toward aggression (e.g., wing-raising accompanied by feather erection and puffing up, like a broody hen).

Electrodes from Group 3 which elicit agonistic responses of one particular kind (e.g., defense) may nevertheless show variability in the specific behavioral items which fall in this category. An attempt can be made to specify this variability by counting the number of different kinds of agonistic

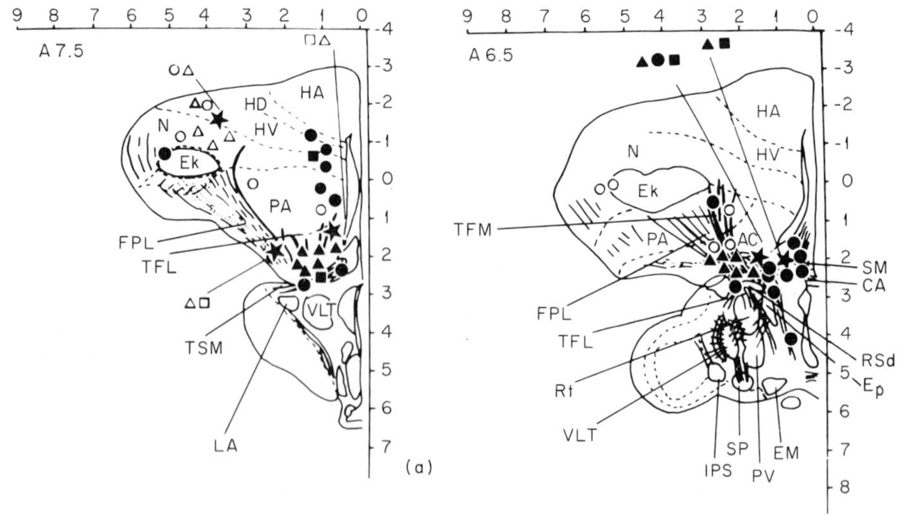

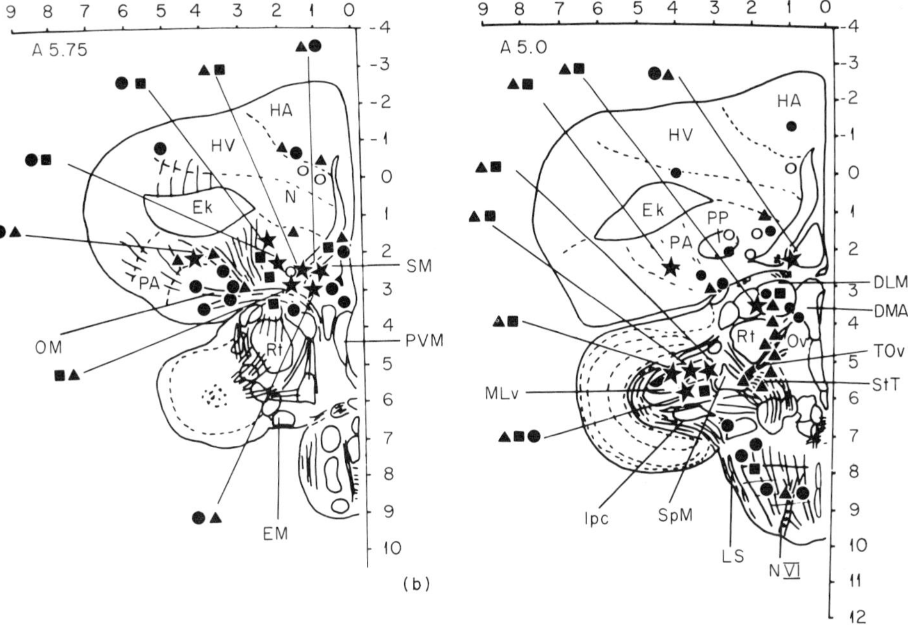

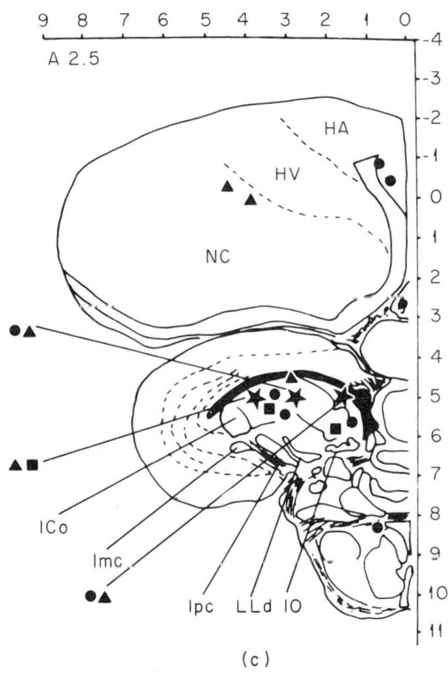

Fig. 9. Location of electrodes which produced defensive kinds of agonistic responses in the Barbary dove. The hollow symbols represent birds from Group 3 and the solid symbols from Group 1. (○, ●), fearful behavior; (■), aggressive behavior; (△ . ▲), defensive behavior; (★), combinations of agonistic behavior.

Table IV. *The Number of Sites Affecting Different Types of Agonistic Behavior in Different Regions of the Dove Brain*

Location	Fear	Aggression	Defense	Total for Area
Paleostriatum primitivum	15	5	7	27
Archistriatum, and tractus occipito-mesencephalicus	12	5	10	27
Neostriatum	7	1	3	11
Hyperstriatum	14	0	2	16
Septal Area	4	2	4	10
	20	*3*	*10*	*33*
Tractus septo-mesencephalicus	2	1	4	7
Anterior dorsal thalamic nuclei	3	2	3	8
Tractus thalamo-frontalis medialis	5	5	7	17
	8	*7*	*10*	*25*
Intercollicular nucleus and nucleus mesencephalicus lateralis dorsalis	11	8	8	27
Anterior hypothalamic supraoptic and preoptic areas	4	6	10	20
Anterior commissure	5	0	0	5
Mesencephalic reticular areas	6	2	1	9
Tractus tecto-thalamicus dorsalis	3	1	4	8
Others	8	2	5	15
Totals	99	40	68	207

responses that can be elicited from any one electrode. When this is done the greatest variability is found in electrodes in three main areas: (1) the intercollicular nucleus plus the nucleus mesencephalicus lateralis dorsalis; (2) the dorsal anterior thalamic nuclei; and (3) the basal palaeostriatal and septal areas where the occipito-mesencephalic tract, septomesencephalic tract, and the tractus thalmo-frontalis medialis run close together. This distribution is confirmed by the concentration in these areas of electrodes which give rise to two or more kinds of agonistic responses (stars in Fig. 9). Within these areas there are also concentrations of electrodes that individually give rise to only one kind of response, but whose near neighbors give rise to another kind.

1. Comparison with Other Species

Direct comparison with the results of other workers raises some difficulties. First, although the description of behavior is at a molecular level (e.g., crouching, fleeing, escape), these actions are often classified in a molar way for purposes of interpretation (e.g., fear, defense, aggression). Thus, in the present study, crouching is ranked as defensive, whereas in

Putkonen's (1967) and Maley's (1969) results it would be described as fearful. Similarly threat is described by them as aggressive, while in the present study (wing-raising) it is called defensive. Thus, it is necessary to compare the actual behavior described when this is possible. Second, in the present work no evidence could be obtained concerning behavior which is associated with intraspecific activities, whereas other workers tested such behavior.

Despite these qualifications there is nevertheless good agreement with the results for other species as can be seen from Figs. 5–8. More fearful responses were associated with the septal areas in the dove, as already indicated, and more aggressive and defensive behavior with the anteriodorsal thalamic nuclei. As in Putkonen's (1967) work there was no evidence that the archistriatum or the OM were biased toward aggression or fear. Potash (1970), Maley (1969), Delius (1970), and Brown (see this volume) have reported that in the region of the intercollicular nucleus agonistic behavior produced by ESB is often accompanied by appropriate vocalizations. In only two cases were vocalizations obtained in the present study: low and indignant squawks, of a most atypical nature.

Many points eliciting "panic" have also been described by workers for the mesencephalic region. No such concentration of loci were obtained in the dove, although some similar effects were observed. It is not clear whether frantic fleeing and flying should be regarded in the same category as agonistic responses to predators or to members of the same species. It might be associated with painful effects of stimulation.

D. Interaction with External Stimuli

Some electrodes produced behaviors in which the subject interacted with the model predators. Whether such interactions occur seems to depend not only upon the general location of the electrode, but also whether its tip resides in a tract or an anatomical nucleus. In Table V the results are classified to show this relationship. Only those loci which two independent observers both rated as having the electrode tip unequivocally in a tract or a nucleus were included.

It can be seen that in the diencephalon and mesencephalon virtually all the electrodes in nuclei showed interactions with the model predators. This is reassuring, since it implies that the models were adequate stimuli. Conversely the electrodes which lay in tracts were equally divided between those which did or did not produce interactions. This does not imply that there are no loci elsewhere in these regions which do or do not produce interaction effects (the reticular system loci in fact do so), bui does suggest the decreased possibility of such interactions when the electrodes are in tracts. These results all refer to Group 3 loci which do and do not interact with external stimuli.

Table V. *Distribution of Electrodes Which Elicit Agonistic Behavior Directly, and Which Differ in Their Interaction with External Stimuli*

	Location			
	Diencephalon and mesencephalon		Tracts between diencephalon and forebrain	Forebrain areas
	nuclei	tracts		
No interaction with external stimuli	1	7	10	7
Interaction	24	6	30	22
Totals	25	13	40	29

When the electrodes are located in the main tracts connecting the telencephalon with the diencephalon and midbrain (OM, FPL, TSM, AC), many more loci are found which permit interaction between ESB and external stimuli. In the wider forebrain areas the majority of loci also permit interactions; there are, of course, many diffuse fibers in the telencephalon and nuclear and tract effects cannot be meaningfully separated. The significance of these findings will be discussed in the final section.

So far, the interactions between ESB and the external stimuli have not been described. These interactions were assessed in three different ways (although unfortunately not all the techniques were used for all electrodes, since the methods were developed over several years).

1. Orientation

When overt agonistic behavior is elicited by ESB alone, the dove commonly adopts a posture as if responding to an intruder on the contralateral side to the electrode. To test if this behavior could be oriented to an external stimulus the model predator was moved about the cage, and two independent observers judged whether appropriate orientation occurred. In many cases correct orientation was shown. This was more frequent when the predator was presented on the contralateral side to the electrode; presumably when ESB alone produced a strong contralateral orientation the bird found it difficult to turn to meet a stimulus on the opsilateral side. Unfortunately, these tests were only done on electrodes in those regions involving the TOM, FPL, TSM, and adjoining areas in the basal forebrain and anterior hypothalamus (further tests for diencephalic and mesencephalic sites are now underway, with an indication of positive results). Of 37 loci tested, 24 showed positive orientation to the external stimulus. When stimulus currents were increased more stereotyped behavior was elicited, which resulted in a failure to show successful orientation to the external stimulus. It is possible that those electrodes which permitted no orientation to an external stimulus even at low currents were placed so that they maximally excited a pathway involved in a motor command. Effects of this sort correspond to the "automatisms" reported by other workers.

2. *Latency*

The group of electrodes just discussed was also examined in terms of the effect of the presence of the model predator on the latency of agonistic responses elicited by ESB. The model predator was always presented contralaterally to the electrode. Only those subjects were used that did *not* show the same responses to the model predator as to ESB. Electrodes that gave a positive effect for the orientation tests also showed a shortening of the latency of responses to ESB when the external stimulus was present. When subthreshold current was used the external stimulus also facilitated the appearance of the agonistic response, which would not otherwise have been shown. Figure 10 shows the relationship latency and stimulus strength. These results must mean that the excitatory effects of ESB and an external stimulus can summate.

3. *Probability of Response*

In these tests the effects of ESB on the responses to the model predators were examined. Subthreshold currents were used for those electrodes which produced overt agonistic behavior with ESB alone. All electrodes were tested. No effects, either facilitatory or inhibitory, were found for Group 2. All electrodes in Group 1 (by definition) and approximately 75% of those in Group 3 produced an increased probability and vigor of agonistic responses. No significant anatomical pattern could be detected in the distribution of positive and neutral loci with the exception of the "tracts versus nuclei" distinction already made. Eleven of the electrodes that showed no interaction with an external stimulus in either the latency or the orientation tests did show facilitation here. Thus some of the "automatisms" described by other workers might be more flexible than previously supposed. It should be noted that the agonistic responses to the external stimulus which were facilitated by ESB were not always the overt responses elicited by ESB alone, although they often fell in the same general category (e.g., defensive).

4. *Laterality of Response*

When carrying out the tests just described, it was found necessary to control and distinguish between responses to presentation of the model predator in the right and the left visual fields. It was discovered (Vowles &

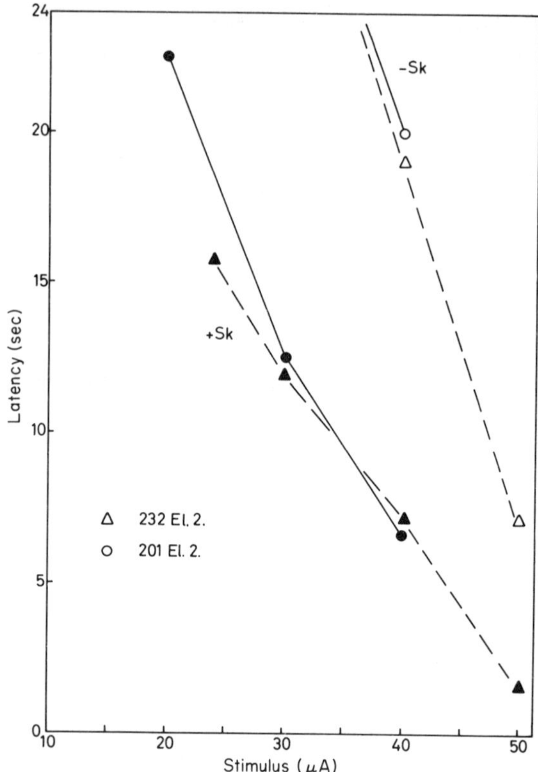

Fig. 10. The relationship between latency and stimulus strength for eliciting defensive wing raising in two Barbary doves. In the hollow symbols no external stimulus was present (−Sk). In the solid symbols a model predator (+Sk) was present. Each point is the mean of four tests.

Tarala, 1971) that facilitation of agonistic responses sometimes occurred when the external stimulus was presented contralaterally to the electrode, sometimes ipsilaterally, and sometimes bilaterally. In some cases the same electrode could increase fearful responses to the model spider presented on one side, and aggressive responses when presented on the other! This sometimes depended upon current strength but not always. Detailed examination of the loci giving ipsilateral, contralateral or bilateral effects showed no obvious anatomical patterns. However, in those cases where ESB alone produced overt agonistic behavior the facilitation produced by subthreshold stimulation was bilateral in the great majority of tests. In Group 1, the three possible effects occurred with fairly equal frequency. In no cases could the different effects be attributed to any anatomical substrate.

E. The Aftereffects of Stimulation

In a previous paper (Harwood & Vowles, 1967) it was shown that stimulation through some electrodes in Group 3 produced fairly long aftereffects. After a few seconds of strong ESB there were either raised agonistic thresholds (following aggressive or defensive responses) or lowered thresholds (following fearful behavior). The threshold changes built up to a maximum over 3–8 min after stimulation, and declined slowly over 10–30 min. These changes were not artifacts, since similar effects could be produced by the vigorous agonistic responses to the model predator. The following procedures all produced effects of a fairly similar sort:

Initial Strong Stimulation	*Test for Threshold Changes*
ESB	ESB
ESB	Model Predator
Attack by model predator	Model Predator
Attack by model predator	ESB

Standard ethological methods were used to measure the thresholds and responsiveness to the model predator.

Only those electrodes used for orientation and latency studies were studied, but similar tests are underway for electrodes in the thalamus and mesencephalon. No aftereffects were found in Group 2.

In Group 1 ESB produced aftereffects but with a different time course. This was tested by presenting the model predator at various intervals after ESB and recording the agonistic responses obtained. Subjects proved to be very variable. Three kinds of responses were obtained as shown in Fig. 11. The same subject could behave in any of the three ways.

In brief, the aftereffect of ESB in the neostriatal and hyperstriatal region is to produce oscillations in emotionality from fear to aggression, with a period of about 2 min for each peak or trough of fear or rage. The phenomenon is reminiscent of robound in physiology. The oscillations usually died away in 4–6 min.

F. Effects of Repeated Stimulation

Most of the observations reported here are qualitative in nature. Nevertheless the data show certain regularities which consistently occur both within and between subjects. These are summarized as follows.

1. In Group 1 repeated stimulation during one experimental session (2–4 min intervals between stimulation) leads to a progressive build up of emotionality. The subject becomes increasingly anxious, and testing has usually to be discontinued after 30–45 min involving 10–15 test stimulations.

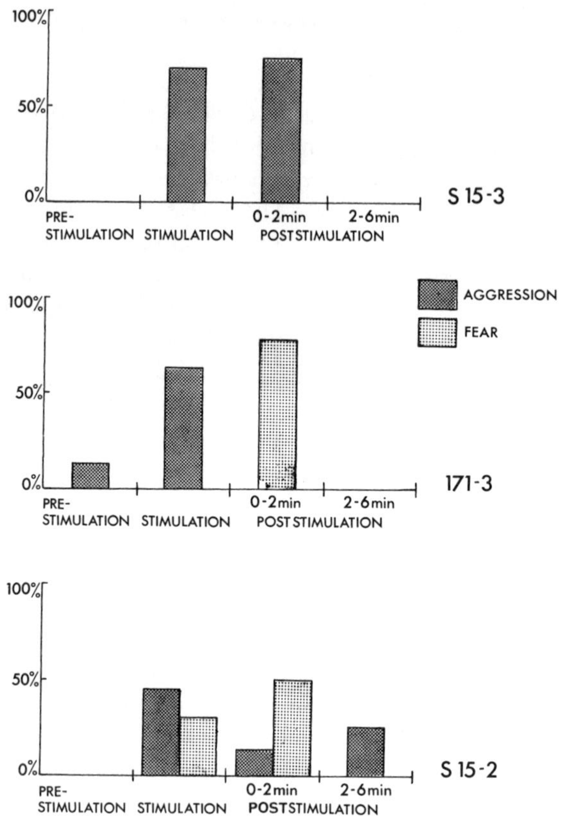

Fig. 11. The number of trials in which agonistic responses to the model spider were produced before, during and after ESB in three different subjects from Group 1.

It seems likely that temporal summation of excitatory states is occurring, which is consistent with the data on aftereffects. If stimulation is too prolonged, the bird becomes unreactive and adopts a sleeplike attitude. This is probably accompanied by some kind of epileptic activity.

2. In Group 3, similar repeated stimulation leads to more unpredictable effects. It is common for adaptation to occur during a period of stimulation; the behavior originally elicited becomes increasingly infrequent over the whole stimulation period, although otherwise the subject appears normal and spontaneously shows other activities. The time course of adaptation and recovery have not been followed in detail, but recovery is certainly complete after 24 hr. Von Holst and von St. Paul (1963) described adaptation and recovery and showed the first to be an active process. In the present study, adaptation was often preceded by facilitation (an increase in responsiveness with repeated stimulation) and by a change in the agonistic-responses displayed (e.g., from defense to flight).

When ESB was repeated over a longer time span—days or weeks—the subjects in Group 3 showed another effect. Although the thresholds for eliciting agonistic responses remained stable, the magnitude of the aftereffects of ESB declined, reaching noise level after 11–14 repetitions.

Although it might have been expected that the subjects would have become emotionally conditioned to the cages and stimulus situations, we have not detected any such signs. Habituation seemed to proceed normally, leading to greater placidity rather than anxiety in the test cage.

G. The Relationship between Agonistic Behavior and Neural Structures

This chapter has reviewed the types of behavioral change which ESB produced in the Barbary dove. The neuroanatomical substrate so far identified has been outlined in the various sections. The areas of the brain and the tracts involved are outlined diagrammatically in Fig. 12. The systems

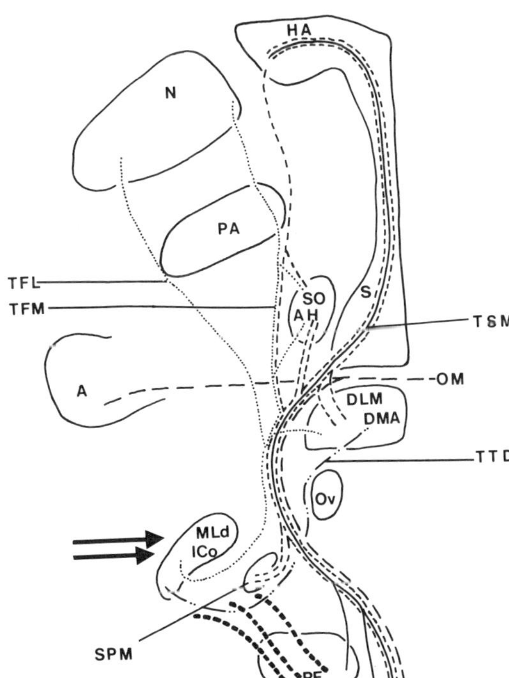

Fig. 12. A diagrammatic representation of the main brain structures involved in agonistic behavior.

outlined make some anatomical sense. Thus in the negative sense the main ascending visual pathways from tectum to rotundus and via the core-component of the lateral forebrain bundle to the ectostriatal regions, and the ascending auditory pathways via the ovoidalis to the telencephalon (Karten & Hodos, 1967; Karten & Revzin, 1966; Revzin, 1969; Biederman-Thorson, 1967, 1970a,b; Harman & Phillips, 1967) do not seem to be involved. In the positive sense Powell and Cowan (1961) have shown that the dorsal anterior thalamic nuclei give connections to the paleostriatum, the septum, hippocampal, and hyperstriatal regions—the fibers running close to those in the septomesencephalic tract, and possibly involving also the tractus thalamofrontalis medialis. Zeier (1971) has shown that parts of the archistriatum involved with emotional behavior in frustration have connections with the opposite side of the brain via the anterior commissure, and with diencephalic and mesencephalic areas via the occipito-mesencephalic tract. Between the tectum, other parts of the midbrain, the diencephalon, and the forebrain there is richness of connections.

In view of such a variety of neural structures we must ask why such complex circuitry is involved. What behavioral functions may correlate with the activity of different brain areas? Why should the circuits be designed in this particular way?

Perhaps we should be cautious when interpreting any of the results of experiments involving ESB. We know neither the effects of nervous tissue of the stimulus, nor even the extent of current spread around the electrode. Maybe it is time we stopped waiting for someone else to determine this empirically. In work on the dove we obtained behavioral effects with currents as low as 5 μA, and commonly work with 20–25 μA. The exposed area of the barrel is much greater than the tip of the electrode so that maximal stimulation should occur near the tip, where current density is greatest. In the diagrams illustrating the electrode loci the circles are drawn to scale with the width of the electrode tracks. We assume that stimulation is quite local. This is supported by the fact that electrodes with tips close together may produce very different results, and by the functional difference we found between stimulating tracts and nuclei. While remembering the need for cautious qualifications we nevertheless put forward the following speculations concerning the behavior functions of different parts of the brain and their interaction to produce an integrated pattern of agonistic behavior.

1. The MLD and ICO

Stimulation in these areas produced variable behavior to ESB alone and also facilitated responses to the model predator. These responses were appropriately oriented. Other workers have firmly established that these nuclei have an auditory function. It also appears likely that they have a visual function, although this supposition needs firmer histological and electrophysiological confirmation. In fish and mammals the collicular area is involved in locating stimuli in space: This applies to both visual and auditory stimuli, and there is some evidence that visual and auditory space are mapped together by a common mechanism.

When detecting and responding to predators, the dove will need at least two mechanisms: first, to identify the dangerous stimulus, which could involve both light and sound; and second, to locate and track the intruder so that the agonistic responses can be appropriately guided. We suggest that when a bird responds to an intruder these two processes are correlated with activity in particular groups of neurons and coded in particular fibers leaving the nuclei. When ESB stimulates cells in such tracts this would send to higher centers information that a potentially dangerous creature is present in a particular region of behavior space. The higher centers may produce a variety of appropriate actions, but since the output fibers of the sensory system has been stimulated by ESB, there could be no interaction with the external stimuli. A similar situation would hold for strong ESB within the nuclei, which would presumably excite cells there to discharge and produce an output. However, with weak ESB the cells in the nuclei would be only mildly facilitated and normal sensory input could interact with the facilitated cells. Thus the effects of the external stimuli will be facilitated by the ESB, increasing the subjects apparent responsiveness.

If these suggestions are correct, we could make four predictions. First, when ESB in these areas facilitates responses to a model predator, it should be effective only for objects in a particular spatial location. Such was shown by von Holst and von St. Paul (1963) in the domestic chicken, and we have seen similar effects in the dove. Second, a particular electrode should facilitate response to particular types of external stimuli; this again was found by von Holst and von St. Paul. Third, ESB in these loci should not by itself produce the aftereffects described for other loci. Fourth, the response produced by ESB should vary with, for example, the drive state and previous history and experience of the bird. All these predictions can be tested.

2. *Basal Forebrain and Anterior Hypothalamic Regions*

These regions were most likely to give rise to full but variable agonistic behavior patterns that interact with external stimuli and show prolonged aftereffects. A clue to their role may be obtained from the work of Hutchinson (1967, 1969, 1970), who showed that in these regions implants of solid testosterone elicited courtship behavior in castrated male Barbary doves. The courtship included aggressive activities such as charging (see Fig. 3). It is particularly noteworthy that a band of sensitive loci were found in the

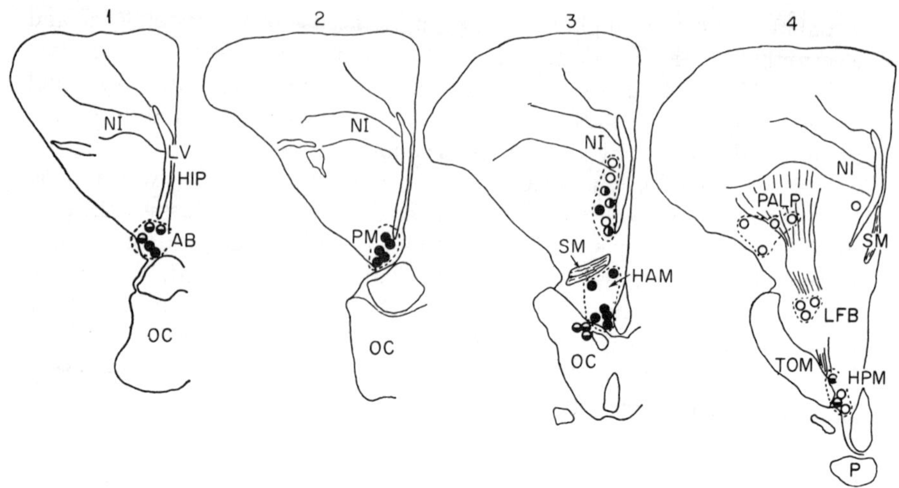

Fig. 13. Regions of the brain in the Barbary dove in which testosterone implants elicited charging (solid black circles and half black circles). Charging is an aggressive sexual activity. (From Hutchison, 1970.)

medial paleostriatum just lateral to the ventricle, an area also involved in a agonistic responses to ESB.

We have also found single units in this region which respond to exogenous hormones such as prolactin and progesterone. These hormones are involved in establishing aggressive and defensive behavior (Vowles & Harwood, 1966; Vowles & Prewitt, 1971). During the breeding season, marked changes in drive occur under the influence of these two hormones, causing changes from fearful to the aggressive and defensive behavior characteristic of courting and broody birds.

We suggest that these areas of the brain are concerned with the relative weighting given to different agonistic behaviors in response to external stimuli; the nature of these responses are determined by hormonal states, acting on cells in this area, to bias their output. The output of particular groups of cells would correspond to particular kinds of agonistic behavior, but at a high level of integration. Thus the output would determine whether behavior should be aggressive, defensive, or fearful, or a mixture of these, but would not determine specific actions, and would normally come into play only when appropriate stimuli occurred in the environment.

These areas would thus determine the basic motivational state of the individual, for in addition to influencing the kind of response to be shown they would also control their threshold and vigor. It is possible that the prolonged aftereffects obtained and detected in these regions may be correlated with this effect; and provide the kind of waxing and waning of behavioral tendencies which Lorenz (1950) subsumed in his early concepts of drive. The function may be biochemical in mechanism and perhaps be correlated with high levels of chemical transmitters in this region (Nauta & Karten, 1970). The specific exhaustion and recovery of behavior following repeated elicitation might result from the temporary exhaustion of transmitter reserves which the cells must later resynthesize.

If the general hypothesis is accepted, then we would expect weak ESB to produce a facilitation of agonistic responses. Strong ESB would have such a strong facilitation that the agonistic pattern might emerge in response to usually negligible stimuli in the environment, as in the so-called vacuum activities. The fact that even with strong ESB the overt agonistic behavior interacts with the external stimulus implies that the output from these forebrain and hypothalamic areas interacts elsewhere with afferent input to produce the properly integrated and oriented behavior.

3. *Septal and Hippocampal Areas and the Hyperstriatum*

Electrical stimulation of the brain in these areas also tended to elicit or facilitate agonistic display but of a generally fearful nature, the behavior interacting with external stimuli. MacPhail (1967) showed that stimulation in these areas is negatively reinforcing. A possible explanation is that these regions are concerned with those emotional aspects of learning which are commonly associated with aversion: the development of conditioned emotional responses, the termination of behavior, and avoidance learning. There is little evidence on this point, although the disruption of timing performance on some operant schedules after hippocampal and hyperstriate lesions might be interpreted in this way.

4. *The Archistriatum*

The archistriatum is also clearly involved in emotional response. Macphail (1967) found that more lateral areas of the forebrain were positively rewarding, and Zeier (this volume) has shown that archistriatal lesions produce effects which may be interpreted as destroying the ability to cause perseveration of behavior. The normal function might therefore be to cause the persistence and repetition of behavior which proves adaptive in a particular situation.

Zeier (1971) also showed an important somesthetic function which could perhaps be involved in sensory–motor skill, for other regions of the archistriatum. It is also possible the ESB effecting somesthetic pathways could produce emotional responses, e.g., anxious and fearful behavior, unconnected with antipredator behavior.

5. *Neostriatum*

Stimulation in this and closely associated areas facilitated rather specific agonistic items of behavior. Many workers have also described simple motor effects from stimulation. A possible interpretation is that in the neostriatum we have behavior represented as a series of rather molecular acts, which can be individually facilitated or inhibited. This would permit short-term modifications of otherwise stereotyped patterns to fit the present environment and long-term refinements of behavioral skills.

6. *Anterior, Dorsal Thalamic Nuclei*

So far, we have considered various parts of the brain in artificial isolation and attributed to them specific types of behavior function: Analysis of the stimulus situation was attributed to the tectal areas; motivational effects, depending upon biochemical factors, implicated the basal forebrain and anterior hypothalamic areas; the reinforcing aspects of motivation were attributed to the septum and archistriatum, and the refinement of motor skills were associated with archistriatum and neostriatum. We may also suppose that the reticular formation will be involved in arousal effects of agonistic drives. Clearly a mechanism is needed which will integrate these various activities. The anterior dorsal thalamic nuclei seem well suited to play this role since they give and receive connections from all the other regions just listed. If this is the case it would be expected that ESB would produce and facilitate various kinds of integrated agonistic behavior from these nuclei, which is indeed the case.

The attribution of functional roles to the various brain structures must be very speculative, but it agrees well with the kind of mechanisms which are required by ethological, hormonal, and learning studies. If we are to understand the physiological mechanisms of behavior, it is necessary to synthesize the results from all three fields, and this can be done only by very detailed study of actual behavioral effects which ESB or lesions may produce. Although the sophisticated methods of learning studies have been well applied in this context, the equally sophisticated ethological methods have seldom been used in a quantitiative way. The combination of both methods should give rise to new developments in the field of functional neuroanatomy. The advantage of the scheme just outlined is that it can be tested by further experiments. Thus a combination of lesions and stimulation, or stimulation through combinations of electrodes enables us to study interaction between brain areas acting together to produce an integral pattern of behavior. At the same time such concepts suggest which physiological activities could be meaningfully analyzed at a neuronal level in order to understand the detailed way in which the mechanism works. The conclusions and theories of the past provide only the foundations for exciting questions in the future.

Abbreviations for Figures 3, 4, 9, 12, and 13

A	Archistriatum
AC	Nucleus accumbens
AH	Anterior hypothalamus
AL	Ansa lenticularis
APH	Area parahippocampalis
AV	Archistriatum, pars ventralis
Bas	Nucleus basalis
CA	Commissura anterior
CB	Cerebellum
DLM	Nucleus dorsolateralis anterior thalami, pars medialis
DLP	Nucleus dorsolaleralis posterior thalami
DMA	Nucleus dorsomedialis anterior thalami
DMP	Nucleus dorsomedialis posterior thalami
DP	Dorsal posterior thalamic area
Ek	Ectostriatum
EM	Nucleus ectomammillaris
Ep	Nucleus reticularis superior, pars ventralis
FDB	Fasciculus diagonalis Brocae
FLM	Fasciculus longitudinalis medialis
FPL	Fasciculus prosencephali lateralis
HA	Hyperstriatum accessorium
HD	Fasciculus dorsale
HP	Hippocampus
HV	Hyperstriatum ventrale
ICo	Nucleus intercollicularis
Imc	Nucleus isthmi, pars magnocellularis
IO	Nucleus isthmo-opticus
Ipc	Nucleus isthmi, para parvocellularis
IPS	Nucleus interstitio pretecto subpretectalis
LA	Nucleus lateralis anterior thalami
LLd	Nucleus lemnisci lateralis, pars dorsalis
LS	Lemniscus spinalis
NLd	Nucleus mesencephalicus lateralis, pars dorsalis
NLv	Nucleus mesencephalicus lateralis, pars ventralis
N	Neostriatum
NC	Neostriatum caudale
NV1	Nervus abducens
n1X-X	Nucleus nervi glossopharyngei et nucleus motorius nervi vagi
OM	Tractus occipito-mesencephalicus
Ov	Nucleus ovoidalis
PA	Paleostriatum augmentum
PD	Nucleus pretectalis diffusus
PLH	Nucleus lateralis hypothalami posterioris
POA	Nucleus preopticus anterior
PP	Paleostriatum primitivum
PT	Nucleus pretectalis
PV	Nucleus posteroventralis thalami
PVM	Nucleus periventricularis magnocellularis
RF	Reticular formation

RSd	Nucleus reticularis superior, pars dorsalis
Rt	Nucleus rotundus
S	Septal area
SM	Nucleus septalis medialis
SO	Nucleus supraopticus
SP	Nucleus subpretectalis
SPL	Nucleus spiriformis lateralis
SIM	Nucleus spiriformis medialis
StT	Tractus spinotectalis
TFL	Tractus thalamo-frontalis lateralis
TFM	Tractus thalmo-frontalis medialis
TOv	Tractus ovoidalis
TPc	Nucleus tegmenti pedunculo-pontinus, pars compacta
TrO	Tractus opticus
TSM	Tractus septomesencephalicus
TT	Tractus tectothalamicus dorsalis
VLT	Nucleus ventrolateralis thalami

References

Akerman, B. Behavioral effects of electrical stimulation in the forebrain of the pigeon. *Behavior*, 1966, I and II, **6**, 323–350.

Andrew, R. J. Intracranial self-stimulation in the chick and the causation of emotional behavior. *Annals of New York Academy Sciences*, 1969, **159**, 625–639.

Barfield, R. J. Induction of aggressive and courtship behavior by intracerebral implants of androgen in capons. *American Zoologist*, 1965, **5**, 203–204.

Barfield, R. J. Activation of copulatory behavior by androgen implanted into the preoptic area of the male fowl. *Hormones and Behavior*, 1969, **1**, 37.

Biederman-Thorson, M. Auditory responses of neurones in the lateral mesencephalic nucleus (inferior colliculus) of the barbary dove. *Journal of Physiology*, 1967, **193**, 695–705.

Biederman-Thorson, M. Auditory evoked responses in the cerebrum (Field L) and the ovoid nucleus of the ring dove. *Brain Research*, 1970, **24**, 235–245. (a)

Biederman-Thorson, M. Auditory responses of units in the ovoid nucleus and cerebrum (Field L) of the ring dove. *Brain Research*, 1970, **24**, 247–256. (b)

Cannon, R. E., & Salzen, E. A. Brain stimulation in newly hatched chicks. *Animal Behaviour*, 1971, **19**, 375–385.

Cohen, D. H., & Pitts, L. H. The hyperstriatal region of the avian forebrain: somatic and antonomic responses to electrical stimulation. *Journal of Comparative Neurology*, 1967, **131** 323–336.

Delius, J. D. Displacement activities and arousal. *Nature*, 1967, **214**, 1259–1260.

Delius, J. D. Neural substrates of vocalizations in Gulls and Pigeons. *Experimental Brain Research*, 1970, **12**, 64–80.

Gardner, J. E., & Fisher, A. E. Induction of mating in male chicks following preoptic implanation of androgen, 1968.

Harman, A. L., & Phillips, R. E. Responses in the vian midbrain, thalamus and forebrain evoked by click stimuli. *Experimental Neurology*, 1967, **18**, 276–286.

Harwood, D. & Vowles, D. M. Defensive behavior and the aftereffects of brain stimulation in the ring dove. *Neuropsychologia*, 1967, **5**, 345–366.

Hodos, W. & Karten, H. J. Brightness and pattern discrimination deficits in the pigeon after lesions of nucleus rotundus. *Experimental Brain Research*, 1966, **2**, 151–167.

Huber, G. C. & Crosby, E. C. The nuclei and fibre paths of the avian diencephalon, with consideration of telencephalic and certain mesencephalic centres and connections. *Journal of Comparative Neurology*, 1929, **48**, 1–225.

Hutchison, J. B. Initiation of courtship by hypothalamic implants of testosterone propionate in castrated doves. *Nature*, 1967, **216**, 591–592.

Hutchison, J. B. Changes in hypothalamic responsiveness to testosterone in male Barbary doves. *Nature*, 1969, **222**, 176.

Hutchison, J. B. Influence of gonadal hormones on the hypothalamic integration of courtship behavior in the Barbary dove. *J. Reprod. Fert: Suppl.* 1970, **11**, 15–41.

Karten, H. J. Hodos, W. A. A stereotaxic atlas of the brain of the pigeon (*Columba livia*). Baltimore, Maryland: John Hopkins Press, 1967.

Karten, H. J. & Revzin, M. The afferent connections of the nucleus rotundus in the pigeon. *Brain Research*, 1966, **2**, 368–377.

Komisarak, B. Effects of local brain implants of progesterone on reproduction behavior in ring doves. *Journal of Comparative & Physiological Psychology*, 1967, **51**, 32–37.

Lorenz, K. The comparative method in studying innate behavior patterns. *Symposium of Society for Experimental Biology* 1950, **4**, 221–268.

Macphail, E. M. Positive and negative reinforcement from intracranial stimulation in Pigeons. *Nature*, 1967, **213**, 947–948.

Macphail, E. M. Self-stimulation in pigeons: tests of two alternative explanations. *Psychonomic Science*, 1968, **11**(1), 3–4.

Maley, M. J. Electrical stimulation of agonistic behavior in the mallard. *Behavior* 1969, **34**, 138–160.

Nauta, W. J. H., & Karten, H. J. A general profile of the vertebrate brain, with sidelights on the ancestry of cerebral cortex. In F. O. Schmitt (Ed.), *The neurosciences, second study program.* New York: Rockefeller Univ. Press, 1970.

Peck, F. W. & Phillips, R. E. Repetitive vocalizations evoked by local electrical stimulation of avian brains (*Gallus gallus*). *Brain, Behavior & Evolution*, 1971, **4**, 417–438.

Phillips, R. E. Approach-withdrawal behavior of peach-faced lovebirds, Agapornis reseicolis, and its modification by brain lesions. *Behavior*, 1968, **31**, 163–184.

Phillips, R. E. & Youngren, O. M. Brain stimulation and species typical behavior: activities evoked by electrical stimulation of the brains of chickens (*Gallus gallus*). *Animal Behavior*, 1971, **19**, 757–779.

Potash, L. M. Vocalizations elicited by electrical brain stimulation in *Coturnix coturnix japonica*. *Behavior*, 1970, **36**, 149–167.

Powell, T. P. S. & Cowan, W. H. The thalamic projection upon the telencephalon in the pigeon (*Columba livia*). *Journal of Anatomy*, 1961, **95**, 78–109.

Putkonen, P. T. S. Electrical stimulation of the Avian brain. *Annales Academiqe Scientarium Fennicae*, 1967, **130**, 1–95.

Revzin, A. M. A specific visual projection area in the hyperstriatum of the pigeon (*Columba livia*) *Brain Research*, 1969, **15**, 246–249.

Tinbergen, N. Comparative studies of the behavior of gulls. *Behavior*, 1959, **15**, 1–70.

Von Holst, E. & Von St. Paul, U. On the functional organization of drives. *Animal Behaviour*, 1963, **11**, 1–20.

Vowles, D. M. & Beazley, L. D. Brain stimulation and assymetrical motivation in the dove. in press (1973).

Vowles, D. M. & Harwood, D. The effect of exogenous hormones on aggressive and defensive behavior in the ring dove (*Streptopelia risoria*). *Journal of Endocrinology*, 1966, **36**, 35–51.

Vowles, D. M. & Prewitt, E. Stimulus and response specificity in the habituation of the anti-predator behavior in the ring dove, (*Streptopelia risoria*). *Animal Behavior*, 1971, **19**, 80–86.

Vowles, D. M. & Tarala, L. The neural substrate of emotional behavior in birds. Paper presented at Second Lashley Memorial Conference, West Virginia Univ. Morgantown, West Virginia, 1971.

Zeier, H. Changes in operant behavior of pigeons following bilateral forebrain lesions. *Journal of Comparative & Physiological Psychology*, 1968, **66**, 198–203.

Zeier, H., DRL—performance and timing behavior of pigeons with archistriatal lesions. *Physiology & Behavior*, 1969, **4**, 189–193.

Zeier, H., Archistriatal lesions and response inhibition in the pigeon. *Brain Research* 1971, **31**, 327–339.

Zeier, H. & Akert, K. Transient changes in operant behavior during bilateral electrical forebrain stimulation. *Physiology & Behavior*, 1968, **3**, 293–296.

Zeier, H. & Karten, H. J. The archistriatum of the pigeon: Organization of afferent and efferent connections. *Brain Research*, 1971, **31**, 313–326.

Zeigler, P. H., Green, H. L., & Karten, H. J. The neural control of feeding behavior in the pigeon. *Psychonomic Science*, 1969, **15**(3), 156–157.

Author Index

Numbers in parentheses indicate pages on which the complete references are listed

A

Abt, J. P., 204, *218*
Adamo, N. J., 49, 57, 59, 64, *68*, 196, 197, *198*
Adams, H. E., 205, *218*
Adamson, L., 46, *69*
Adey, W. R., 179, *198*
Akerman, B., 88, 89, 98, *98*, 113, *130*, 222, 234, *256*
Akert, K., 222, *258*
Albert, D. J., 129, *130*
Allen, W. F., 166, *198*
Allison, T., 134, 135, 136, 137, 138, *152*
Alpern, H. P., 204, 207, 209, *219*
Altman, J., 48, *68*
Altmann, S. A., 4, 5, *12*
Andrew, R. J., 87, 90, *98*, 222, *256*
Andersson, B., 113, *130*
Aprison, M. H., 149, *150*
Armstrong, J., 38, *72*
Aronson, L. R., 6, *12*
Atz, J. W., 22, *24*

B

Bachrach, H., 182, 183, *198*
Baker-Cohen, K. F., 62, *68*
Bang, B. G., 16, *24*
Bamborough, P., 64, *71*
Barfield, R. J., 222, *256*
Barnikol, A., 39, *68*
Barrett, R. J., 206, *220*
Beazley, L. D., 225, *257*
Behrend, E. R., 187, *199*
Beitel, R. E., 204, *218*
Békésy, G. V., 78, *85*
Belekhova, M., G., 64, *68*
Bennetto, K., 63, *69*
Berger, H., 134, *150*
Berger, R. J., 134, 136, 137, 138, 139, *150*, *152*
Bertler, A., 62, *68*
Biederman-Thorson, M., 52, 63, *68*, 85, *85*, 250, *256*
Bitterman, M. E., 168, 187, 195, *198*, *199*
Bivens, L. W., 209, *220*
Blough, D. S., 171, *198*
Bock, W. J., 20, *24*
Boguch, S., 173, *198*
Bohdanecky, A., 204, *218*, *219*
Bohdanecky, M., 204, *218*
Bolles, R. C., 111, *130*
Bonbright, J. C., Jr., 177, *199*
Boord, R. L., 51, 53, *68*, *69*, 79, 85, *85*
Botezat, E., 38, *69*
Bower, G. H., 168, *199*
Boyden, A., 21, *24*
Brady, J. V., 179, *198*
Brémond, J. C., 77, *85*
Bresnahan, E., 208, *218*
Brodal, A., 36, *69*

Brown, J. L., 87, 88, 90, 91, 98, *98*, 140, 145, *150*
Brunelli, M., 145, 146, *150*
Brunner, R. L., 204, 208, *218*
Buchwald, N. A., 139, *150*
Bures, J., 204, *218*
Buresova, O., 204, *218*
Bush, G., 207, *220*
Butler, R. A., 166, *198*

C

Cajal, S. R., 30, 46, 51, 52, *69*
Campbell, C. B. G., 20, 22, *24*
Cannon, R. E., 222, 234, *256*
Carpenter, M. B., 48, *68*
Catania, A. C., 154, *163*
Cate, J. ten, 145, *150*
Chase, M., 129, *132*
Chavez-Ibarra, G., 149, *150*
Cherkin, A., 209, 210, 211, *218*
Chevalier, J. A., 204, 206, *218*
Chorover, S. L., 204, 205, 209, *218, 220*
Church, S. M., 137, 138, 146, *150*
Cizek, L. J., 107, *130*
Clemente, C. D., 138, 139, 141, 145, *151*
Cobb, S., 16, *24*, 50, *69*
Cohen, D. H., 39, 40, 41, 42, 43, 49, 64, 65, *69, 71*, 147, 168, 173, 175, 189, 190, *198, 201*, 222, *256*
Collier, G., 108, 111, *130*
Coons, E. E., 204, *218*
Cooper, R. M., 206, *218*
Cowan, W. M., 30, 34, 46, 48, *69, 71, 72*, 231, 250, *257*
Cramer, H., 133, *151*
Crosby, E. C., 30, 34, 59, *69, 70*, 231, *257*
Cuenod, M., 191, 193, *198, 200*

D

Dane, B., *5, 12*
Delius, J. D., 63, 64, *69, 72*, 87, 98, *98*, 222, 243, *256*
Dement, W., 137, *150*
Dethier, V. G., 128, *130*
Deutsch, J. A., 204, *218*
Diamond, I. T., 23, *24*, 48, *69*, 166, *198*
Donner, K. O., 16, *24*
Dooley, S., 49, 64, *69*
Dooling, R. J., 83, *85*
Dorward, P. L., 38, 39, *69*
Doty, R. W., 204, *219*
Dow, R. S., 30, *69*
Dubbeldam, J., 52, 57, 58, 59, *69, 71*, 150, *151*
Durkovic, R. G., 168, *198*

E

Ebbesson, S. O. E., 45, *69*
Edinger, L., 55, 56, 57, 65, *70*
Ehman, G. K., 129, *130*
Elliot, D. H., 166, *198*
Elterman, M., 139, *151*
Epstein, A. N., 123, 129, *130, 131*
Erulkar, S. D., 50, 52, 63, *70*, 165, *198*
Essman, W. B., 204, *218*

F

Fabricius, E., 113, *130*
Falck, B., 62, *68*
Falls, J. B., 77, 85, *85*
Farner, D. S., 102, *130*
Feldstein, S., 105, 112, 128, *132*
Ferrier, D., 145, *150*
Ferster, C. B., 153, *163*
Fink, R. P., 158, *163*
Fisher, A. E., 222, *256*
Flexner, J. B., 204, *218*
Flexner, L. B., 204, *218*
Flourens, P., 55, *70*, 113, *130*
Fox, C. A., 62, *70*
Frishkopf, L. S., 85, *86*
Fulcher, J. K., 206, *218*
Fuxe, K., 62, *70*, 149, *150*

G

Galambos, R., 78, *86*
Gardner, J. E., 222, *256*
Gellerman, L. W., 155, *163*
Gerbrant, L. K., 156, *163*
Ghiselin, M. T., 21, 22, *24*
Giaquinto, S., 211, *219*
Glickman, S. E., 204, 205, 207, *218*
Gloor, P., 162, *163*
Goddard, C. J., 204, 208, *218*
Goddard, G. V., 162, *163*
Goldberg, J. M., 166, *199*
Goldstein, M. M., Jr., 85, *86*

Gonzales, R. C., 187, *199*
Goodman, I. J., 90, 98, *98*, 133, 136, 137, 138, 140, 141, 143, 145, 146, 147, *150*, *151*, 211, 212, *219*
Gossette, M. F., 195, *199*
Gossette, R. L., 187, 195, 196, *199*
Gotoh, J., 133, 135, 136, *151*
Gottfries, C. G., 62, *68*
Green, H. L., 55, *73*, 102, 105, 106, 107, 108, 109, 110, 114, *132*, 159, *164*, 222, *258*
Greenough, W. T., 206, *218*
Grossman, S. P., 129, *130*

H

Haartsen, A. B., 58, *70*
Haas, O., 21, *24*
Haefelfinger, H. R., 30, 61, 65, *70*
Hall, W. C., 23, *24*, 48, *69*
Hamburg, M. D., 204, *218*
Harman, A. L., 52, *70*, 250, *256*
Harrison, R. H., 170, *199*
Harwood, D., 87, 98, *98*, 113, *130*, 140, *150*, 222, 224, 226, 247, 252, *256*, *257*
Hays, K. J., 204, *219*
Hecht, M. K., 19, *25*
Heimer, L., 158, *163*
Heinecke, P., 92, *98*
Heise, G. A., 170, *199*
Hernandez-Peon, R., 149, *150*
Herrmann, A., 66, *70*
Herz, M. J., 204, *220*
Hess, W. R., 139, *151*
Heuser, G., 139, *150*
Hilgard, E. R., 168, *199*
Hill, H. H., 112, *131*
Hillman, D. E., 62, *70*
Hinde, R. A., 101, *130*
Hishikawa, Y., 133, 137, 138, *151*
Hodos, W., 20, 22, 23, *24*, *25*, 31, 32, 37, 47, 49, 64, *70*, *71*, 102, *131*, 145, 151, 165, 175, 176, 177, 178, 181, 182, 189, *199*, 214, 215, 216, *219*, 231, 250, *257*
Holden, A. L., 64, *71*
Holman, G. L., 123, *130*
Holmes, G., 57, *70*
Horvath, F. E., 162, *163*
Howard, E., 11, *12*
Huber, G. C., 30, 34, 37, 38, 40, 43, 59, *70*, 231, *257*
Hudspeth, W. J., 205, *219*
Hughes, R. J., 78, *86*
Hunt, E. B., 207, *220*
Hutchison, J. B., 222, 251, 252, *257*

I

Ilyichev, V. D., 50, *70*
Isomura, G., 39, *73*

J

Jacobs, B. L., 204, *219*
Jansen, J., 36, *69*
Jarvick, M. E., 204, 207, *218*, *219*, *220*
Jason, K. M., 211, 212, *219*
Jenkins, H. M., 170, *199*
Johnstone, B. M., 85, *86*
Johnstone, J. R., 85, *86*
Jouvet, M., 133, 139, *151*
Jungherr, E. L., 30, *70*
Juorio, A. V., 62, *70*, 150, *151*
Jurgens, U., 89, 98, *98*

K

Kaiser, L., 39, *70*
Kalischer, O., 49, 56, 57, *70*
Kappers, C. U. A., 30, 31, 36, 38, 49, 59, 61, 62, *70*
Karten, H. J., 16, 23, *25*, 30, 31, 32, 33, 37, 43, 44, 45, 46, 47, 48, 49, 50, 51, 52, 54, 55, 57, 58, 59, 61, 62, 63, 64, 65, 66, 67, 68, *69*, *70*, *71*, *72*, *73*, 85, *86*, 102, 110, 114, 116, 117, 118, 122, *130*, *131*, *132*, 145, 147, 150, *151*, 154, 156, 157, 159, 161, 162, *163*, 165, 175, 176, 177, 178, 179, 181, 182, *199*, *200*, 214, 215, 216, *219*, *220*, 222, 231, 250, 253, *256*, *258*
Kasper, R., 204, *219*
Kawakami, M., 139, *151*
Kesner, R. P., 204, 205, 208, 209, *219*
Key, B. J., 148, *151*
Kiang, N. Y. S., 82, *86*
Kimberly, R. P., 64, *71*
King, R. A., 209, *220*
Kitchell, R. L., 123, *131*
Klein, M., 133, 137, 138, *151*

Kleitman, N., 137, *150*
Knauss, T., 138, *151*
Koelle, G. B., 62, *71*
Kohlenberg, R., 206, *219*
Komisarak, B., 222, *257*
Konishi, M., 77, *86*
Kopp, R., 204, 207, *219*
Koppenall, R. J., 206, *218*
Kramer, E., 92, 97, *98*
Kramer, T. J., 161, *163*
Krasnegor, N. A., 178, *199*
Kreinick, C. J., 208, *219*
Kristiansen, K., 36, *69*
Kruger, L., 34, *72*
Krushinsky, L. V., 197, *199*
Kuhlo, W., 133, *151*
Kuo, Z. Y., 6, *12*

L

Langley, J. N., 40, *71*
Larsell, O., 30, 36, *71*
Larsson, B., 113, *130*
Lashley, K. S., 15, 24, *25*, 29, *71*, 101, 130, *131*
Lavail, J. H., 30, *71*
Layman, J. D., 185, *200*
Lee-Teng, E., 209, 210, 211, 215, *218, 219*
Lehmann, D., 138, *151*
Lehrer, R., 105, 106, *132*
Lehrman, D. S., 6, *12*
Le Magnen, J., 105, *131*
Leonard, R. B., 41, 42, 43, *71*
Lettvin, J. Y., 122, *131*
Levin, J., 191, *199*
Levine, M. S., 208, *219*
Lewis, D. J., 205, *218, 219*
Lidsky, T. I., 208, *219*
Limpo, A. J., 155, *164*
Ljunggren, L., 62, *68*, 149, *150*
Lo Presti, R. W., 133, 136, 141, 143, *151*
Lorenz, K., 5, 6, 8, 9, *12*, 253, *257*
Lovejoy, A. O., 20, *25*
Luttges, M. Q., 206, *219*

M

Macdonald, R. L., 39, 40, *69, 71*
Mackintosh, N. J., 187, 189, *200, 201*
MacPhail, E. M., 155, 156, *163*, 183, 184, 186, 187, *200*, 222, 253, *257*
Madsen, M. C., 204, *219*
Magni, F., 145, *150*
Magnus, J. G., 211, *219*
Maher, B. A., 205, *219*
Mahut, H., 204, 207, 208, *219*
Maier, V., 193, *200*
Maiorana, V., 7, *12*
Maley, M. J., 98, *98*
Malinovsky, L., 39, *71*
Marley, E., 148, 149, *151*
Marshall, J. F., *129*
Martin, G., 207, *220*
Maturana, H. R., 122, *131*
Matyniak, K., 171, 187, *200, 201*
Mayor, S. J., 173, *200*
Mayr, E., 8, *12*
McBride, G., 7, *12*
McCulloch, W., 122, *131*
McCollum, R. H., 211, 212, 213, 215, 217, *219*
McDonough, J. H., 204, 205, 208, 209, *219*
McFarland, D., 107, *131*
McGaugh, J. L., 204, 205, 206, 207, 209, *219*
McIntyre, A. K., 38, *69*
Mead, W. R., 183, *200*
Medway, Lord, 50, *71*
Megibow, M., 108, 111, *131*
Meier, R. E., 192, 193, *200*
Mello, N. K., 191, 192, 193, 194, *200*
Meyer, D. R., 166, *200*
Meyers, R. E., 191, *200*
Michel, F., 133, *151*
Miller, A. J., 206, *220*
Miller, J. D., 83, *85*
Miller, N. E., 111, *131*, 204, *218*
Miller, R. R., 205, *219*
Misanin, J. R., 205, *219*
Morgane, P. J., 129, *131*, 149, *150*
Moruzzi, G., 30, *69*, 145, 146, *150, 151*
Muller, G. E., 205, *220*
Mulligan, J. A., 83, *85*
Mundinger, P., 5, *12*
Münzer, E., 46, *71*
Murphey, R. K., 87, *99*
Musumeci, D., 145, *150*
Myers, T. T., 38, *73*

N

Nauta, W. J. H., 33, 47, 48, 49, 64, 66, 68, *71*, 154, 162, *163*, 165, *199, 200*, 215, *220*, 253, *257*
Neff, W. D., 166, *198, 199*
Nelson, G. J., 17, *25*
Newman, J. D., 87, *99*
Nielson, J. C., 204, 206, *220*
Nieuwenhuys, R., 30, *72*
Nistico, G., 149, *151*
Northcutt, R. G., 183, *200*
Nottebohm, F., 87, *99*
Novick, A., 50, *72*

O

Olds, J., 179, *200*
Orvig, T., 19, *25*
Oscarsson, O., 43, *72*
Owen, R., 21, *25*
Okuma, T., 139, *150*
Olmos, N., 139, *151*
Ookawa, T., 133, 135, 136, *151*

P

Pagano, R. R., 206, *220*
Papez, J. W., 30, *72*
Parker, D. M., 64, *72*
Paulson, G., 134, *151*
Payne, R. S., 50, *72*
Pearlman, C. A., 204, *220*
Pearson, R., 30, 38, 45, 50, *72*
Peek, F. W., 87, 97, *99*, 222, *257*
Peeke, H. V. S., 204, *220*
Pellegrino, L., 162, *164*, 204, 208, *220*
Penaloza-Rojas, J. H., 139, *151*
Peters, J. J., 133, *151*
Pfingst, B. A., 209, *220*
Phillips, R. E., 52, 70, 87, 88, 92, 97, 98, *99*, 113, *131*, 156, *164*, 197, *200*, 222, 227, 237, 250, *256, 257*
Pilzecker, A., 205, *220*
Pitts, L. H., 39, *69*, 122, *131*, 189, *198*, 222, *256*
Ploog, D., 89, 98, *98*
Porter, P. B., 204, *218*
Portmann, A., 30, *72*, 196,, *200*
Potash, L. M., 87, 97, *99*, 222, 243, *257*
Powell, T. P. S., 34, 46, *69, 72*, 231, 250, *257*
Prewitt, E., 252, *258*
Pritz, M. B., 183, *200*
Pryer, R. S., 204, *220*
Pumphrey, R. J., 82, *86*
Putkonen, P. T., 98, *99*, 113, *131*, 140, 145, *151*, 221, 222, 227, 231, 234, 236, 243, *257*

Q

Quilliam, T. A., 38, *72*
Quinton, E. E., 204, *220*

R

Rasmussen, G. L., 51, *69*, 79, *85*
Ray, O. S., 206, 209, *220*
Reynolds, G. S., 154, 155, *163, 164*
Revzin, A. M., 23, *25*, 47, 49, 64, *71, 72*, 250, *256, 257*, 165, *199, 200*
Riddell, W., 195, *199*
Rilling, M., 161, *163*
Rodgers, W. L., 129, *131*
Rogers, F. T., 55, *72*, 113, *131*
Rojas-Ramirez, J., 134, 136, 137, 138, *151*
Romer, A. S., 16, 19, *25*
Rose, J. E., 78, *86*
Rose, M., 51, 63, *72*
Rosen, I., 43, *72*
Rosenblatt, J., 6, *12*
Rosengren, E., 62, *68*
Rossi, G. F., 139, *151*
Rossi, R. R., 204, 208, *218*
Roth, T. C., 204, *218*
Routtenberg, A., 208, *218*
Ruckebusch, Y., 138, *151*
Ruskin, R. S., 147, *151*

S

Salzen, E. A., 222, 234, *256*
Salzman, A., 179, *201*
Sanders, E. B., 50, *72*
Saunders, J. C., 85, *86*
Sawyer, C. H., 139, *151*
Schade-Powers, A., 185, 186, *200*
Schiller, P. H., 205, 209, *218*
Schjelderup-Ebbe, T., 7, *12*
Schleidt, W. M., 4, 5, 6, 7, *12, 13*

Schmid, D., 133, *151*
Schnall, A. M., 39, 40, *69*
Schneider, A. M., 204, 209, *220*
Schneider, G. E., 23, *25*, 48, *72*
Schrader, M. E. G., 145, *151*
Schultz, W., 185, 190, *201*
Schuknecht, H. F., 82, *86*
Schwartzbaum, J. S., 208, *219*
Schwartzkopf, J., 50, *72*
Schwitzgebel, R. L., 206, *218*
Sether, L. A., 62, *70*
Shalter, M., 5, *13*
Sharpless, S. K., 204, *220*
Sherman, W., 204, 209, 210, *220*
Shimp, C. P., 153, *164*
Sidman, M., 160, *164*
Siegel, J., 107, 108, 109, 110, *132*
Siegesmund, K. A., 62, *70*
Sillman, A., 45, *72*
Silver, R., 118, 119, 120, 122, 129, *131, 132*
Simpson, G. G., 16, 21, *25*
Skinner, B. F., 11, *13,* 153, *163, 164*
Smith, C. A., 50, *72*
Sorenson, C. A., 204, *219*
Sperry, R. W., 191, *200*
Spooner, C. E., 148, 149, *151*
Ssinelnikov, R., 40, *72*
Stamm, J. S., 161, *164*
Steen, L., 113, *130*
Stein, D. C., 204, *220*
Stellar, E., 112, *131,* 204, *218*
Sterman, M. B., 138, 139, 141, 145, *151*
Stettner, L. J., 171, 185, 187, 190, 192, *200, 201*
Stingelin, W., 30, 52, *72*, 114, *131*, 196, *200*
Stopp, P. E., 51, *72*
Storlien, L. H., 129, *130*
Streeter, G. L., 37, *72*
Ström, L., 123, *131*
Strominger, J. L., 107, *131*
Stutz, R. M., 204, *218*
Sutherland, N. S., 189, *201*
Svensson, L., 113, *130*

T

Takahashi, R., 149, *150*
Takasaka, T., 50, *72*
Tauber, E. S., 134, 136, 137, 138, *151*
Teitelbaum, P., 123, 129, *131*
Terni, T., 40, *72*
Thompson, C. W., 205, *219*
Thompson, R. F., 166, *201*, 204, 207, *220*
Timo-Iara, C., 149, *150*
Tinbergen, N., 5, *13*, 239, *257*
Tobach, E., 6, *12*
Tokizane, T., 179, *198*
Trabasso, T., 206, *219*
Tradardi, V., 133, 138, *152*
Trahiotis, C., 166, *198*
Turner, B. H., *129*

U

Uddenberg, N., 43, *72*
Ursin, H., 162, *164*

V

Van der Kloot, W. G., 5, *12*
Van Den Akker, L. M., 30, 41, 42, 43, *72*
Van Twyver, H., 134, 135, 136, 137, 138, *152*
Verhaart, W. J. C., 58, *70*
Vogt, M., 62, *70*, 150, *151*
Vonderahe, A. R., 133, *151*
Von Holst, E., 97, *99*, 140, *152*, 221, 222, 239, 248, 251, *257*
Von St. Paul, U., 92, *98, 99*, 97, 140, *150, 152*, 221, 222, 239, 248, 251, *257*
Vowles, D. M., 87, 98, *98*, 113, *130*, 140, *150*, 222, 224, 226, 247, 248, 252, *256, 257, 258*

W

Walker, J. M., 134, 136, 137, 138, 139, *150, 152*
Wallenberg, A, 43, 50, 54, 57, 63, *70, 73,* 114, *131*
Watanabe, T., 39, *73*
Weissman, A., 204, 206, *220*
Welty, J. C., 16, *25*
Wenzel, B. M., 179, *201*
Wheeler, G., 187, *200*
Whitfield, I. C., 51, *72*
Wiener, H., 46, *71*
Williston, J. S., 204, *220*
Winkelmann, R. K., 38, *73*
Winter, P., 10, *13,* 50, *73*

Winters, W. D., 148, 149, *151*
Witkovsky, P., 52, *73*, 102, 119, 120, 122, 123, 129, *131*, *132*
Wolin, B. R., 102, *132*
Wood, D. J., 129, *130*
Woodburne, R. T., 59, *69*, 114, *132*
Woolsey, C. N., 166, *200*
Wrightman, W. P. D., 20, *25*
Wyers, E. J., 139, *150*, 204, 208, 209, 213, *220*
Wyrwicka, W., 129, *132*

Y

Yasuda, M., 39, *73*
Youngren, O. M., 87, 92, 97, *99*, 222, 237, *257*

Z

Zecha, A., 43, 57, 58, 59, 66, *73*
Zeier, H., 43, 54, 57, 59, 63, 65, 66, *73*, 117, 118, *132*, 147, *152*, 156, 157, 158, 159, 161, 162, *164*, 191, *198*, 215, *220*, 222, 250, 253, *258*
Zeigler, H. P., 52, 55, 63, *73*, 102, 105, 106, 107, 108, 109, 110, 111, 112, 114, 116, 117, 118, 119, 120, 122, 127, 128, 129, *131*, *132*, 156, 159, 162, *164*, 170, 181, 182, 183, 185, 190, *201*, 222, *258*
Zinkin, S., 206, *220*
Zotterman, Y., 123, *131*
Zubin, J., 206, *220*

Subject Index

A

Abducens nerve, 255
AC, *see* Anterior commissure
Acetylcholine, 146, 149
Acetylcholinesterase, 62, 162
ACH, *see* Acetylcholine
Acoustic stria, 51
Adenosine derivatives, 149
Adrenaline, 148, 149
Aggressive behavior, *see* Agonistic behavior
Agonistic behavior, 88–90, 221–254
Alarm responses, 4, 226, 229–230
Alphamethylnoradrenaline, 149
Amphetamine, 148–149
Amnesia, retrograde, 204–217
Amygdala, 65, 156, 161–162, 208, 215
Anesthesia, halothane, 209
Ansa lenticularis, 55, 58–60, 62, 66, 114, 147, 157, 162, 255
Anterior commissure, 55, 66, 91, 94, 157–158, 191–192, 240, 242, 244, 250, 255
Anterior dorsal thalamic nuclei, *see* Nucleus dorsolateralis anterior thalami
Anterior hypothalamus, 251, 255
Anticholinesterase, 204
Aphagia, 113, 115, 118, 122–123, 126–129
Appetitive behavior, 111
Archistriatum, 32–33, 54, 57–59, 65, 67, 88, 118, 156–159, 160–162, 181–182, 197, 212–217, 231, 234, 242–243, 250, 253–255
Area mesencephalicus lateralis, *see* Mesencephalon
Area parahippocampalis, *see* Parahippocampal area
Ataxia, 148
Attack behavior, 90, 134, 239
Audiometric curve, 82–84
Auditory cortex, 166–167
Auditory discrimination, 179
Auditory lemniscal system, *see* Lemniscal pathways
Auditory nerve, 55
Auditory pathway, 16, 34, 85, 87, 165, 178, 250
Automatism, *see* Stereotyped behaviors
Autonomic nervous system, 39–41
Avoidance behavior, 4, 90, 134, 208, 226–229, 237, 239–240, 243, 246–247, 249, 252–253

B

Barbary dove (Columbidae), 232, 234–235, 241, 246, 249, 251–252
Basal ganglia, 60
Basilar membrane, 77–79, 82
Bat (Choroptera), 10, 22
Blackbird (*Agelaius phoeniceus*), 88, 98
Blowfly (Calliphoridae), 128
Bobwhite quail (*Colinus virginianus*), 185, 187, 189, 195–197
Brachial plexus, 37
Brain temperature, 136–138
Buccal cavity, 104
Bulbus olfactorius, *see* Olfactory bulbs
Bullfrog (*Rana catesbeiana*), 85
Bundle of Bagley, 58, 66, 158
Burrowing owl (*Speotylo cunieularia*), 23, 45, 48–49, 52, 134, 136–139

C

Canary (*Serinus canarius*), 82–83
Capsula interna occipitalis, 181
Carbachol, 149
Cat (*Felis catus*), 82, 141, 145, 166, 193
Catecholamines, 62, 149
Cauda equina, 37
Caudate nucleus, 139, 207, 208
Caudate–putamenal complex, 62–63
Caudoputamen, 162
Cells of Waldeyer, 41
Cerebellum, 30–32, 36, 55, 61–62, 255
Cerebral cortex, 31, 113, 139, 145
Cervical enlargement, 40
Cheetah (*Acinonyx jubata*), 10–11
Chiasma opticum, *see* Optic chiasma
Chicken (*Gallus domesticus*), 39, 52, 87–88, 90, 92, 98, 113, 133, 135–138, 145, 148–149, 195–197, 209–212, 217, 222, 234, 236–237, 251
Chimpanzee (*Pan troglodytes*), 22
Cobefrin, 148
Cochlea, 51–52, 77–79, 81–82
Collicular area, 251
Color discrimination, 11, 176–177, 179, 182–183, 185, 194
Columella, 50
Column of Clarke, 41–44
Column of Terni, 40, 42
Column of von Lenhossèk, 41
Commissura anterior, *see* Anterior commissure
Consolidation process, in memory, 205–206, 208–211, 217
"Consummatory" behaviors, 111
Corpus callosum, 191, 193
Corpus striatum, 65, 162, 165
Corpus trapezoideum, 51
Cortex piriformis, prepiriformis, 181
Cortical ablation, 197
Corticoid areas, 182–183
Corticoidea dorsolateralis, 157, 181
Coturnix quail (*Coturnix coturnix japonica*), 5, 87, 90, 98, 222
Courtship behavior, 5–6, 9, 89, 134, 222, 224–225, 251, *see also* Sexual behavior
Cranial nerve I, *see* Olfactory nerve
Cranial nerve III, *see* Oculomotor nerve
Cranial nerve IV, *see* Trochlear nerve
Cranial nerve V, *see* Trigeminal nerve
Cranial nerve VI, *see* Abducens nerve
Cranial nerve VIII, *see* Auditory nerve
Cranial nerve X, *see* Vagus nerve
Cricket (*Teleogryllus commodus*), 85
Crow (*Corvus brachyrhynchos*), 22
Cuneate nuclei, *see* Nucleus cuneatus

D

Defecation, 225
Defensive behavior, 226, 229, 240–241, 243, 247, 249, 252, *see* also Nest defense
Diencephalon, 34, 36, 47, 54–55, 89, 114, 140, 145, 147, 176, 231, 243–244, 250
Discrimination learning, 155, 166–197
DLA, *see* Nucleus dorsolateralis anterior thalami
Dog (*Canus familiaris*), 197
DOPA, 149
Dopamine, 62, 148–150, 162
Dorsal magnocellular cell column, 41, 43
Dove (*Columba palumbus*), 113, 222, 224, 231, 239, 243, 250–251
Drinking behavior, 90, 102, 115, 126, 129, 146
  interaction with feeding behavior, 107–108
Drowsiness, 135–135, 141, *see also* Sleep

DSO, *see* Supraoptic decussation
Ducks (*Anas* spp.), 8-10, 98, 102, 113, 156, 222, 234, 238

E

ECS, *see* Electroconvulsive shock
Ectomammillary nucleus, *see* Nucleus ectomammillaris
Ectostriatum, 32–33, 47–49, 64, 67, 175–177, 181, 194, 215, 231, 250, 255
Edinger–Westphal nucleus, 53
EEG, *see* Electroencephalogram
EKG, *see* Electrocardiogram
Electrical stimulation of the brain, 87–98, 113, 129, 139–141, 145–146, 173, 221–230, 239, 243–254, *see also* Intracranial stimulation
Electrocardiogram, 145–146, *see also* Heart rate
Electroconvulsive shock, 204–217
Electrocorticograms, 211
Electroencephalogram, 134–137, 141, 143, 146, 148–149
Electromyogram, 134, 137, 141, 143
Electrooculogram, 134, 137, 141, 143
EMG, *see* Electromyogram
Emotional behavior, 162
Entorhinal area, 31
EOG, *see* Electrooculogram
ESB, *see* Electrical stimulation of the brain
Escape behavior, *see* Avoidance behavior
Ethology, 3, 10, 90, 247, 254
Evoked behaviors, 90, 95, 140, 143

F

Falcon (Falconinae), 134–138
Fasciculus prosencephali lateralis, *see* Lateral forebrain bundle
Fasciculus prosencephali medialis, *see* Medial forebrain bundle
Fear response, *see* Avoidance behavior
Feeding behavior, 3, 55, 90, 101–106, 112–115, 118, 121, 124, 127–129, 134, 140, 146–147, 159, 222, *see also* Mandibulation
  interaction with drinking behavior, 107–108
Feedometer, 104–105, 111, 115, 121, 127–128
Field L, 51–52, 60, 63
Finches (Fringillidae), 5
Fixed action pattern, *see* Stereotyped behaviors
Flight behavior, *see* Avoidance behavior
Flocculus, 36
Flurothyl, 210–211
Forebrain, 139, 158, 175, 179, 181, 193, 197, 230, 251, *see also* Telencephalon
Fornix, 59, 208
FPL, *see* Lateral forebrain bundle
Frog (*Rana* spp.), 122
Fronto-archistriate tract, *see* Tractus frontoarchistriaticus
Funiculus
  dorsal, 42–43, 59–60
  lateral, 37, 41–44
  ventral, 37

G

Gallus, see Chicken
Geese (Anserinae), 8–9
Geniculostriate system, 47, 64–65
Globus pallidus, 62, 162
Glycogen body, 31, 37
Golgi material, 40
Grandry corpuscles, 38–39, 52
Grooming behavior, 134
Gulls (Larinae), 87, 98, 222, 239

H

HA, *see* Hyperstriatum accessorium
Habenular nuclei, 53
Hallucination, 227
Haloperidol, 149
Hawk, *see* Falcon
HD, *see* Hyperstriatum dorsale
Heart rate, 134, 136–138, 168, 184, *see also* Electrocardiogram
Herbst lamellar corpuscles, 38–39, 52
Hippocampus, 31, 33, 134, 139, 155, 157, 181, 185, 187, 208, 240, 250, 253, 255
HIS, *see* Hyperstriatum intercalatus superior
Homology, 10, 15, 17, 20–23, 62–63, 65, 88, 118, 162, 215

Homoplasy, 21–23
House sparrow (*Passer domesticus*), 22, 81, 84
5HT, 146
Hummingbirds (*Trochilidae*), 102
HV, *see* Hyperstriatum ventrale
Hyperstriatum, 32, 64, 145, 154–156, 174, 179, 181, 189–190, 193–194, 197, 231, 234, 240, 242, 247, 250, 253
  accessorium, 32–33, 55, 58, 63, 155, 177, 181–183, 186–187, 193, 255, *see also* Wulst
  dorsale, 33, 49, 55, 63, 155, 181, 183, 186, *see also* Wulst
  intercalatus superior, 63, 181, 186, *see also* Wulst
  ventrale, 32, 55, 58–60, 155, 157, 181–184, 186, 230, 255
Hypodipsia, 115
Hypophagia, 115, 122, 124, 127
Hypothalamus, 33–34, 59, 65, 88–89, 92–95, 97–98, 113–114, 128–129, 149, 157, 215, 223, 239, 244

I

ICO, *see* Nucleus intercollicularis
ICS, *see* Intracranial stimulation
IHA, *see* Nucleus intercalatus
Incubation, 90
Inferior colliculus, 16, 51–52
Intercollicular nucleus, *see* Nucleus intercollicularis
Interstitiospinal tract, 43–44
Intertectal commissure, 53
Intracranial stimulation, 204, 207–209, 212–217; *see also* Electrical stimulation of the brain
Isoprenaline, 148–149

K

Kinocilium, 51
Krause end bulbs, 38

L

Lamina archistriatalis dorsalis, 157
Lamina frontàlis, 32, 181
Lamina hyperstriatica, 32, 145, 181
Lamina medullaris dorsalis, 33, 145–146, 157, 181
Lateral anterior thalamic nucleus, *see* Nucleus lateralis anterior thalami
Lateral forebrain bundle, 55, 114, 141, 147, 157, 197, 231, 234, 244, 250, 255
Lateral geniculate body, 49
Lateral lemniscus, 35, 51, 53, *see also* Lemniscal pathways
Lateralis anterior, 49
Lemniscal pathways, 34, 45, 51, 57, 63, 65, *see also* Lateral lemniscus
Lentiform nuclei, 45, 47
Limbic system, 179
Lingula (cerebellar), 36
Lissencephalic, 31–32
Lobes of Lachi, 37–38, 41
Lobus parolfactorius, *see* Parolfactory lobe
Locus coeruleus, 58
Lovebird (*Agapornis* spp.), 196–197
Lumbosacral enlargement, 37, 40

M

Macula lagena, 51, 53
Magpie (*Pica* spp.), 195–196
Mandibulation, 103–104, 115, 118, 121–122, 124, 127–128, *see also* Feeding behavior
Mebanazine, 149
Medial forebrain bundle, 59, 65
Medulla oblongata, 43–44, 78, 88, 161
Meissner corpuscles, 38
Merkel discs, 38
Mesencephalon, 16, 34, 36, 44, 55, 88–98, 175, 231, 243–244, 247, 250
Methyl anthranilate, 209
Methyl phenidate, 148
Microsmatic, 11, 31
Midbrain, *see* Mesencephalon
MLD, *see* Nucleus mesencephalicus lateralis pars dorsalis
Model predators, 224–227, 229, 243–251
Monkey (*Macacca* spp.), 22–23, 162, 197
Monoamine oxidase, 149
Morphotype, 17–18, 20
Mouse (*Mus musculus*), 206, 209
Mynah bird (*Acridotheres* spp.), 196

N

NB, *see* Nucleus basalis
Neocortex, 24, 57, 60, 66–68, 154, 161, 165
Neostriatum, 32–33, 51, 53, 55, 58–60, 63–64, 67, 88, 117, 140, 145, 154–156, 159, 178, 181–183, 212–216, 229–231, 234, 242, 247, 254–255
Nest building, 5, 90
Nest defense, 224, *see also* Defensive behavior
Nodulus (cerebellar), 36
Noradrenaline, 148–149
NREM, *see* Sleep
Nucleus
 abducens, 53
 accumbens, 146, 181, 255
 angularis, 51–52, 78-79, 81
 ansae lenticularis, 60
 basalis, 54, 114, 116–117, 119–123, 159, 181, 255, *see also* Nucleus prosencephali trigeminalis
 cuneatus, 43, 57–59
 dorsalis intermedius posterior thalami, 60
 dorsolateralis anterior thalami, 48, 231, 240–243, 250, 254–255
 dorsolateralis posterior thalami, 63, 255
 ectomammillaris, 36, 45, 55, 255
 geniculatus lateralis, pars ventralis, 157
 gracilis, 43, 57–59
 intercalatus, 49, 59, 146, 177, 181
 intercollicularis, 231, 240, 242–243, 250, 255
 intrapeduncularis, 32, 55, 59–60
 isthmi, pars magnocellularis, 34, 46, 53, 255
 isthmi, pars parvocellularis, 34–35, 46, 53, 255
 Isthmo-opticus, 34–36, 46, 255
 Laminaris, 51, 53
 lateralis anterior thalami, 141, 255
 magnocellularis, 51, 53, 78–79, 81
 mesencephalicus lateralis pars dorsalis, 34–35, 46, 53, 89, 93, 231, 240, 242, 250, 255
 mesencephalicus lateralis pars ventralis, 88, 255
 opticus principalis thalami, 177–178, 181
 ovoidalis, 34, 51–53, 63, 89, 231, 250, 255
 pretectalis, 36, 45, 59, 255
 principalis nervi trigemini, 55
 principalis precommissuralis, 57
 prosencephali trigeminalis, 54, 56, 63, 114, 116, 119–123, 128, 159, *see also* Nucleus basalis
 quadrangularis parvocellularis, 51
 reticularis, 58, 141, 146, 255, 256
 reuniens, 51
 rotundus, 33–34, 47–48, 53, 55, 140, 146, 157, 175–178, 194, 231, 250, 256
 semilunaris, 36, 53
 septalis lateralis, 146, 181
 spirifomis medialis, 57, 256
 striae terminalis, 158
 subcoeruleus, 58
 subpretectalis, 36, 255
 subrotundus, 57, 147
 subtrigeminalis, 58
 superficialis parvocellularis, 63
 suprarotundus, 49
 taeniae, 157
 tegmenti pedunculopontinus pars compacta, 53, 60, 62, 256
 tractus ascendens, nervi trigeminus, 52
 tractus descendens, nervi trigeminus, 54–55, 58
 trigeminalis, *see* Nucleus prosencephali trigeminalis
 ventrolateralis thalami, 141, 146–147, 256
NVI, *see* Abducens nerve

O

Occipitomesencephalic tract, *see* Tractus occipitomesencephalicus
Oculomotor nerve, 35–36
Oculomotor nucleus, 53
Olfaction, 11, 16, 123
Olfactory bulbs, 16, 31, 179, 181
Olfactory cortex, 54
Olfactory nerve, 179
Olivary nucleus, 37
OM, *see* Tractus occipitomesencephalicus
OMT, *see* Tractus occipitomesencephalicus
OPT, *see* Nucleus opticus principalis

Optic chiasma, 157, 181, 191
thalami
Optic lobes, *see* Optic tectum
Optic tectum, 30, 32, 34–36, 45–48, 50, 55, 57–59, 64–65, 158, 161, 175, 191, 194, 250
Organ of Corti, 50
Ostrich, (*Struthio camelus*), 37
Owl, *see* Burrowing owl

P

Pacinian corpuscle, 38–39
Pair bond, 5, 90
Paleostriatum, 32, 88, 162, 212–214, 229–230, 240, 250, 252
augmentatum, 32–33, 48, 55, 58–59, 62–63, 146–147, 149, 157, 162, 181, 231, 255
primitivum, 32–33, 54–55, 58–59, 62, 66, 114, 146–147, 157, 162, 181, 234, 242, 255
Pallium, 158
Panic responses, *see* Avoidance behavior
Panting, 225, *see also* Respiration
Papilla basilaris, 50–51
Parafloccular lobes of the cerebellum, 36
Paragriseal cell column, 37–38, 41–42
Parahippocampal area, 157, 181, 185, 255
Parahippocampal cortex, 33
Paramedian nuclei, 47
Paravertebral sympathetic chain, 31
Parolfactory lobe, 32, 59, 65, 114, 146, 181
Parrot (Psittacidae), 56, 102, 195–196
Pecking behavior, 103–104, 118, 121, 146, 155, 158, 161–163, 169, 211, 225–226
Pelicans (*Pelecanus* spp.), 102
Periectostriatal belt, 47, 50, 59, 64, 67
Phenlyephrine, 148
Phenmetrazine, 148
Phyletic development, *see* Phylogeny
Phylogenetic tree, *see* Phylogeny
Phylogeny, 8, 23, 29, 31
Physostigmine, 204
Pigeon (*Columba livia*), 30–33, 37, 40, 42, 44–45, 48–49, 52, 55–56, 79, 87–89, 101–110, 113–119, 121–124, 129, 133–138, 140–141, 143, 145–147, 154–163, 169, 176–178, 181–185, 189–193, 197, 209–213, 215, 222, 226
Pipadrol, 148
Pons, 44
Pontine nuclei, 57–59, 65, 118, 158
Posterior commissure, 55, 192
Posterodorsal pretectal nucleus, *see* Nucleus pretectalis
Preening, 3–4, 146
Prefrontal cortex, 162
Preoptic area, 89, 139, 231, 234, 242
Pretectal area, 45, 47, *see also* Nucleus pretectalis
Pretectal nuclear complex, *see* Nucleus pretectalis
Proactive inhibition, 205
Progesterone, 252
Prolactin, 252
Proprioception, 39, 104, 122
PrV, *see* Nucleus prosencephali trigeminalis
Pulse duration, *see* Stimulus parameters
Pulse frequency, *see* Stimulus parameters
Purkinje shift, 11
Puromycin, 173, 204
Pyramidal tract, 57–58, 66, 118, 158, 217

Q

QFT, *see* Tractus quintofrontalis
Quinto-frontal structures, 116–118, 122, 126, 129

R

Rami communicantes, 40
Rat (*Rattus norvegicus*), 10–11, 102, 105, 112, 123, 128–129, 206–207, 224
Raven (*Corvus corax*), 196–197
Red nucleus, 36, 57, 59–60
Red-winged blackbird (*Agelaius phoeniceus*), 91–96
REM, *see* Sleep
Reproductive behavior, *see* Courtship behavior
Respiration, 136–138, 168, 184, *see also* Panting
Reticular activating system, *see* Reticular formation

Reticular formation, 37, 44, 46, 57–59, 65, 139, 158, 161, 207–208, 231, 240–243, 254, 255
Reticulospinal pathway, *see* Spinoreticular tract
Retina, 16, 21, 30, 36, 45–46, 48, 176, 191, 195
Rhombencephalon, 36
Rhomboid sinus, 31, 37–38, 42
Ringdove (*Streptopelia risoria*), 85, 98, 140
Rotundal–ectostriatal pathway, 182
Rubrospinal tract, 43–44, 60

S

Self-stimulation, 90
Septal area, 139, 187, 234, 240, 242–243, 253, 256
Septomensencephalic tract, *see* Tractus mesencephalicus
Septum, 32, 208, 230, 250, 254
Serotonin, 149
Sexual behavior, 222, 252, *see also* Courtship behavior
Slate-colored junco (*Junco hyemalis*), 84
Sleep, 133–150, 225, *see also* Drowsiness
  sleeplike states, 162
  spindles, 137
Species–specific behavior, 6, 10, 101
Spinal cord, 30–31, 37, 39–44, 57–58, 60–61, 65, 158, 161
Spinal nerves, 37
Spinocerebellar pathway, 43–44
Spino-olivary pathway, 44
Spinoreticular tract, 43–44
Spinotectal pathway, *see* Tractus spinotectalis
Spinothalamic tract, 44, 63
Spiriformis medialis and lateralis, 36, 59–60
Squirrel monkey (*Saimiri* spp.), 98
ST, *see* Stria terminalis
Starling (*Sturnus vulgaris*), 82
Steller's jay (*Cyanocitta stelleri*), 92–98
Stereotyped behaviors, 5, 88, 103–104, 163, 226–227, 245, 254
Stimulus frequency, *see* Stimulus parameters
Stimulus parameters, 90, 97, 141, 146, 213, 223–224
  pulse frequency, 93–94, 97–98, 141
  pulse duration, 92–95
Stimulus spread, 91–92
Stratum opticum, 46
Stria medullaris, 181
Stria terminalis, 59, 65–66, 157
Striatal complex, 31–32, 89
Striate cortex (mammalian), 49–50
"Striatum," 61
Strutting, *see* Courtship behavior
Succinic dehydrogenase, 62
Superior colliculus, 46–47
Superior olivary nucleus, 51, 53
Suprochiasmatic nucleus, 45
Supraoptic area, 231, 234, 242
Supraoptic commissure, 49
Supraoptic decussation, 53, 93, 192–194
Swallowing, 103–104, 121, 146
Swift (*Chaetura* spp.), 10–11

T

Tameness, 158, 161
Taste, 123
Taxon-specific behaviors, 10
Tectal commissures, 191–192, 194
Tectal gray, 36
Tectal ventricle, 46
Tectofugal pathway, 23–24, 34, 46–48, 50, 64–65, 176–177
Tectospinal tract, *see* Tractus spinotectalis
Tectum, *see* Optic tectum
Tegmentum, 57, 158, 162, 207
Tegmental reticular structures, 36
Telencephalization index, 196
Telencephalon, 23, 30–34, 44, 47–49, 54–55, 57, 59–60, 63–65, 67, 113–114, 140, 145, 147, 154, 156, 176, 177, 191, 231, 244, 250, *see also* Forebrain
Territorial calls, 85
Testosterone, 251–252
TFM, *see* Tractus frontothalamicus medialis
Thalamic nuclei, 207, 231
Thalamic reticular nucleus, 147
Thalamofugal pathway, 23–24, 34, 47–48, 50, 64–65, 176–177, 189
Thalamus, 33–34, 45, 47–49, 51, 54, 57, 64, 67, 139–141, 146, 158, 165, 175–177, 181, 191, 193, 223, 247
Thoracolumbar system, 40

Threat behavior, *see* Agonistic behavior
Tonotopic organization, 78, 81–82
Torus externus, 88–89, 93, 95–96
Torus semicircularis, 51, 88
Tractus
  archistratalis dorsalis, 157–158
  cortico-habenularis et cortico-septalis, 181
  dorsoarchistriaticus, 54
  frontoarchistriaticus, 54, 63, 66, 117, 157–158, 181
  frontothalamicus medialis, 229, 231, 234, 240, 242, 250, 256
  occipitomesencephalicus, 43, 57–59, 65–66, 88, 91, 118, 122, 157–158, 217, 234, 242–244, 250, 255
  opticus, 157, 256
  oviodalis, 53, 256
  quintofrontalis, 54–55, 114–116, 121, 157, 162, 181
  septomesencephalicus, 53, 55, 58–59, 91, 141, 181, 197, 231, 234, 240, 242, 244, 256
  spinotectalis, 44, 47, 256
  tectothalamicus dorsalis, 231, 242, 256
  thalamo-frontalis medialis, *see* Tractus frontothalamicus medialis
Trapezoid body, 46
Trigeminal nerve, 52, 114, 123
Trigeminal nucleus, *see* Nucleus prosencephali trigeminalis
Trigeminal system, 39, 52, 54–56, 63, 114
Trochlear nerve, 36
Trophotrope syndrome, 140
TSM, *see* Tractus septomesencephalicus
Turkey (*Meleagris gallopavo*), 4–6
Tympanic membrane, 50

U

Uvula (cerebellar), 36

V

Vacuum activities, 253
Vagas nerve, 39
Vallecula, 181
Ventricle, fourth, 78
Ventricle, third, 149
Ventral amygdalofugal pathways, 59
Ventral funiculus, 41–42, 44
Ventral geniculate nucleus, 45, 47, 59, 64
Ventriculus, 157, 181
Ventrolateral thalamic nucleus, *see* Nucleus ventrolateralis thalami
Vestibular nucleus, 53, 55
Vestibulospinal pathways, 43
Visual pathways, 23, 31, 165, 175, 193
Vocal frequencies, 83–85
Vocalization, 222, 225, 243
Vocalization systems, 87–90

W

Wakefulness, 133–137
Wulst, 32–33, 48–50, 59, 61, 63, 66, 174, 177, 179, 181–186, 189–190, 196–197, *see also* hyperstriatum dorsale, Intercalatus superior, Hyperstriatum accessorium

Z

Zone of Lissaucr, 42, 44

A 4
B 5
C 6
D 7
E 8
F 9
G 0
H 1
I 2
J 3